Management in Engineering

Management in Engineering

Principles and Practice

Gail Freeman-Bell
James Balkwill

Prentice Hall
New York London Toronto Sydney Tokyo Singapore

First published 1993 by
Prentice Hall International (UK) Ltd
Campus 400, Maylands Avenue
Hemel Hempstead
Hertfordshire HP2 7EZ
A division of
Simon & Schuster International Group

Typeset in 10/12 pt Garamond
by MHL Typesetting Ltd, Coventry

Printed and bound in Great Britain at the Cambridge University Press

Library of Congress Cataloging-in-Publication Data

Freeman-Bell, Gail.
Management in engineering: principles and practice / Gail Freeman-Bell and James Balkwill.
p. cm.
Includes bibliographic references and index.
ISBN 0-13-554023-2 (pbk.)
1. Engineering—Management. I. Balkwill, James. II. Title.
TA190.F74 1993
620'.0068—dc20 92-1603
CIP

British Library Cataloguing in Publication Data

A catalogue record for this book is available from the British Library

ISBN 0-13-554023-2

1 2 3 4 5 97 96 95 94 93

To

Martyn, Emily,

Elizabeth,

Mint and Willie

Contents

Foreword

There is always a question about whether a perception is actually formed from elements of reality or, in fact, becomes a self-fulfilling prophecy.

My earliest mental image of an engineer was that of a rather dishevelled, introverted individual characterised by the protrusion of a slide-rule from the top pocket of his crumpled jacket. The chances are that, these days, the younger generation of engineers may not even recognise a slide-rule, let alone carry one.

One of the longstanding complaints of engineers from all generations is the lack of public recognition that they enjoy. Overseas, our engineers are highly regarded but in the United Kingdom tend to be viewed as subordinate to other professions such as medicine and the law — a prophet is not without honour, except in his own country?

I believe that this can be due, in part, to the public perception of the engineer fuelled by the engineer's own dedication to a specialism. For too long, engineers have tended to concentrate their efforts within the — often narrow — band which is their sphere of expertise, to the exclusion of involvement in the general enterprise. Personnel, commercial and finance issues have been left to the departments considered appropriate, with the result that management and business issues have been settled without an engineering input. Furthermore, engineers working in a company environment have often felt their creativity constrained by the specialists from other company areas. Representatives from personnel, finance, sales and marketing each have their own, particular styles which can hinder effective team play.

The need for broadly based engineers who can communicate intelligently and knowledgeably on the 'home ground' of all company functions has long been recognised. Considerable effort has been expended in providing the means for this educational process. Gone are the days when an engineer would be promoted beyond his level of competence as a manager purely because he was a good engineer.

We are now beginning to witness the advent of the engineer who not only has an understanding of the alternative disciplines but whose thought processes actually emulate those of the indigenous experts. These skills which the engineers embrace, in addition to their core skills, are those which equip them to be effective and successful managers. It is this level of overall understanding and competence which has provided the environment demanded for effective teamworking.

There are three major factors upon which every manager must focus:

- Operational processes.
- Management style.
- Organisational structure.

All other business topics will rest comfortably beneath these generic titles. The declared priority is crucial.

This book considers all the business elements that an engineer who desires to manage effectively will require and I am glad to see it. Understanding the elements described will provide a sound foundation from which engineers will be able to establish a knowledge of how those elements are applied through discussions with the fact-holders from within their own organisations. It is imperative that the readers are not satisfied with the theory alone, for therein lies sterility, but actually seek this practical information at an operational level.

Nothing, apart from bankruptcy, comes easily. The tools for engineers in management are known and available — use them, tailor them to your circumstances and style, and manage well!

John Towers
Group Managing Director
Rover Group Limited

Preface

Commercial success comes from being able to meet the needs of customers. For engineers this involves the creative application of engineering principles. The challenge of engineering also includes effective management of resources to produce cost-effective solutions.

This book addresses the management needs of engineers. It is written by professional engineers, and can be used by both student and practitioner alike. The learning objectives are:

- To be able to apply management topics as tools with which to achieve engineering solutions.
- To show how effective management is achieved in engineering organisations.

These learning objectives are met primarily by the way in which the content has been determined, structured and presented. In addition, we have included case studies that illustrate the techniques described, as well as many worked examples and revision questions. A Teacher's Manual to accompany this book is available free of charge from the Publishers to lecturers who use it in their teaching.

The book is totally biased towards engineering and the examples are drawn from the electro-mechanical engineering industry. We have also given consideration to those topics which have specific relevance to engineers, such as design and manufacturing. All the chapters are self-contained which allows them to be used on their own. However, the book leads the reader through all the appropriate management topics in a logical progression, culminating in a case study of a large engineering project which shows how all the topics in the book come together in practice.

Acknowledgements

Many different people have played an important role in the creation of this book and before we begin it is appropriate to spend a few lines acknowledging them. We wish to thank the companies that have allowed us to produce case-study material: Oxford Lasers Ltd, Edwards High Vacuum, Rover Group, Warwick Power Washers, Research Machines, Seagate Technology International and JP Engineering. We also wish to thank those engineers who have allowed us to use their personal experiences: Hilary Briggs, Martin Twycross, Matthew Hobbs, Simon Tu, Rhys Lewis, Richard Benfield, John Boaler, and Steff Inns. Thanks must also go to Mr John Towers for preparing the foreword. There are many others, too many to name, who have helped and advised us during the process of writing and we acknowledge their help and support. However, we must thank the School of Engineering at Oxford Polytechnic for providing us with the opportunity that allowed us to write this book.

Lastly we wish to acknowledge the incalculable contribution made to humanity by Monsieurs Alain Voge and Pierre Ricard. They have produced, without doubt, the finest white and red wine (respectively) on the planet. 'Saint Peray — Cuvée Boisée', and 'Vacqueyras — Domaine le Couroulu' are wines beyond compare. That they have played a vital part in the creation of this text there can be no doubt, and we hereby salute their greater art — as should you.

Chapter 1
Engineers as managers

The most important skill of a creative mind is the ability to ask good questions. Before we embark on the body of this textbook there is a question that should be asked, and we must provide an answer to it before we can sensibly proceed with the study of 'Management in Engineering'. The question is 'why?' Why are we bothering, as engineers, to study management and what relevance does it have? The answer to this question, however, rests on another one: 'What do engineers actually do?' If they have a need for management then the book is worth study; if they don't then it should be thrown away now.

Many dictionaries define an engineer as 'one who drives engines' or 'one who designs and builds engines'. Few professional engineers actually are engine drivers and only a tiny fraction design and build engines. Popular understanding is often worse, with many people thinking the words 'engineer' and 'mechanic' are interchangeable. They are not. The word 'engineer' has two roots: one means 'the creative one' or 'the one who uses ingenuity'; the other means 'to bring about' or 'to make manifest'. These definitions are much closer to the truth, but they still do not fully explain what an engineer does and whether engineers have a need for management.

The engineering profession is like other professions, such as law or medicine, in that it has a regulating body controlling qualifications, there is a code of ethics, professional status is distinct and the law regulates the practice of engineering. It is therefore not surprising that the period of study prior to achieving membership of a professional engineering body is a lengthy and testing one. In the United Kingdom it takes a minimum of five years, and is often much longer. To examine whether engineers need to study management or not we will look at the two distinct periods of an engineer's life — training and professional practice.

During the training period required to become a qualified professional engineer the student is continually under pressure to acquire technical knowledge and skills. A mechanical engineer's training places emphasis on thermodynamics, stress analysis and mechanical manufacturing processes. Electronics and electrical engineers will be putting their efforts into analogue and digital circuit design, data processing and electronics manufacturing techniques. The emphasis is on technical excellence, on the application of advanced theory and the ability to describe the world with mathematics and then to manipulate it to meet desired needs. Management is a curious thing to study at this time. Whenever was a student of applied mathematics helped by knowing how to hold a performance-related appraisal? For engineering undergraduates to study 'management' may seem not only unreasonable in terms of the extra effort required but also it may seem unconnected with undergraduate engineering.

But what about life *after* qualifying? Of course, most undergraduates are aware that life after qualification may involve different skills and that management may yet prove to be useful. However, what reasons are there for thinking this?

Perhaps it is worth opening the dictionary again, this time to see what it has to say about 'management'. The *Oxford Concise Dictionary* (Clarendon Press, 1990) gives a number of definitions of management. These include 'the process or an instance of managing or being managed' — certainly this doesn't tell us very much as we don't understand the process of managing. Another definition is 'trickery; deceit', not a definition we would use! The best definition for the term as it is used in engineering is 'the professional administration of business activities'.

How does this affect practising engineers, and what are 'business activities'? In fact, business activities are anything that relate to the organisation and operation of a business. Engineers certainly play a major role in the organisation and operation of an engineering business. More than this, it is the engineering that defines the very structure of such a company: a jet engine manufacturer is constrained by the limits imposed by the design and manufacture of the jets. Engineers wrestle with the design of the engines, striving to get more power, greater reliability, more profit and greater safety. It is engineers who state the form the product will actually take when it is made. All other aspects of the company's management must derive from this, not the other way round. The question to ask now is: 'What sort of jobs do engineering organisations need their engineers to do?' The following are extracted from advertisements that have appeared in recent professional engineering journals:

> Principal Mechanical Engineer . . . provide technical and commercial direction within the engineering field. Matching business acumen with a detailed understanding of thermal and acoustics . . .

> Design Engineers . . . With responsibility for design, manufacturing, specification, installation, commissioning and the provision of operational support, you will be encouraged to pursue the creative application of new technology.
>
> Design/Development Engineer . . . The successful applicant will be engaged in product concepts generation, detail design and analysis, development testing and client liaison.

In the jobs advertised there is no explicit reference to 'the professional administration of business activities', but let us look more carefully at each job in turn.

In the first of the advertised jobs there is reference to commercial direction and business acumen, as well as to technical issues. 'Direction' implies that this job involves getting others to do something, showing them the way. For this job then one could assume that in addition to the technical knowledge required you have to know something about marketing, finance and directing others.

The second job looks much more technical, except where it says 'with responsibility for . . .'. There are many technical issues involved in this job but it is clearly the intention that the successful applicant will be responsible for others who will assist in the work. There is simply too much for one person to do. The third job is similar.

Clearly the technical issues so familiar to undergraduate engineers are important here but so is something else. In all of these jobs, the advertisers are looking for specific technical knowledge *and something else*. The phrase 'the professional administration of business activities' may not trip off the tongue but it does describe the 'something else' that the advertisers of these jobs want. The 'something else' is, of course, management skill. These advertisements show that there is need for a combination of skills, both technical and management, and that the two skills go together.

How can an engineer's training period best provide for post-qualification life, in both technical skills and managerial ones? Clearly it is important that during the qualifying period an engineer should gain a sound technical knowledge and learn enough of the physical principles so that when a new situation is encountered it may be analysed. The engineer will then be able to identify the knowledge needed to solve the problem. The academic environment is ideal for this and most engineering courses are characterised by intense study.

But what of the management? What important management principles may be learnt at this time? It is clearly not possible to study and experience management in undergraduate life in the way that it will be experienced in postgraduate life, so what is required is a combination of technical and managerial courses. The undergraduate must be sufficiently well prepared for the business environment so that the transition is not traumatic or unexpected but is beneficial and the graduate can quickly absorb the new environment.

From a management training point of view it is clear that in order to prepare for life in engineering business, a training is required that provides the important management skills as they are needed by engineers. The training should make clear how the topics relate specifically to engineering. Management topics that do not help engineers should be omitted. Those of use to engineers, but often omitted from other management courses, should be included. The material should be written for engineers by engineers, and it should explain how an individual's engineering abilities can be greatly extended by good management, for that is the essence of management. This book provides the material for such a training.

We need a structure for such a large undertaking and this book has been carefully constructed to present the important aspects of engineering management in a logical and readable format. To begin with, we study the environment of engineering practice and then examine the needs that this environment creates. These needs are studied as separate issues since this makes their understanding easier. Of course, in reality the issues are not separate and occur together. For this reason we have cross-referenced material and used engineering examples to show its relevance. More importantly, we have provided a large case study of a real engineering project which includes a description of its technical side. This case study brings together the separate themes of the book. The case study shows the essence of engineering, it illustrates vividly the equal importance of management and technical skills. It serves as an excellent illustration of how the important principles explained in the book are used in practice.

We will now explain in a little more detail the contents of each chapter.

In Chapter 2 we start by examining the environment of business. We explain the legal establishment, the structure and the purpose of organisations. This provides the context for our work as engineers. In treating these subjects the chapter explains why the organisation exists, the way that the structure assists the organisation in its purpose and the legal identity of the organisation. Once we have this picture of an engineering organisation we can examine the various tasks that such an organisation will need to execute. These tasks are called 'functions' to make clear the active role they take in the success of the organisation. Chapter 3 is called 'Functions of Organisations' and is a detailed explanation of the tasks that must be managed if the objective of the organisation and purpose of its existence are to be met.

After completing Chapters 2 and 3 the reader should have a complete picture of the engineering environment. We have explained why the organisation exists, what it is trying to achieve and the tasks, or functions, that it must accomplish in order to be successful. From this point on the nature of the book is different. It now explains the management tools that are required in order for the company's personnel to be successful in their aims. The most important issue is treated first in Chapter 4, 'Personnel Management'.

If you visit a company on a Sunday you will more than likely find the whole place at a standstill. There is nothing wrong with the organisation — all the things that are there during the week are still there on a Sunday, with one exception: the people are missing. No other single resource has such a dramatic effect on an organisation. In Chapter 4 we look at the specific issues associated with employing and managing people. The contents of Chapter 4 will give the reader a formidable and comprehensive battery of personnel skills that will serve well for life in professional practice.

The emphasis in Chapter 5 is still on personnel management but it is a very specific aspect, that of teamworking. Successful teamworking brings a staggering capacity for output. In Chapter 5 we study how this comes about in order that we may be able to control and enhance the success of teams to assist the organisation in achieving its goals. Engineers are not the only professionals to recognise the benefits of successful teamworking and much of the material in this chapter is of use to other professions.

Engineers must be creative and able to work on more than one task at a time. However, if engineers are to be able to orchestrate technical work and deliver results on time then they must be well organised. Personal organisation is not an attribute of our genetic coding; we are not born tidy or messy people. We can give ourselves good personal organisation by working at it. Chapter 6 tells you how. Clearly those who have good personal organisation will be of more interest to the originators of the advertisements above than those who do not.

In Chapter 7 we consider the management of the greatest resource after personnel — money. In this chapter we look at the role of money in the operation of a company and specifically at the financial issues that involve engineers. The management of money is a large issue in its own right and some people never work with anything else. As engineers we have a great responsibility and it is important that we understand issues such as product costing and financial control. Chapter 7 provides an explanation of the financial issues that are important to engineers.

Engineers often undertake work in project groups, either as team member or team manager. Chapter 8 looks specifically at the management of projects, their definition, planning and monitoring. It describes techniques that can be used for planning and considers the aspects that have to be investigated when deciding whether or not to proceed with a project, such as the cost. There is also a case study which shows how one company tackled a project that had a significant impact on its business and profitability.

Engineers are processors of information; they gather it from a wide range of sources. If engineers are to be competent and enjoy the challenge of creative engineering, they must be good at gathering information and then disseminating it. It is clear that one skill required by all of the above aspects of engineering and management has not been addressed: it is not possible to be successful in any of these areas if one is not able to communicate. Of course, we can all communicate at some level but professional engineers, along with other professionals, have the most demanding needs for clarity and excellence in communication. Chapter 9 explains this need and provides a training in communications that forms the basis of a separate undergraduate course in many engineering programmes.

Finally, in Chapter 10 all the different skills are brought together in a case study which clearly shows the balance of technical and managerial skills that a practising engineer actually uses. In the case study the boundaries between the separate management subjects we have introduced have gone. In practice they are never there. The technical and managerial issues are solved simultaneously and engineers use many of the skills at once. Just as a good piece of music is composed of different instruments playing together, each making exactly the ideal contribution for the given instant, so is good management practised.

It is this last chapter in which the very spirit of engineering management is encapsulated. All of these issues explained in the previous chapters are brought together and used in conjunction with the commensurate technical skills of the qualified engineer. It is certainly important to dissect and examine the individual tools of good management as we have in the previous chapters, but it is important to see how they fit back together again and to understand how the skills are used in practice. This textbook therefore expounds a complete understanding of engineering management. It first makes clear the environment in which engineering management is practised. It then dissects each of the important tools of management that an engineer needs. Finally, it brings them all back together, explaining, with a real example, how successful management by engineers may be achieved. Figure 1.1 shows this structure diagrammatically and includes a very short summary of the contents of each chapter.

In the first paragraph of this introduction we asked: 'Why are we bothering, as engineers, to study management?' As you can see, it is impossible to be a professional engineer without this knowledge.

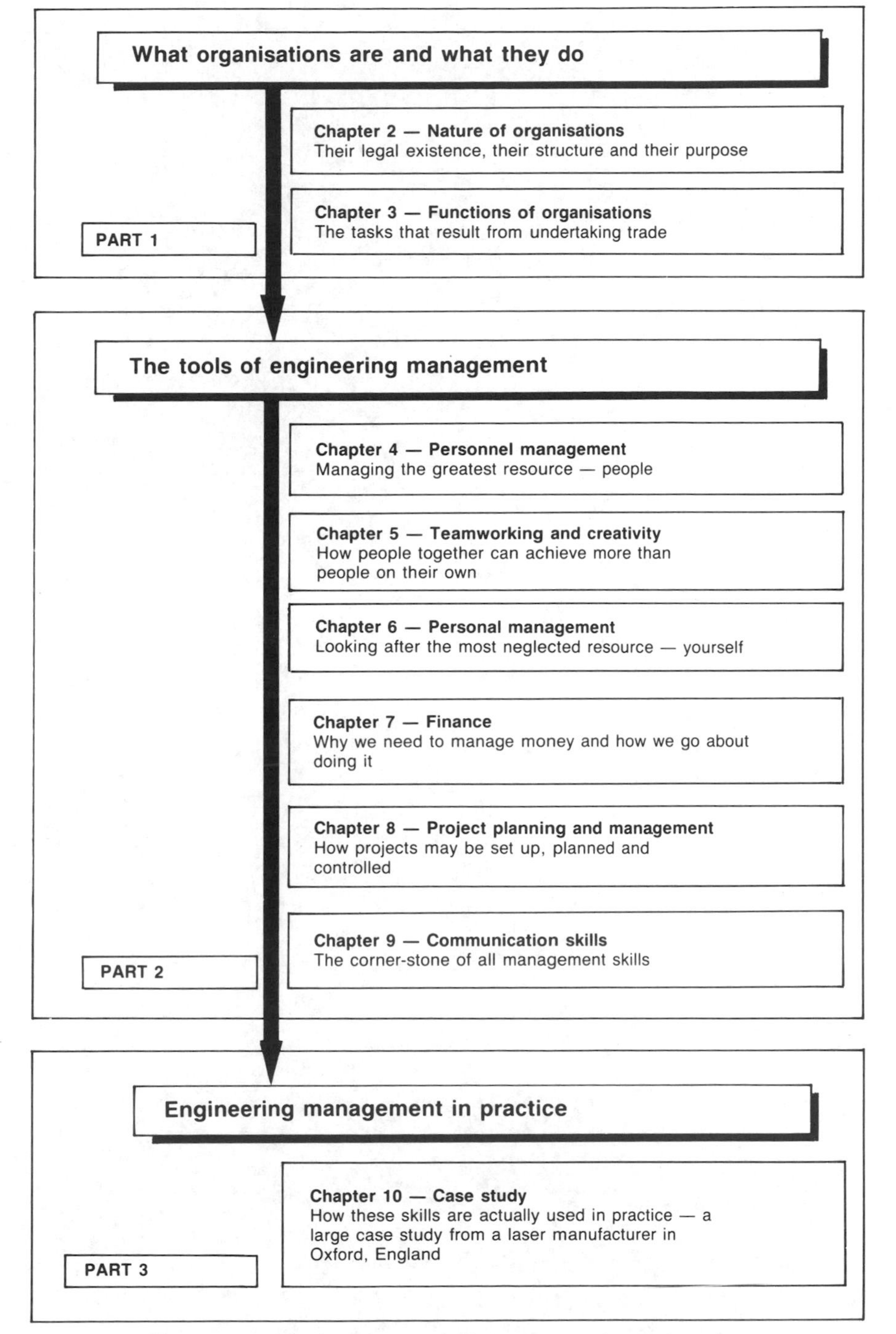

Figure 1.1 The structure of *Management in engineering*

Part 1

What Organisations Are and What They Do

Chapter 2
Nature of organisations

Overview

In this chapter we look at the management of organisations. We consider the legal forms of organisations in the United Kingdom, including a review of some of the legislative and other factors that affect the way in which organisations are able to transact business.

Following establishment, companies have to be organised and people deployed amongst the functions that are required in order to meet the objectives of the business. We will examine overall organisation structure, and consider the factors that determine the structure adopted. We will also describe some of the structures commonly found in organisations.

In order to determine the optimum structure for the business and to direct the employees towards achieving the aims of the business, there must be a clear statement of these aims and objectives. In this chapter we look at the role of corporate objectives and the way in which they define the operation of the company.

Finally, we will look at an engineering company and the way that it has been established and operates.

2.1 Introduction

Many qualified engineers work in the engineering industry in manufacturing companies, some work in design consultancies, others in public organisations. There are also many people who qualify as engineers who take a different career route, moving away from engineering but using their numerical and managerial skills in areas such as finance, computing and management services, or administration. All of the engineers entering these areas will need the management knowledge and skills described in this book. They will also all work in organisations.

Organisations vary enormously from the two- or three-person design consultancy, where everyone is a partner, to a multinational conglomerate where an engineer could be working in a division with two or three hundred other engineers. However, if the engineer is to manage a project, finance or personnel, effectively then an understanding of the organisation is necessary to put the management requirement in context. The engineer should know what the goals of the organisation are, and should be able to assess the contribution of the engineering activity to meeting these goals. In addition an engineer should be able to examine how a company is organised and be able to assess the communication routes, determine who has authority for particular aspects of the organisation, and evaluate the effectiveness of the structure for meeting the company's goals. This latter aspect of understanding organisations is particularly important for engineers who, as agents of change, are responsible for implementing new technology or working practices.

As well as the organisation structure and its objectives, the operation of a company will be constrained by the way in which it has been formed, for example the way in which it can raise capital for expansion will be greatly affected by its legal form. In this chapter we will examine the way in which businesses can be legally established in the United Kingdom and how they can be structured. We also look at the way in which organisations set objectives and how they develop strategies for meeting their objectives.

The learning objectives for this chapter are as follows:

1. To understand why organisations are formed and to see how they can achieve their aims.
2. To consider different ways that businesses can be structured and understand the implications of such different structures.
3. To look at the types of business that can be formed in the private sector in the United Kingdom.

2.2 The legal establishment of business

In this section we will examine five ways in which businesses in the private sector in the United Kingdom can be established. These are sole traders, partnerships, co-operatives, and public and private companies. We will also consider the use of franchises. These types of organisations will also exist in countries outside the United Kingdom but the legal forms and legislation that affect them will vary from country to country.

2.2.1 Sole traders

The establishment of a business with a single owner — a sole trader — is the simplest form of business and is therefore very common, particularly in the high street. Sole traders run small shops, newsagents, work as electricians, plumbers, builders and consultants. This form of business is also the simplest to dissolve.

If you want to start up a business as a sole trader, it is quite simple — you just have to start trading. The business has no separate legal identity from the owner which means that all debts, liabilities — and profits — belong to the owner. You do not have to use your own name for the business, although this is usual, but if you do use a different name you have to be careful not to use another company's trade name. Operating as a sole trader doesn't mean that you have to work on your own, you can still employ other people.

Some features of sole trader businesses are listed below.

1. The accounts do not have to be disclosed publicly, and are therefore not available to competitors.
2. The owner makes all the decisions, and therefore has total control of the business. This can be a great advantage, particularly because of the speed with which decisions can be made and implemented. It can also be a disadvantage as the owner's competence as a manager and businessperson can restrict the business. In addition the sole trader may have no one with a vested interest in the business with whom to discuss ideas.
3. A sole trader can offer a personal service which may be valued more highly by some customers.
4. All the profits and the debts belong to the owner. This can lead to hardship if a business does badly, as many sole traders use their homes as security for their business. In any case, if debts are excessive it can lead to the forced sale of personal possessions; liability for debts is unlimited.
5. Often the only form of security a sole trader will have is their own home and it can therefore be very difficult to get extra capital for development or expansion of the business.
6. Working for yourself, particularly with all the administration and bookkeeping required to run a business, can mean that you work very long hours. It may also be some time before you are able to pay yourself an appropriate salary. Similarly, taking holidays can be a

problem unless there is someone that can be trusted to run the business while you are away.

7. It can be a high-cost enterprise because the sole trader rarely benefits from economies of scale. However, some businesses get over this problem by forming a sort of collective. This is particularly common in the retail trade, where a collective can purchase larger quantities from wholesalers and manufacturers, thus getting a better price.

2.2.2 Partnerships

Partnerships exist when there are a number of people involved who are part-owners of the business. This type of business can be set up without any formality but it is usually advisable to draw up a deed of partnership which covers such matters as the way in which the profit will be split between the partners.

Under English law the maximum number of partners that is allowed is twenty. There are two exceptions to the limit: in banking the maximum number is ten, and in certain professional partnerships where companies are not allowed to be formed by law, such as for solicitors and accountants, there may be more than twenty.

In a partnership each partner is jointly liable for any contract made by any other partner, irrespective of prior knowledge. This means that if, for example, one partner agrees to a contract to supply a piece of equipment within a certain time and another partner does not agree to this, the contract is nevertheless binding on all the partners. If a partner leaves the business then all previous contacts have to be told of the situation and made aware of the fact that the person will not be liable for future contracts. The ex-partner's name should also be removed from the business stationery as soon as possible.

The main features of operating as a partnership are listed below.

1. The liability and decision-making is shared. This can be an advantage when the skills and experience of the partners complement each other and the management of the business is carried out effectively. However, problems can ensue because liability as a partner exists even if the partner did not know of, or agree with, the contracts made. It may also be more difficult to reach decisions since they will all require consultation and agreement. This problem can be alleviated by having very clear lines of demarcation for the various aspects of management and operation of the business.
2. There is less formality and expense than is involved when forming a company.
3. Accounts do not have to be publicly disclosed to competitors.

4. Because you have a number of people with resources that can be drawn upon, it may be possible to raise large amounts of capital initially. However, extra capital may be difficult to raise due to limited additional security — many people use their homes as security for a loan.
5. The partnership could be dissolved suddenly if one partner dies or goes bankrupt, unless this event is covered in the deed of partnership.
6. Holidays and illness can be covered by the other partners.

2.2.3 Co-operatives

Co-operatives involve a voluntary association of people, called *members*, who operate an enterprise collectively. The aim of a co-operative may not necessarily be to make a profit. Co-operatives are run on the basis of one member, one vote, irrespective of how much each individual has invested. The managers are also democratically elected on this basis. In the United Kingdom limited liability for the members is achieved by registering with the Register of Friendly Societies.

The main advantages of co-operative operation are that the co-operatives give support to their members, there is little asset stripping (where assets are sold off to provide profits), and because of the essence of membership you should have good industrial relations.

Five principles are involved and form the main features of running a co-operative. These are listed below.

1. There is open and voluntary membership for all who work in the organisation.
2. There is one member, one vote, the management being elected by the membership.
3. Interest paid on capital is limited.
4. The business is conducted for the mutual benefit of the members and all profits are shared. In particular, people are paid on the basis of what they need, rather than on the basis of 'market rate for the job'.
5. The business must be socially aware and act responsibly to other businesses, customers, suppliers and the local community; it is not necessarily profit-making.

2.2.4 Companies

A company has a separate legal identity to its owners and will continue to operate even if the owners and managers change. It is owned by shareholders who each have a liability which is limited to the nominal value of their shares. This nominal value is the amount originally promised in return for the allotment of shares. The nominal value is usually different from the price of shares being traded on the stock market. The trading price reflects the value placed on the company by investors.

In the United Kingdom companies, both public and private, are registered with the Registrar of Companies. We will now look at how UK companies are established, and consider the main features of operating a company.

The creation of a company

The process of creating a company is called *incorporation*. The process requires that two documents are drawn up — the Memorandum of Association and the Articles of Association. The Memorandum relates to the external affairs of the company, giving information such as the company name, its registered office, the purpose (Objects) of the company and its share capital. The Articles relate to the internal operation of the company and include such things as the rights of its shareholders and the powers of its directors.

These two documents and others defining the amount of capital and number of directors are submitted, with a fee, to the Registrar of Companies, who is then able to issue a certificate of incorporation. Following incorporation the company details are made available for public inspection and a private company can start to trade immediately. A public company may not trade until the Registrar is satisfied that the necessary share capital has been allotted.

The regulation of UK companies is achieved through the Companies Act (1981). This Act states that the company must regularly prepare and publish information regarding its activities. The company is also only allowed to carry out the activities defined in the Objects clause of its Memorandum of Association, although this clause can be amended in accordance with the Act.

The annual report

Ownership of limited liability companies belongs to the shareholders and they delegate the operation of a company to the directors. It is the directors who are responsible for producing, in the correct form, annual reports and accounts as required by law. The annual report must contain both the balance sheet and the profit and loss account (these are discussed in detail in Chapter 7).

Large companies also have to include in the annual report other items of financial and general information such as a chairman's review and the auditors' report. A statement of source of funds and the way in which they are used is also included, giving a full record of the financial transactions undertaken during the accounting period.

Closing a company

If a company experiences financial difficulties and cannot pay its debts it may be closed by court order. The way in which a company is wound up is called *liquidation*, and it can be done voluntarily or compulsorily. Voluntary liquidation occurs when the company agrees to sell its assets and pay off any debts. Compulsory liquidation is when the company is given no choice and the court appoints an Official Receiver who will be put into the company to sell off the assets and pay off debts. In some cases the receiver may be able to sell the company as a going concern, which means that it continues to trade.

The main features of operating as a company are listed below.

1. Shareholders have limited liability, which is limited to the nominal value of their shares.
2. The company has separate legal identity from its owners and managers.
3. Extra capital can be raised in public companies by issuing more shares. Shares in public companies can be freely transferred without consulting other shareholders.
4. Once formed, a company has perpetual succession since the company management is separate from its ownership.
5. The accounts have to be disclosed publicly, and therefore it is possible to monitor closely what is happening in a competing company.
6. Company activities are limited by the Objects clause.
7. The company is more closely controlled and regulated by outsiders than sole traders or partnerships, because of the Companies Act (1981).

Public companies

Public companies tend to be larger than private companies and can be distinguished from private companies in that their names end in the letters PLC (Public Limited Company).

In order to become public the company must have a minimum defined amount of capital. The company must also have at least one-quarter of each share actually paid for. It can offer shares to the public and transfer them freely, subject to rulings by the Monopolies Commission. This share dealing is done through the stock exchange. The minimum membership of a public company is two, the maximum is defined by the number of shares. The company must have at least two directors and its accounts have to be disclosed to the public.

Private companies

Private companies can be distinguished because their names must contain the word 'Limited', abbreviated to 'Ltd'. The company is owned by shareholders but unlike the public company the shares cannot be offered to the public, unless this is done under agreements made with the Inland Revenue. In this case shares can be offered to company employees. Private companies do not have to disclose as much information in their accounts as public companies; they also only require one director but must have a company secretary.

2.2.5 Franchising

Franchising is an increasingly popular way of setting up a business. A person wanting to go into business will pay a sum of money to a large organisation for a franchise. The franchise is a licence to sell the franchisor's product or service; this form of business is very common in the fast-food industry. In return for the franchise fee and other ongoing payments the business receives the support of the franchisor including a brand name or label under which they are licensed to trade and full support for marketing and training. In return for this support the franchisee is exclusively bound to the franchisor for the period of the franchise. In addition to paying the franchise fee the franchisee has to find premises and equipment in the house style. Sole traders, partners and companies can operate as franchises.

The main features of operating a franchise are listed below.

1. There is limited outlay and risk due to the support of the franchisor; however, there may be problems if the support provided is not adequate.
2. The product has been tested, marketed and there is a well-known image, and in addition the franchisor may provide large-scale advertising.
3. The franchisee may not be allowed to manage the product in the best way for their own business because it is not in the franchisor's best interest.

2.3 Structure of organisations

In this section we will examine three ways in which organisations can be structured. We will then consider the factors that affect the structure adopted by the organisation. In order to provide a graphical representation of a company, organisation charts are used and so we will start the section by looking at how these are drawn and what they show.

2.3.1 Organisation charts

Organisation charts provide a picture of the organisation that can be readily understood by people internal and external to it. The most common form of organisation chart is the line chart, as shown by the manufacturing company chart in Figure 2.1.

The amount of detail shown on the chart will vary with the audience for which it has been prepared. It may show an overview of the company's departments or it may show the personnel deployment in a particular department or group.

The manufacturing company chart, Figure 2.1, shows the way in which the various functions within the company are addressed. For instance, it shows that the function of storekeeping falls within the remit of the Production Director. The chart also shows the number of people working in each area.

The chart shows the company hierarchy, line management authority, routes for consultation and the span of control of each manager. For a further example of this, let us consider the chart shown in Figure 2.2.

The 'span of control' defines the number of people reporting to a particular person. Obviously the span of control directly affects the task of each manager. However, it is important to recognise that the number of people involved is not the only factor that has to be considered when looking at the effective control of groups. One needs to consider the types of task involved, the ability of those carrying out the tasks, and the

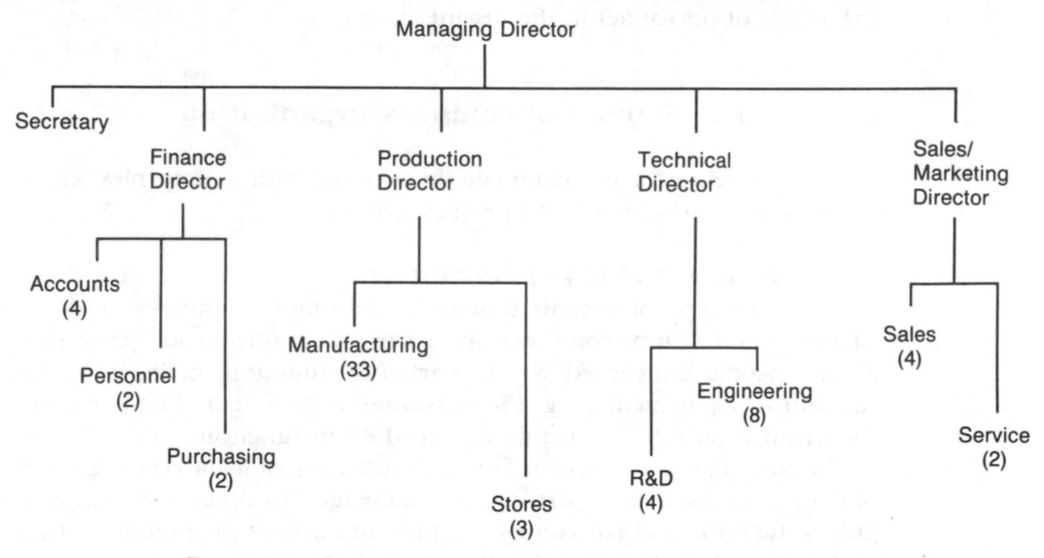

Figure 2.1 Organisation chart for a manufacturing company

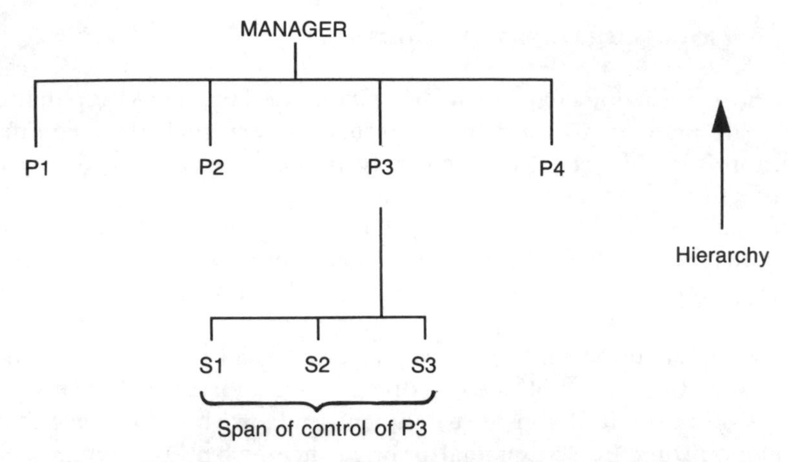

Figure 2.2 Organisation chart showing personnel deployment

complexity of the task. For example, an Engineering Manager with six project engineers reporting to him/her might be overwhelmed, whereas a Production Supervisor responsible for twenty operators working on a production line may cope very easily.

Authority and reporting is shown by the vertical lines. In Figure 2.2 employee S2 reports to employee P3 who in turn reports to the manager. If P2 wanted S2 to carry out a particular task then the correct route would be for P2 to consult with P3 and then for P3 to authorise the work. The horizontal lines indicate lines of consultation. The hierarchy indicates the levels of personnel in the organisation.

2.3.2 Methods of company organisation

There are three common methods of organising companies. These are functional, divisional and project structures.

Functional organisation

This type of organisation is the one most commonly used in smaller manufacturing companies. In this organisation you group together all the people concerned with a particular function, giving separate functional departments, e.g. the personnel department. The company shown in Figure 2.1 is organised according to function.

The advantages of this type of organisation are that it allows integration of people with similar expertise and knowledge. However, as a company grows, functional organisations can present barriers to communication with people in a different area because of the size of each area. As a company grows it may be necessary to carry out subdivision of the areas.

Divisional organisation

Many large companies organise themselves into divisions. In these cases each division will contain a number of functional areas, and consequently some functions will be found in more than one division. There are a number of reasons for using this approach, two of which are listed below.

Division by product or service: When a company has a range of very different products, or products addressing different markets, it may make more sense to use separate divisions to allow more appropriate manufacturing and marketing techniques. It would also allow more accurate costing of products by clearly differentiating the resources used for manufacturing each type of product. In addition it may be difficult for a company to make a name for itself in a new market segment if it has a reputation in a different one. For example, if a company has an established reputation as a producer of high-volume low-cost goods it might be difficult to persuade potential customers that it can also make a range of high-quality, high-cost goods even if this were for a different product range.

Geographical division: Geographical division makes organisation and administration easier when divisions are far apart or when operating to different laws and regulations. It also eases communication and transport problems. Geographical division usually means that a company operates on different sites, which may be in different counties or even in different countries. It is particularly common to find geographical division when the product must be close to the market, for example perishable goods, or with consultants who serve a particular locale and therefore need to be close to their clients.

Some companies mix both functional and divisional structure by retaining overall control of key functional areas such as personnel and finance. In this case each division would pay a contribution for the cost of these services.

Project organisation

When a company organises itself such that distinct groups of people address particular projects or pieces of work this is 'project organisation'. This type of organisation is commonly found in the construction and civil engineering industry. In manufacturing industry project organisation tends to be of a more temporary nature, with people being brought together to tackle specific jobs on secondment from their usual job in the function or division. However, this form of organisation is used widely in product development (see Section 3.6).

Although we can now look at a company organisation chart, as in Figure

2.1, and see how it arranges its functional areas we have not yet considered how people may be deployed within those areas. In Chapter 3 we will look at each functional area and the ways in which they can be organised; however, there are some general rules that relate to the methods of organising people, and we shall now look at these.

2.3.3 Deployment of personnel

Any organisation must be concerned with fitting into its structure people of varying skills, abilities and personalities, in order that they can be used to maximum effect within the organisational system. The organisation must also recognise that when it buys the skills, energy and abilities of people it must accept the individuality of people and their feelings of uniqueness and importance.

By dividing the workforce one can build a hierarchy in which people have other people reporting to them, and for whom they are responsible. The purpose of division is to enable people to manage the company more effectively and to give a sense of belonging to a discrete group.

Consider the structure shown in Figure 2.3. In this structure the Managing Director (MD) has a span of control of sixteen. With so many people to look after, the task of management becomes very difficult, if not impossible. The MD will not have sufficient time to carry out the planning of work and be able to spend the time necessary to motivate, organise, control, direct and provide feedback to all sixteen people, let alone undertake all the strategic tasks that this senior role requires.

The other disadvantage with this type of structure is that as there is little, or no, hierarchy the possibilities for promotion and advancement are very low. This will inevitably be a big demotivating factor for many people.

A structure that alleviates these problems and provides more opportunity for advancement is shown in Figure 2.4. This structure also has the advantage that for the people involved in the lower levels it gives a much more well-defined and clear route to authority. In the previous

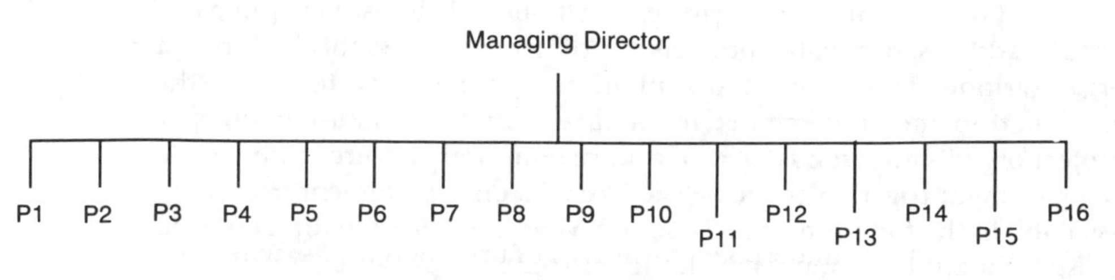

Figure 2.3 Single-level organisation structure

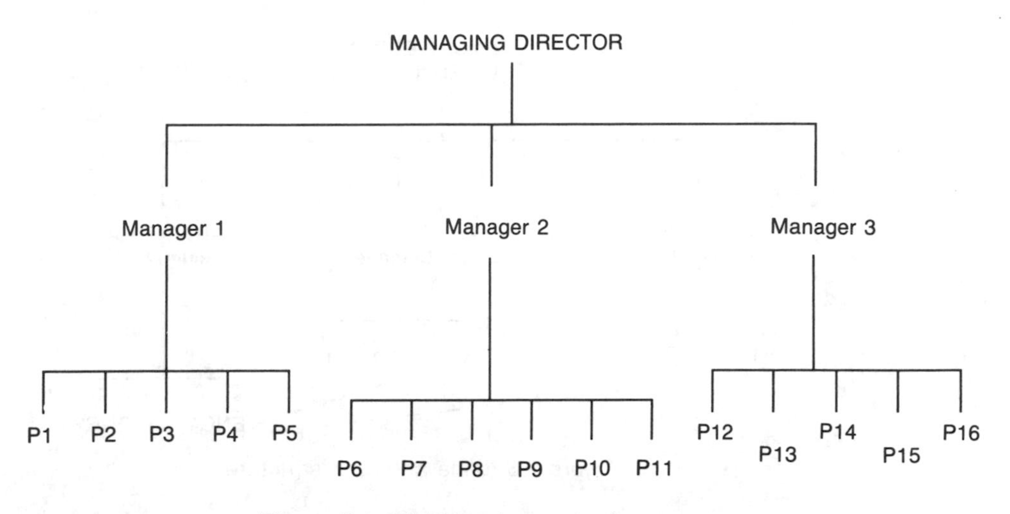

Figure 2.4 Multi-level organisation structure

structure it appears that the reporting route to the MD was very short and clear-cut. Unfortunately, in organisations that operate in this way the people involved tend to form their own hierarchy which, firstly, does not necessarily meet the needs of the organisation and, secondly, causes confusion because it is not generally known.

Hierarchical structures can also be problematic and care must always be taken to ensure that people at the bottom of the organisation are not left feeling remote from the company management. One problem with imposing a very formal and hierarchical structure is that it can make communication very difficult for people at these lower levels. In particular they can lose sight of the company's overall objectives. A consequence of this is that informal structures tend to be adopted by the workforce. This can be very good — it gives the workers a sense of belonging because the communication links across the company improve. However, it can be quite disadvantageous for the company. An informal structure implies that the management have no control of it. It can lead to an excessive burden being placed upon the people who comply with the requirements of this arrangement and are used as advisers or mentors, in place of the managers who should really have the responsibility for this.

The structure shown in Figure 2.4 is the commonly used *line structure*. In this structure everyone has a well-defined manager and there is clear definition of the routes of authority and communication. The advantage of this type of structure is that everyone knows where they stand, in terms of responsibility, authority and opportunities for advancement. The disadvantage is that this type of structure requires consultation across the organisation to be undertaken formally at fairly high levels, with the consequent loss of benefits this might have.

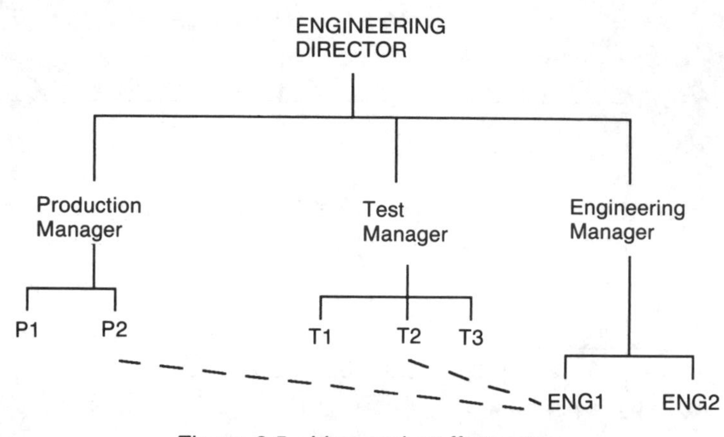

Figure 2.5 Line and staff structure

Another type of structure is 'staff structure', also known as 'matrix organisation'. In this type of structure there are no line managers — each manager has specialist functional responsibility and a pool of staff which can be drawn upon for a specific project. The problem with this type of structure is that people at lower levels in the organisation, and some at higher levels, find that working for more than one boss is both inefficient and demotivating. It also requires people in the organisation to be fairly adept at time management and diplomacy. For the people at higher levels it can be very frustrating not having total control of a particular resource.

The most common organisational structure is a compromise of line and staff, with the level of compromise being appropriate to the organisation involved. An example is shown in Figure 2.5.

In the line and staff structure shown in this figure, the production personnel, P1 and P2, have a line manager, the Production Manager. Similarly for the test personnel and the engineers. However, the engineers also have some technical responsibility which means that they work directly with the people in the other functional areas when required. For example, the engineer ENG1 has technical responsibility for a particular product which is to be manufactured by P2 and tested by T2. P2 will be authorised by the Production Manager to work on that product. P2 will also receive line management from the Production Manager. So the Production Manager will produce work schedules, be responsible for approving holidays and overtime, and have responsibility for any other line management tasks. However, during the period of working on the product P2 will work closely with ENG1, perhaps receiving detailed work schedules or discussing manufacture problems.

This line and staff structure can be very effective. It ensures that the managers of each functional area are able to manage their areas and they know what is happening in terms of work schedules. They are in control. However, for the technical input there is a quick and easy route directly to the engineers. The only real disadvantage of this type of structure is for the engineer, who may have considerable technical responsibility and may have responsibility for meeting project deadlines. However, the engineer has no direct authority over the resources needed to do the job. This type of organisation is very common for companies operating 'project offices', and the project engineers need very good interpersonal skills in order to meet their objectives.

2.3.4 Factors that affect company organisation

We have considered some of the organisation structures that are widely used. We will now consider the factors that affect the way in which a company is organised, and the people deployed within it. These are:

1. Product and manufacturing system
2. Functions and expertise
3. Size
4. Responsibility and authority
5. Decentralised/centralised control
6. Communication

Product and manufacturing system

The type of product and the way in which it is manufactured will have the greatest effect on the organisation of the business. In particular it will dictate the emphasis put on areas such as research, sales, purchasing, and product development. However, the manufacturing system will also dictate the way in which people are deployed within the manufacturing area. For example, it will define whether they work in groups on a complete product, or individually on discrete parts.

The state of development of the company's products will also affect the organisation. If the company has a stable product range and plans very little development, then it lends itself to the functional organisation. However, if the product range is changing considerably and the company requires flexibility to be able to react to that change, then a matrix organisation may be more appropriate.

Functions and expertise

Depending upon the nature of the business in which the company is involved, there will be various functions that have to be carried out, for example manufacturing, testing, exporting. The company must have expertise in these functional areas and the organisation of the company will reflect the distribution of that expertise. In small companies particularly this can lead to some unexpected combinations of functions due to limited personnel with appropriate expertise. For example, personnel and finance may be combined because the company accountant has some previous personnel experience.

Size

The size of the company will have great influence on the way in which it is organised because of the need to establish a level where the number of people being managed is low enough to allow effective management and delegation, but high enough to make sure that there are sufficient people 'doing' rather than 'organising'. If there are too many managers the company will have a large financial burden that has to be charged against profits, and which is not directly contributing to those profits. If there are too few managers it is likely that they are spending the majority of their time scheduling or carrying out day-to-day management activities. The company suffers because no one has the time to look forward and consider company development.

Responsibility and authority

The amount of responsibility and authority that an organisation gives to its managers will affect the structure because of the amount of resource that can be committed to any particular area. For example, the managers will have to consider how many people they are prepared to commit to one area and whether line management, functional or technical responsibility can be given to a particular person.

Centralisation and decentralisation

Whether or not an organisation operates in a centralised or decentralised manner is largely dependent upon the style of the Chief Executive or Managing Director. Both types of organistion have benefits which should be considered. Centralisation should produce strong leadership and direction, and defined strategies. However, decentralisation supports the worker, therefore morale should be higher, communications should be improved and decision-making should be brought closer to the situation it is addressing.

Communication

Good communication is essential in any company and the lines of communication — both essential and desired — should be considered, as an inappropriate structure can easily lead to poor communication which acts to the detriment of the company as a whole. An hierarchical structure, for example, can inhibit communication between people at very different levels — this may be unacceptable in a company where there is a lot of technical knowledge at the 'top' which has to be accessible by people at the 'bottom'.

2.4 Corporate objectives

2.4.1 The need for corporate objectives

It is usual for newly qualified engineers to join organisations and start practising the skills they have learnt. As their competence advances they are able to make gainful use of assistance. Before long, the successful engineer earns managerial responsibility and is provided with constant assistance from subordinates. This process continues with senior managers often having several layers of management beneath them.

With such large numbers of people, simply deciding what everybody should be doing can be a major problem. There are always many more constructive tasks available than it is practical to complete. Consequently, successfully managing such numbers involves choosing the most effective options for each person. In short the professional manager will want to know, not only that everybody is doing something constructive, but also that it is the most worthwhile task that they could be doing.

It is very much in the interests of the organisation to ensure this. People engaged in less than optimal tasks represent time wasted since they could contribute more effectively by doing something else. For instance, a design engineer clearly knows that any material or labour savings that do not compromise performance, are desirable. However, unless the engineer knows how much the organisation stands to benefit from the project, no rational choice can be made between this work and any alternatives.

It is also of benefit to the individual to be certain that the work in progress is the most beneficial alternative for the company. Such knowledge is very motivational since it makes clear the value of the work. We are all motivated by doing useful work and de-motivated by pointless labour. In Section 6.3.1, we explain how objectives are needed from a personal, psychological point of view. People need to know what they are supposed to be doing in order to be happy; one cannot be fulfilled if one has no direction to one's efforts. Having an important objective for which to aim, answers this need directly and understanding how our objective fits into and assists the aims of our organisation is part of being convinced that what we are doing is worthwhile. Successful organisations therefore ensure that everyone is not only engaged in constructive tasks but also can see the benefit of the contribution they are making and so are motivated to achieve their aims.

How can these important aims be achieved? In order for everyone to know what tasks are best and for all managers to be able correctly to identify optimal projects for those who report to them, everybody must have access to some underpinning theme to which all efforts are directed. Such an aim is clearly going to be of great importance to the organisation and its selection must be a matter for senior management since if it is chosen poorly, the whole organisation will be directing efforts into activities that provide less than the best return. Once a good underpinning theme has been chosen, a method must be identified by which everybody in the organisation can have access to, and be guided by, this central aim.

We will now introduce the two management concepts that engineering organisations use to achieve this organisation-wide co-ordination of effort. The first step is choosing the underpinning theme, and this is dealt with in the following section. The next section explains how the efforts of every member of the organisation can be guided by the central aim.

2.4.2 The mission statement

Imagine a company, perhaps a small pump manufacturer. The business employs, say, thirty people. What is the purpose of the business? Does it exist to push back the frontiers of pump design? Probably not, although the company might conceivably be engaged in doing so. Does it exist to provide wealth for its employees? Certainly the owners of the business may well be anxious to provide fair pay for their workforce but this is unlikely to be the reason the company was founded. Does the company exist to make money? Many companies produce less profit than would be provided by investing the assets elsewhere, but still they carry on trading; money alone is therefore not enough to explain a company's existence.

There may be any number of different reasons for which the company exists and each reason brings its own strategy. For example, if the pump manufacturer does actually exist to lead the world in pump technology, then much research and development would be appropriate. If, on the other hand, the purpose of the company is long-term survival, then perhaps moderate research and development coupled with the establishment of alternative business lines would be best. If the aim is to meet all the pump needs of a particular market sector, then forging special links with that market and excluding competitors from it would be best.

From this demonstration of how the purpose of the company affects its management we can conclude that before any rational business decisions can be made, the company must decide once and for all — at the top — what it is trying to achieve. You cannot steer a course without knowing where you are heading.

Well-managed companies have such a statement, made at the highest level, explaining what they are trying to achieve — this is known as a 'mission statement'. It endeavours to be a lasting summary of the whole spirit of the company and to explain succinctly what the company is about. It is this statement that sets the direction and thrust of the whole organisation. The statement is an objective and as such should have the properties of a well-written objective such as are described in Section 6.3, which explains the mechanics of good objective writing. It details why they must have these properties and it should be read before preparing any objectives or mission statements oneself.

A salutory lesson in the importance of choosing the right mission statement may be taken from the example of the American railway companies early this century. One such company might have had a mission statement such as:

> To provide a quarter of all the railway miles sold in the USA over the next decade.

This statement seems a perfectly sensible goal — the organisation is engaged in making money from the sale of rail journeys. We cannot tell from such a statement whether it is a realistic target for the particular company but it certainly does sound like a worthy goal for the organisation to head for. In fact many such companies at the time used exactly such mission statements and yet they ultimately went out of business. In the first years of this century, at a place called Kittyhawk in the USA, two brothers spent their days making flying machines. A few years later one of them actually flew in one of these machines and history was made as the Wright brothers became the first humans to fly in a machine weighing more than the air it displaced. What relevance had this to the mission statement of the rail companies? In their view, it had none and the rail companies turned a blind eye towards this new development. However, after only thirty years, mankind had put the aeroplane to all manner of uses and cargo flights were routine; the new machines carried freight with unchallenged speed. The railways were relegated to transporting goods to ports where they were loaded onto ships to cross the Atlantic but the advent of transatlantic flight took even this from them. The error of the rail companies was very simple: they chose the wrong mission statement. Imagine how different things would have been if the rail companies had chosen the following mission statement:

> To provide a quarter of all the transport miles sold in the USA over the next decade.

There is only one word that has been changed in this new mission statement: 'railway' has become 'transport'. A rail company with this mission statement would have turned a keen eye towards developing new transport technologies, and such a company would have been the first to add air transport to its portfolio of products. A rail company that was successful in this goal would have ended up owning the airlines instead of going out of business because of them — all because one word was incorrectly chosen. To use an archaeological metaphor, you will never find buried treasure if you are digging in the wrong place.

2.4.3 Managing by objectives

Once a mission statement has been selected and agreed upon, the management of the organisation must communicate it to all members within the organisation. This, however, means far more than simply making the words of the statement available. To arrive at tasks for all members of the organisation that contribute to this central task clearly involves a great deal of subdivision and is far more than one person, or even a small team, is likely to be able to do. The process starts with an examination of the implications of the mission statement and, from these, subordinate tasks can be identified. Each of these tasks may itself have component tasks and may be further subdivided. Consider the example of the pump manufacturer, whose mission statement might be of the form:

> To return to a profit of 14 per cent of turnover
> by supplying centrifugal water pumps to
> European customers over the next calendar year.

From this central aim the senior management can use their technical knowledge and skills to produce supporting tasks. By considering the sales price and the profit margin for all the different models of pump, target numbers for each model to be sold can be prepared. The organisation will then have to manufacture the pumps, and they will have to be sold and shipped. Using their knowledge of the business and the company, the directors can now start to make objectives for each department. Indeed, even the very presence of the departments is governed by the mission statement — the manufacturing department of the pump manufacturer exists to meet the manufacturing needs of the mission statement, not the other way round. The sales department might be given the objective of logging fifty orders a month. The manufacturing department might be given the objective of meeting every order within twenty-two days, with an average cost of manufacture of less than 65 per cent sales price. And so on for the other departments. All these departmental objectives are chosen to support the mission statement. If the mission had been different, the departmental objectives would also have been different.

Once this first level of objectives has been agreed, individuals are made responsible for achieving them. Normally in a company it will be the directors who take on objectives at this level.

Once departmental objectives have been set, the objectives for the groups of individuals within the departments can be set. In the example above the manufacturing manager might know that the current material and labour cost is too high to meet the departmental objective and so might set up a group of engineers to solve the problem. The importance of the departmental objectives can now be seen. For instance the manufacturing department may feel that the amount of subcontract labour is too high or that commonality of parts between models should be increased. The departmental objectives are used to decide which tasks are actually important, and only those which support the departmental objectives, and therefore the mission statement, are selected.

The number of times objectives are subdivided depends only on the size of the organisation. The important thing is that by this progressive subdivision of the main task, the whole organisation is directed to acheive the same overall goal. No effort is wasted in undesirable directions and no contradictory programmes of activity occur. The objective setting extends as far as necessary throughout the organisation. Through the use of appraisals, it is common to find management by objectives extending to every single member of staff in the organisation.

Finally, we have seen the need for objectives, we have seen the great effect mission statements can have on organisations and we have seen how objectives may be used to manage the activities of everybody in an organisation. People tend to think of objectives, both at the corporate and personal level, as being things that don't change. This is wrong — times do change and business opportunities change with them. For this reason the senior managers of the organisation should meet regularly to review progress towards each of their objectives and to re-assess the value and sensibility of the mission statement and its consequent objectives.

Corporate objectives are not immovable and they need maintaining to keep them appropriate. It is clearly not advantageous for organisations to stick doggedly to objectives which have become outdated, and so miss opportunities. Conversely there is no benefit in constantly changing objectives in response to every external factor. The most testing assessment of a management team is how well they maintain their corporate objectives in order that they steer the organisation between these two pitfalls.

2.5 Case study — JP Engineering

JP Engineering is a subcontract machining company based in Oxfordshire. Its business is to produce parts to other people's drawings using a variety of machine tools, including Computer Numerical Control (CNC). The business was formed six years ago by John and Phil. At that time they were working together for a company but decided that they wanted to work for themselves. They are both fully skilled machine operators, having served apprenticeships.

When they decided to go into business John and Phil wanted to stay with an area that they knew well, i.e. machining. However, their goal was to become the best subcontract machining company in the region, in terms of quality and reliability — they did not necessarily want to be the biggest. JP Engineering was established as a partnership, with a partnership agreement signed by both men.

One of the first jobs for the partners was to raise capital to start the business and in order to do this they had to produce a business plan describing how they thought the company would trade and grow. Most importantly the plan gave a projection of expected incomes and outgoings. That first plan made a number of bold statements. For the first six months of trading the business would employ only John and Phil; it would then increase by one employee. The partners projected that after twelve months the business would be employing a total of four or five people. They planned the purchase of their first CNC machine after twenty-four months and made an estimate of turnover for the first year. With their business plan the partners were able to secure funding, although some of this was secured on their homes.

During those early days John and Phil worked as machinists, building up a client base of local companies. The job of running the company was done at weekends or in the evening. However, their goals were realised and the business went well. In the first six months they employed a further two machinists, an inspector and a part-time secretary; they also bought a CNC machine. The first year's turnover was two and a half times greater than their target.

Over the six years the business has continued to grow and has survived a recession, although this forced the partners to make four people redundant. The business now employs fifteen people in total, has moved to premises four times greater, and turnover has increased almost five-fold over the first-year level.

The changes in the business operation have been subtle. John and Phil are still working many hours but now 75 per cent of their time is spent running the business. The management tasks, such as scheduling the machine shop, are divided evenly between the partners. There has also been an increase in non-production staff, since the company's growth has necessitated the employment of someone to be responsible for purchasing and sales. The partners' goal to be best has led them down the road of BS5750 Quality Systems.

The business is still run as a partnership although this is regularly reviewed, and as soon as it becomes feasible and beneficial to change status to a limited company, this will be done. The small, local client base is maintained and the partners aim to balance the work between customers to avoid the risks of having too much reliance on a single source. The relationship with clients is informal and the partners feel this is important to their business operation. It is also a necessity when trying to match the needs of the client with the business objectives, trading delivery dates so that quality of the product is not jeopardised.

At this time there are no plans for further expansion of the premises but the partners do plan to update machinery which will lead to improvements in productivity and quality when linked to plans to introduce shift-working and to increase the number of machinists by one or two. ■

2.6 Summary of Chapter 2

In this chapter we have seen that in the United Kingdom there are five ways to establish a business legally. The method chosen will have implications for the way the organisation can transact business, the way in which it can be financed, and the amount of information that has to be publicly divulged. The five forms of business are:

Sole trader
Partnership
Co-operative
Public company
Private company

In Section 2.3 we looked at the different ways in which businesses can be organised and we saw that there are a number of factors that affect the final choice. The greatest influence comes from the type of product and manufacturing system used.

In Section 2.4 we saw the benefits that can be achieved by companies who use corporate objectives to manage their business. The various points are summarised below.

1. A mission statement is a statement of the purpose of the organisation.
2. Corporate objectives are derived from the mission statement and are used to ensure that everyone in the company works to meet the company's goal.
3. The same rules of good objective preparation apply to corporate objectives as apply to personal objectives.
4. Once the corporate objectives have been set, other objectives for each department or group may be made from them. These are divided in turn until every member of the organisation has their own personal objectives.
5. Corporate objectives are the only way of ensuring that everybody in the organisation is undertaking tasks constructive to the organisation as a whole.

6. Corporate objectives are motivational in that they permit each individual in the organisation to see how their own work areas contribute to the success of the organisation.

Finally in Section 2.5 we looked at an engineering company, at the way in which it was established and at the changes that have taken place as it has grown.

2.7 Revision questions

1. Write down a mission statement for the following:

(i) A private company you know something of.
(ii) The American space agency NASA.
(iii) The department responsible for defence.
(iv) A public transport system.
(v) The next five years of your life.

2. Write your own summary of why corporate objectives are important. Include the benefits they bring and any disadvantges or extra work you can think of.

3. Investigate your organisation to discover its mission statement. How does this mission statement relate to you? Write about the effects of the mission statement, listing areas where the behaviour of the institution seems to agree with the mission statement and those where it disagrees. (If your institution does not have a mission statement compose one yourself.)

4. Consider each of the summary points relating to corporate objectives in Section 2.6 and choose the most important, stating why you have chosen it.

5. What are the advantages of operating as a partnership rather than as a sole trader?

6. As an engineer, what possibilities exist for you to operate a franchise? What are the disadvantages of franchising rather than developing your own product?

7. What factors affect the way a company is organised? How might the emphasis on each factor change as a company grows?

8. Compare the operation of a partnership and a limited company.

2.8 Further reading

Argenti, J. (1986), *Systematic Corporate Planning*, Wokingham: Van Nostrand Reinhold (UK) Co. Ltd (ISBN 0-442-30741-1).

A comprehensive book about corporate planning, explaining why it is needed and how it may be achieved. The book includes chapters on corporate objectives and strategy. The second half of the book deals with planning and executing corporate plans.

Dale, E. and Michelon, L.C. (1986), *Modern Management Method*, Harmondsworth: Penguin Books (ISBN 0-14-009111-4).

A readable and thorough guide to many issues in management including managing and communicating by objectives, managerial decision-making, decision-making and payoff tables and critical path analysis. Based on the wide experience of the two famous management consultant authors.

Handy, C.B. (1985), *Understanding Organizations*, 3rd ed., London: Penguin Business.

Kadar, A., Hoyle, K. and Whitehead, G. (1987), *Business Law Made Simple*, London: Made Simple Books (Heinemann).

Ohmae, K. (1983), *The Mind of the Strategist*, New York: Penguin Group (ISBN 0-14-00-9128-9).

In his native Japan, Kenichi Ohmae has been described as 'Mr Strategy'. His abilities as an adviser to top management have earned him considerable respect as an authority on his subject. In this book Ohmae places strategic thinking at the centre of success. The book is in three parts: in Part 1 the art of strategic thinking is expounded; in Part 2 the techniques for building successful strategies are examined; and in Part 3 the realities of modern strategic planning are explained.

Chapter 3
Functions of organisations

Overview

In order to design a device an engineer must have an understanding of the use to which it will be put. The requirement for the device gives the context. Before we can manage people or resources we need to know in what context we are working and understand why management is required. Knowing how businesses come together and operate gives that context — unless we have an understanding of the whole, it makes it very hard to put the detail in perspective.

In Chapter 2 we looked at the way in which companies can be formed and then structured to meet objectives. We saw that in order to meet objectives various functions had to be performed. In this chapter we examine those functions of business considering how they can be organised and managed. We also present some case studies that show how these functions interact in real companies and the role that engineers can have.

3.1 Introduction

In any manufacturing organisation there will be various functions that have to be carried out in order for the business to operate. To illustrate this we will consider the example of the manufacture of electric kettles.

We will start with the assumption that research has been done to ensure that the product is viable and that the company is established and has adequate money to finance the operation. Bearing this in mind, we will consider how the business will operate.

In order to make a profitable business out of manufacturing kettles we will have to consider a number of important areas; the minimum that we would have to consider is shown in Figure 3.1. We need to procure

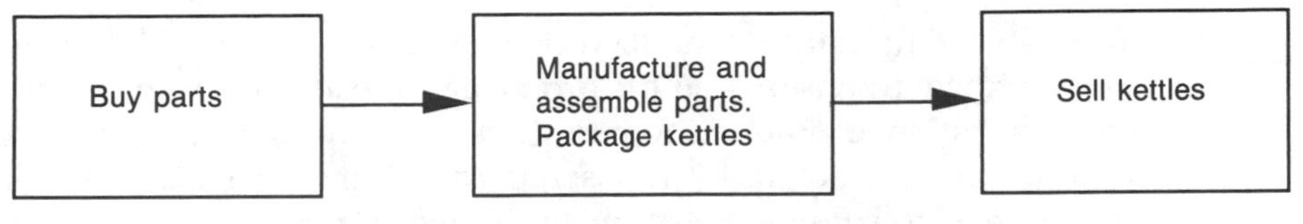

Figure 3.1 The kettle business

(buy) some raw materials, we need to transform those raw materials into a saleable product, and then we need to sell the product at a price high enough to provide a profit. So we have three main functions of business: buying, making and selling.

As part of the business we have identified that parts will have to be bought in and kettles will have to be manufactured. Therefore there needs to be somewhere to keep all these items that can be used both as a means of safe storage and as an aid to controlling the receipt and use of parts. This is usually achieved by having a stores system, which is often part of the purchasing function's remit.

By saying that we can manufacture the kettle, we have assumed that we know what it looks like, that the parts that go into it have been defined and that manufacturing processes have been specified. All of these activities are carried out by some form of product development function which will include design and may include research. If expansion, growth and development of the company are to be considered then how the company and its products will develop over a period of time in order to achieve a desired market share must be determined. Companies can achieve this by using research which is then allied to design activity. In the kettle business the research activity might involve trying to develop a new type of heat exchanger mechanism; the design activity would then take the new mechanism and consider the design for manufacturability, producing drawings and parts lists, and also ensuring compatibility with other parts in the kettle.

In order to sustain itself the business must be profitable, and for this to happen we need to ensure that the cost of buying the raw materials and making a kettle is less than the amount for which we can sell the finished kettle. This introduces two further functions: finance and marketing. The finance function will be involved in calculating the costs to make the kettles, including the costs of buying the raw materials. Amongst many other activities, it will also be involved in making sure that there is enough cash to keep the business operating whilst waiting for customers to pay their bills. Marketing means ensuring that a match is found between customers' needs and the product's attributes — at a profit — and this function will be involved in finding out what the customers want and what the competition is offering.

This leads us to conclude that six areas have to be addressed in order to set up and operate a manufacturing company. However, there is also a requirement to ensure that the products made are always of a sufficient standard to meet the customers' expectations and that they meet appropriate product and safety standards. Companies can address this area by having some form of quality function which is used as part of the management of the other functions in order to ensure that standards and customer requirements are met. In the kettle business 'quality' would involve all aspects of the business including ensuring that the design process addresses the appropriate safety standards and making sure that there are effective measures to ensure that people in the company know what the standards are and that they have the ability to meet them.

Finally, people will be needed to work in the company in order to carry out all the tasks required. Their deployment and welfare, initial recruitment and selection, as well as payment and conditions of employment, will have to be considered, and these areas are addressed by the personnel function.

So we can conclude that in order for a company to manufacture electric kettles, or indeed any product, it needs to address several different areas. These areas (functions) are as follows:

1. The buying and safe storage of materials (purchasing and stores).
2. The manufacture of products (operations).
3. The selling of products (sales).
4. The pricing of the product and its competition (marketing).
5. The financing of the business and the cost of manufacture (finance).
6. The design and development of the product for manufacture (product development).
7. Consistent product standards (quality).
8. The people employed by the business (personnel).

In this chapter we shall consider each of the functions listed above. For each of them we will look at the way in which they might be organised in an engineering company and consider some of the activities that have to be carried out. Towards the end of the chapter we will look at some examples of the way in which engineers are employed in engineering companies and how they interact with each of the functions described.

Because this book is about engineers and the management aspects that they will meet at work, this section is broad, although there is greater emphasis on the three functions that engineers will be most involved with — product development, manufacturing and quality. The 'management' topics of personnel and finance are dealt with briefly in this chapter in order to give an idea of how the functions are managed in an organisation. However, because this is a management book the techniques and theories relating to these two areas are covered in much more detail in Chapters 4 and 7.

The learning objectives for this chapter are as follows:

1. To understand the way in which an engineering company operates.
2. To consider the way in which engineers are required to interact with all departments in a company.
3. To consider how the 'management' topics of personnel and finance relate to the engineering industry.
4. To look at some examples of how engineers work in engineering companies.

3.2 Purchasing and Stores

A company will have to purchase goods and services to feed all aspects of its operation, from special parts required for research to paper goods for the wash-room. In all purchasing activities it is necessary to clarify what is required, obtain supplies and ensure that any business objectives are met. Meeting business objectives may relate to quality requirements or budget controls; it will certainly include making sure that the customer for the good or service gets what is required at the time that it is required.

The purchasing function is complicated by the fact that within a company there will be a diversity of goods and services to be purchased, and there will be different types of customers requiring different levels of service from the purchasing function. Companies therefore have to address the function in a way that is appropriate to their business. In a very large organisation it may be appropriate to have a large purchasing department which handles all requirements and has a number of buyers, each with responsibility for a particular type of good or service. In a small company it may be more appropriate to allow different departments to do their own purchasing. The organisation of purchasing will also be affected by the quality system, if the company has one; the controls imposed by a parent company, if a subsidiary; and the business relationship with suppliers, for example if the company has a direct computer link to its suppliers.

In this section we will look at the job of the purchasing function, the way in which purchasing can be organised, and we will consider some of the activities carried out by a purchasing department. In addition we shall look at the operation of Stores, which has a responsibility for control and preservation of goods for manufacture and which is a part of the purchasing function in many companies.

3.2.1 The role of the purchasing function

In an engineering company the majority of purchases will be used in some way to form the final product — this can lead to purchased goods accounting for as much as half of the total expenditure. Purchasing might involve buying parts for production, negotiating contracts for temporary staff, assessing suppliers or negotiating discounts for bulk purchases. It *always* involves having to ensure the supply of goods of the right quality, in the right quantity, at the right time and at the best price for the level of service provided.

The purchasing function, however, is much more than just buying — it exists to meet the needs of its customers inside the organisation whether in design, manufacturing, personnel or finance. The needs of such a disparate group will be very different and the purchasing function must address them in the most appropriate and effective way.

3.2.2 Organisation of the purchasing function

Often in large companies a centralised purchasing department is established which has control over all aspects of procurement, stock-holding and supply (the department may also be known as 'materials management'). In smaller companies these activities are more likely to be distributed, with different departments having responsibility for buying the goods and services that they require. However, there should still be an overall central purchasing policy which, as a minimum, allows cash-flow control and will define the procedures by which suppliers will be selected and then paid.

When a company employs many buyers in a centralised department they may be given particular areas to cover. For example, there may be buyers with specific responsibilities for procurement of food, furniture, stationery, production parts and so on. One advantage of this type of operation is that each buyer can develop very specialised knowledge of the market and provide even greater economies. Other advantages of centralisation include being able to make bulk purchases with consequent price-saving, minimising the cost of placing orders and avoiding duplication of effort. Against this, if decentralised purchasing is used instead, it can mean that a buyer is able to react more quickly to a problem and that there is likely to be a better understanding of the local needs.

Figure 3.2 shows how the purchasing function might be organised in a medium-sized engineering company.

Some companies allow certain departments or groups the freedom to place purchase orders with a minimum of centralised control — this is found typically in research and product development activities where

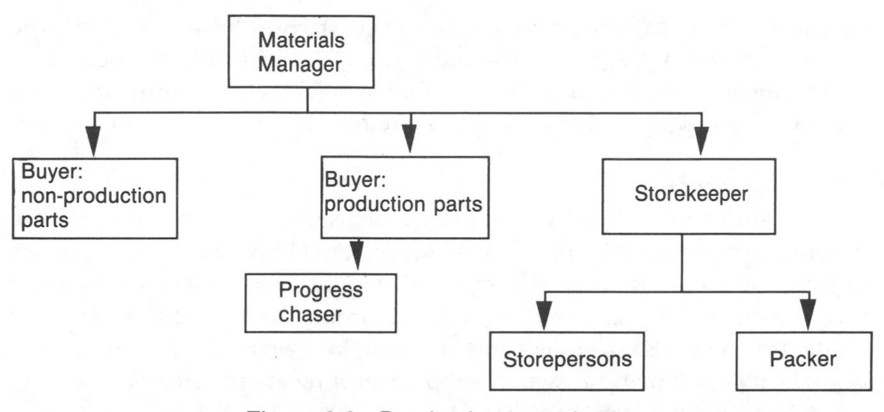

Figure 3.2 Purchasing organisation

it is important that the designers and engineers can communicate with suppliers easily and when they require goods on an *ad hoc* and quick turn-around basis. This can work very effectively particularly if the alternative of using a central buyer would delay the orders or would be very bureaucratic. However, these problems do not have to occur if the purchasing function is properly organised and there are appropriate ways of dealing with different sorts of orders. There is a very large potential problem with this *ad hoc* system which relates to the level of control exerted by central management. This has therefore to be accompanied by strict budget controls and a formalisation of purchasing procedures as a device moves from research prototyping to manufacture. There may also be a need to control the suppliers that can be used because of quality requirements or their ability to supply to production demands.

3.2.3 Activities in purchasing

The most important activities that will be undertaken within the purchasing function will be establishing purchasing policy, sourcing, buying and stock control. In addition the purchasing function must work with other departments to ensure that their needs are understood and met, and that quality is maintained.

Establishing purchasing policy

A purchasing policy will define the rules by which the purchasing function will operate. It may define the criteria on which suppliers are selected, or the geographical location of suppliers. It may define cash limits or specify the limits on the amount of business that can be put to any one supplier.

Purchasing policies will be developed over a period of time but one of the main management issues involved in purchasing will be the implementation of the policy and monitoring to check effectiveness. This will be most important, and most difficult to achieve, in companies that operate a decentralised purchasing system.

Sourcing

Sourcing is about finding suppliers that can meet the company requirements for goods and services. Sourcing will invariably require close consultation with the department requiring the good or service, in order that the buyer fully understands the requirements. It will then involve contacting potential suppliers and collating information that can be used to make a decision about which supplier best meets the needs. Sourcing can take up a lot of time and effort and it can be more cost-effective to let a specialist in a particular area carry out sourcing. For example, in the software field it might be more economically viable to have a software consultant carry out sourcing rather than someone who has only a general knowledge of the field and the terms that are likely to apply.

Many companies also carry out second sourcing, in which an alternative supplier for a particular good or service is found. This ensures that there is some security of supply for the company in the event of a primary supplier not being able to meet the requirements of the company and that a competitive price is secured at all times.

The criteria for selection of suppliers should be defined in terms of quality, location, delivery reliability, financial stability (of suppliers) and any other factors that are appropriate to the company objectives. In companies operating quality systems there will also be a need to ensure that any supplier can provide assurance that the required quality standards can be met. This may involve the purchasing department in carrying out supplier appraisal visits and will always involve monitoring of supplier performance in terms of goods supplied, delivery and quality of after-sales service. This is described in more detail in Section 3.7.

Buying

There are several ways of buying and the methods chosen will depend on the company and on the particular types of parts, items or services being bought. The criteria used for buying should be defined in the purchasing policy. For companies operating a fairly stable production system, buying production parts will be very much a case of buying from a list indicating the parts required and where they can be obtained from. More specialist buying will require more detailed knowledge about the actual goods required and an understanding of the commercial implications for any contract set up.

Many companies hold stock of frequently used items in Stores. For these items buying new stock is done when a specified re-order level is reached.

The quantity of new parts will also be specified — this will be the re-order quantity and will already have been calculated, taking into account purchase price, discounts and stockholding costs. Companies buying for stock may use a *call-off* schedule with a supplier. In this type of buying one order is placed and it covers all the parts that will be required for a period, although they may not all be required at the same time — the company then 'calls off' the parts at the times when it does want them. The advantage to the company is that administration is reduced significantly and yet the parts are delivered as needed, so there is no need to hold stock of one large order.

A vital aspect of purchasing for manufacturing is the ability to predict *lead-time*. The lead-time is the time required to get the goods into Stores and available for manufacture, following the placing of an order. The lead-time offered by suppliers will affect the lead-time that a company is able to offer its customers and will affect the amount of stock that a company must hold. The effect on stockholding can be quite significant because of the company's need to keep enough parts to cover fluctuations in demand and any problems with supply. Because of the importance of lead-time many companies will specify lead-times for goods and the ability to meet these lead-times should be monitored as part of any supplier assessment scheme.

It is important, particularly for production parts, that the buyers have good communication links with the manufacturing department so that they can act quickly in cases of urgency, or advise in cases of any difficulties with supply.

Part of the buying role includes progress chasing. This is following up an order to ensure that the goods are going to be delivered on time and in the right quantity. Some large companies will employ people whose only job is to keep in touch with suppliers' progress on orders.

3.2.4 Stores and stock control

The primary function of a stock control system is to ensure that the right parts are available for manufacture at the right time. It therefore involves identifying requirements, buying, and making the parts available for manufacture. The system will also have objectives relating to cost, quality and levels of stockholding.

The heart of most stock control systems is the 'Stores', where stock is received, held and issued, and where the records of these transactions are generated and maintained. It should be noted that with the use of more sophisticated production systems, such as 'Just-in-Time', there is a move away from having a Stores in the company and a move to using the supplier as a Store. (This affects a relatively small number of companies at present and will not be dealt with here; however, further reading is indicated at the end of the chapter.)

There are three types of stock that are commonly defined:

1. Raw materials, parts and components.
2. Work in progress.
3. Finished goods.

Raw materials, parts and components are all the materials that are to be used for manufacture of the product, in their original form. This might include items such as sheet steel, parts made to the company's design, and proprietary parts, i.e. those made to someone else's specification. These parts are usually bought according to a planned purchase scheme and are kept in the Stores until required for processing. Assemblies that have been manufactured and then returned to Stores prior to further processing will also be included in this category.

Work in progress is the material that is on the shop-floor either being processed or awaiting processing. This will typically include parts that have been released from Stores to make assemblies, and part-built assemblies.

Finished goods are those that have been completely processed and are in a state fit for sale; they are then stored in this state, until despatch, in a finished goods store (FGS).

The purpose of the Stores is to provide a safe and secure place for holding goods and to make those goods available when required by the appropriate company departments. Security is required in order to prevent loss due to pilferage or misuse. The Stores should also preserve the quality of the goods by providing adequate packaging, space and handling. The layout of the Stores must consider the types of parts that are to be stored and the access that has to be provided; it should also consider the way in which parts can be stored, for example whether or not they can be stacked. In addition consideration must be given to any packaging or special storage requirements.

Considering that the parts must be easily accessible in Stores some companies design the layout based on part number, on groups or families of parts, the frequency with which access is required, or on the size of the parts. Independent of how the Stores are laid out it is essential that all parts within the Stores are identifiable and that their inspection status is clear, i.e. whether or not they are available for manufacture or sale.

Some Stores have to have specially defined areas, sometimes with restricted access. These are usually called *bonded* and *quarantine* areas. The bonded area is used to hold parts that have high value, or that are required for special jobs. The quarantine area is the place where the goods that cannot be used are held pending decisions being made on their disposition — this would include parts that had failed inspection.

Depending upon the company and on the types of parts used, there may also be other special areas defined, such as static-free for electronic components and a lockable store for flammable materials.

The Storekeeper is the person who has responsibility for maintaining the Stores, and the goods held there, in good order. The Storekeeper will also be responsible for maintaining stock data accurately, and for carrying out the four important tasks of the Stores function, which are goods-in, goods-out, traceability and stock-taking. These are considered below.

Goods-in

The goods-in operation is carried out to ensure that parts received from suppliers are correct, in the right quantity, and made available to manufacturing as soon as possible. The operation involves checking goods against the purchase order when they are received. Typically this will involve signing a delivery note to confirm receipt and visually checking packaging for signs of damage. Further checking will usually be done at a fairly low level, confirming that the parts are those that were ordered, that the quantities are correct and that there is no visible damage. Further checking, in the form of detailed inspection, will be done after this initial check.

When the goods have been cleared for use they will then be booked into Stores. This is done by updating a stock record, which may be on computer, and then physically locating the goods in Stores. It may also involve some repackaging in order to preserve the goods whilst in Stores.

The Storekeeper will be responsible for ensuring that the purchasing and production departments are informed of the receipt of the goods. This assists production scheduling, prevents further order progressing and gives clearance for payment to the supplier. It may also fall to the Storekeeper to be responsible for dealing with reject and return goods. These can arise from faulty or incorrect goods being delivered, goods having been found to be faulty at a later stage in the production process, or simply delivered at the wrong time. In all cases the stock records will have to be adjusted.

Goods-out

The goods-out task is the process of sending parts out of stores for manufacture ('issuing') or for despatch to suppliers or customers. Goods are issued from Stores either by ***requisition*** or by ***kit issue***.

Requisition: This is used when a person comes to the Stores with a specific requirement. Usually the person will have to fill out a form to say what goods are needed, and in what quantity, also indicating what product the goods are to be used on. The stock record is debited as the parts are taken away. Some companies allow the requisition of free-issue or consumable items. These are items that are of low value and are used in quantity, they are not usually shown on the parts lists for products (*bills of material*, see Section 3.3) and are often kept in bins on the shop-floor. Items that have been specially ordered, but not for use in production, are also usually issued by requisition.

Kit issue: In this method the parts required for a particular assembly or product are marshalled into one area, sometimes into a *kitting bin*. These parts are defined by the bill of material and the number of units being built. All the parts are issued from the Stores at once and the stock is debited for that complete unit or batch, thus removing the need to process many requisitions. If kits are made up too far in advance of requirement and more pressing requests for particular items come in, the kits end up being 'robbed'. The kit then has to be remade at a later date when the required parts come in. This is a time-consuming business and can cause stock records to become inaccurate — consequently kitting should be done as and when required.

Traceability

In certain industries it is necessary to have *traceability*. Traceability refers to being able to trace back all the parts that were put into a particular product or batch of products. The level of traceability will depend upon the product and the industry but may involve requiring the serial numbers of parts to be logged or a record to be kept of which particular batch of raw material was used in an item. Traceability records are often established in the Stores area through the issue records.

Stock-taking

In order for the Stores to operate effectively it is necessary to know what goods are held there and in spite of stock records being updated regularly discrepancies can occur; therefore stock-taking is used. It is also required in order to value the assets of the company for financial accounts. Two forms of stock-take can be done. The most common method is still the annual stock-take — this usually takes place at the end of the financial year and requires the Stores to shut its doors for business during the period of the stock-take. The way this is achieved is to warn that any goods that might be required during the stock-take period should be booked out prior to that date. Suppliers can also be notified that no goods will be accepted during the period, although it is usual to just stack the incoming goods in a temporary bonded area so that they cannot be used. For the stock-take the Stores stock is manually counted and checked against the records.

Perpetual stock-taking can be done as an alternative to the annual stock-take and does not require the Stores to be closed. However, before being able to use this as a method for accounting it has to be approved by the auditors. The technique requires small amounts of stock to be checked on a sample basis with the requirement that all the goods must be checked so many times in a given period.

3.3 Operations

In using the term 'operations' we mean the central business activity of the company. In a manufacturing company, therefore, operations are the activities that transform raw materials into finished product. The majority of engineering companies are involved with manufacturing to a greater or lesser extent. Even in those companies that do not manufacture, designers have to be aware of the constraints imposed on product development by manufacturing processes and organisation.

In this section we shall examine how manufacturing is planned and organised in order to meet company objectives, and look at some of the activities that are carried out within the operations function.

3.3.1 Organisation of manufacturing

Manufacturing takes raw materials, parts and components and transforms them into finished product. This can involve a few or many processes and these can vary from a simple machining process to a complex assembly and test procedure. When considering the organisation of manufacturing there is a need to know what processes are being used and the number and types of products. This informtion will allow an examination of the organisation of the factory in terms of the machines and operators.

Irrespective of specific process there are four production methods used in manufacturing: job, batch, flow and group. Each of these has implications for factory layout, labour and material requirements, and planning.

Job production

In job production products are made singly or in small batches, usually in response to a customer order. This type of production usually requires a great deal of planning and a lot or organisational flexibility because the product is built by either a group of people, where the group could be the whole factory, or by one person who does everything. This type of production applies to any product that is only ever present in the organisation in very small numbers.

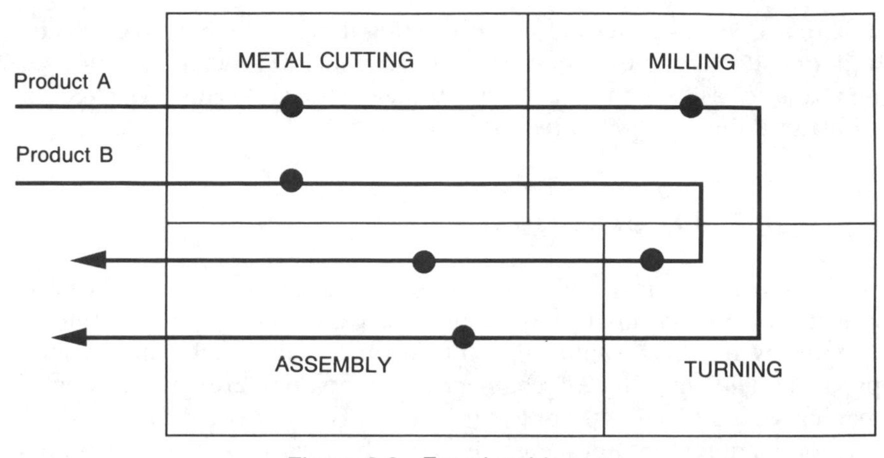

Figure 3.3 Functional layout

Batch production

In batch production the product is built in a series of steps and in large numbers, but not sufficiently large to warrant the use of flow production. The batch moves through a sequential series of operations, usually with every unit in the batch completing each operation before the whole batch moves on to the next. This method of production is particularly suitable if the manufacturer has to support and produce a wide range of products, although it can be used on a small product range as well. This type of production requires very careful planning in order to ensure that waste is minimised, since it entails high work-in-progress stocks. Companies using batch production typically use a functional factory layout where similar processes and functions are grouped together and all work requiring those processes is scheduled into that area. An example of a functional layout for batch manufacture is shown in Figure 3.3.

Flow production

Flow production is the process of manufacturing the product in a number of discrete steps but with no waiting at each stage for a batch of components to be produced. In this case the products pass through a production line as if they were batches of one. In order for the flow production system to operate effectively there should be no waiting time between operations — therefore each operation should take the same amount of time. The method used to calculate the times for successive operations and determine the production rate is called 'line balancing'. Also, because one product will be following after another, the line cannot be allowed to break down at any point; if this occurs product will rapidly build up at the break and this will bring the whole line to a standstill.

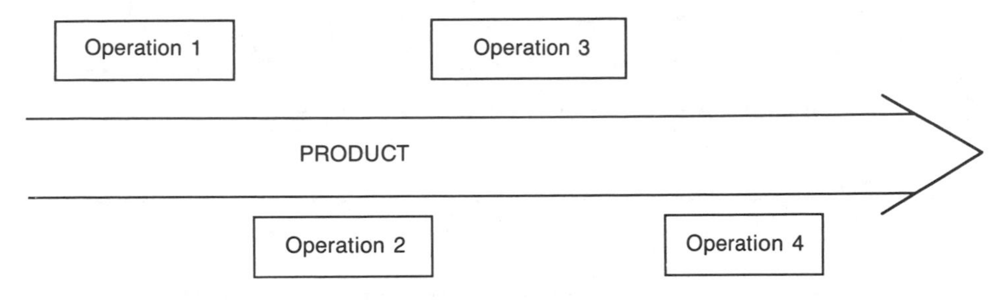

Figure 3.4 Flow production

This type of production can be used where there is a fairly constant demand for the product, and the product is standardised. All operations for the processing of parts must be clearly defined and the materials for the different stages of manufacture must be available promptly at all times. Companies using flow production have a factory layout based upon the product, where all the processes required for the product are grouped around a central line along which the product travels, as illustrated in Figure 3.4.

Group technology

Group technology is an extension of batch production in which the factory is organised with the aim of putting all the processes required for a particular product, or family of products, into one area. The area is called a *cell*. The principle of group technology is that it achieves the benefits of flow production and high standardisation, i.e. low work in progress and minimised handling costs, for a situation where there is small-scale batch production and a range of products. It is illustrated in Figure 3.5.

The capital costs of group technology can be high because of low machine utilisation, therefore it is important that the groups of parts are selected carefully and that people are deployed effectively. Multi-skilling of the workforce is a feature of group technology which enables the workforce to be used to operate many different processes.

3.3.2 Production planning and control

Production planning involves the planning of people, machines and materials in order to ensure that the output targets set for the operations area can be met. The factors that affect this are the lead-times for manufacturing and purchasing, the production system used and its level of complexity, the size of the operations function and the objectives set for the operations area.

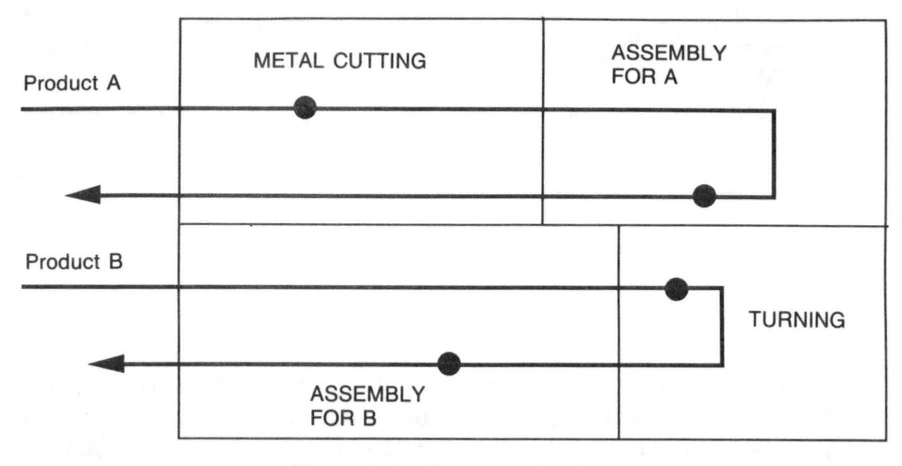

Figure 3.5 Group technology

The activities included in production planning will vary from long-term forecasting, as required for business planning, to daily or weekly shop-floor scheduling. These activities are discussed later, with an overview of some specific production planning techniques that are widely used in industry. However, before production can be planned there needs to be data on which the plans can be based. The data needed falls into two categories: operational data, including stock data, and product information. In each of these categories there will be data that is set up when the system is established and data that is developed during operation and has to be continually monitored and fed back to the user.

Operational data

This is the data concerned with the people and machine resources available. It will include a catalogue of the types of resource available and the capacity associated with each resource. It must also include information on the limitations of the resource — for instance, people may not be available on bank holidays, machines may have programmed maintenance periods. All this type of data needs to be set up in a database.

The operational data that will have to be monitored is that which indicates actual performance; for instance, this will include monitoring the numbers of parts that have been processed on a machine in a given time.

Collection of operational data during the production process is still carried out in many companies by using manual recording methods that are then fed back to a central processing station. Typically, in a machine shop operating a batch production process a ticket will be attached to the batch of parts, and this ticket will indicate what the parts are and what processes are to be carried out. When the batch of parts has been processed the machinist who carried out the work will complete the ticket, indicating the amount of parts processed, the amount that met specification and how long the processing took. The ticket will be passed to the finance department who can use the information to cost the job. The production control department will use the process times indicated on the ticket to update process capacity information.

Product information

This is the information that is used to define the product and the processes that are used in its manufacture. The information falls into three areas: part specifications, bill of materials and routings.

Part specifications: These are required for every part used in the manufacturing process. The parts are usually given a unique part code, or number, which is then used to identify each part. Against this part code there will be part definition information — this will include a clear description of the part, which may be achieved by reference to a drawing number; there will also be supplier information for those parts that are brought from external suppliers, including the supplier's/manufacturer's part code, the cost, and delivery lead-time. Part specifications will have to be set up at the start of any production control system; they will then have to be monitored, and adjusted accordingly, to reflect changes in cost and lead-time, as well as any regrading of the specification.

Bill of materials: These are sometimes referrred to as 'product structures'. The bill of material defines the parts, and quantities, that go into any subassembly or product, and usually reflects the way in which the product is manufactured. Bills of material are usually multi-level — for example, consider a kettle, as shown in Figure 3.6.

To understand how the product is made you work up from the bottom

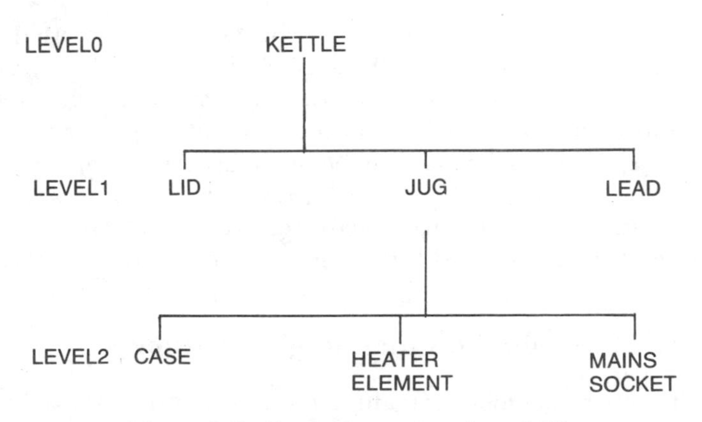

Figure 3.6 Product structure for a kettle

Level	Part no.	Part	Quantity	Supplier code	Lead-time	Stock
0	J100	Jug	1			
1	JJ239	Case	1	D32	24	200
1	EP210	Heater element	1	E65	3	46
1	EP554	Mains socket	1	E65	3	72

Figure 3.7 Bill of material

Part no. DL9003 Description: Base flange						Issue No. 1 Batch size: 50
Operation no.	Description	Dept	Machine			Drawing No.
			No.	Type	SMs	
10	Grind face	M	305		15	DL9003
20	Drill ten 5mm holes	M	306		8	DL9003
30	Transfer to Stores	SI				

Figure 3.8 Route card

— the level 2 parts have to be assembled before the level 1 parts. It is usual to describe any part that has its own bill of material as a subassembly, unless it is sold in its own right as a product. It is also the case that when looking at subassemblies the level numbers will change from those indicated in the product structure; the top level of any bill of material is usually level 0.

Bills of material can also be prepared on the basis of the time that it will take to process the product, so for example each level could relate to all the parts that you would use in one week.

The bill of material for the jug is shown in Figure 3.7.

Because they define the product, bills of material need to be controlled to ensure that any changes to the physical item are reflected.

Routings: These define the processes that are used in the manufacture of the product and the sequence in which they occur. For parts that are made regularly the routings might be given to the manufacturing department in the form of a standard part schedule. For other parts special route cards may be made up. An example of a route card is shown in Figure 3.8.

For all data used in the manufacturing process effective control is imperative. This is discussed further in Section 3.7.4.

3.3.3 Production planning techniques

There are a number of techniques that are used for manipulating the data discussed above in order to plan production so that output targets are met. Two techniques that are widely used are Materials Requirements Planning (MRP) and Just-in-Time (JIT). These are described briefly below. Other techniques include Manufacturing Resources Planning (MRP2) and Optimised Production Technology (OPT).

Materials Requirements Planning

MRP looks at the materials that are needed in order to meet the production requirements. It uses lead-time information to calculate when parts are required on the factory floor, thus allowing effective management of stock, and work-in-progress, and providing cash-flow information. The heart of MRP is the processing of the bill of materials. Figure 3.9 shows the operation of MRP.

Output targets and/or sales orders are the input to the production system, providing demand. The Master Production Schedule (MPS) is the

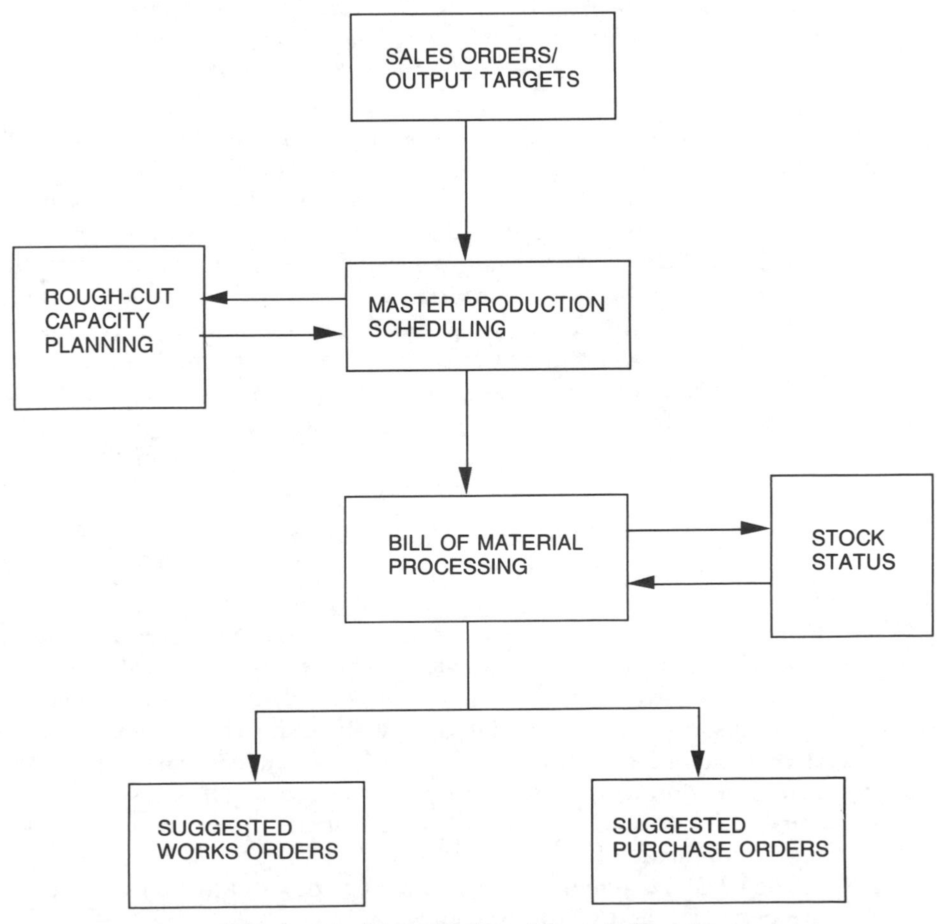

Figure 3.9 Materials Requirements Planning

list used to log all the demands made on the production system; in its most basic form it will record what products are required, in what quantity, and when. An MPS will be updated at regular intervals, appropriate to the product lead-time.

When the MPS is produced some rough-cut capacity planning is done. This is not detailed planning but looks broadly at overall capacity requirements and available capacity for that period. The MPS may be amended following this level of planning.

The bill of material processing will look at all the bills of material for every product required to meet the MPS. It will analyse the bills in order to determine all the parts that are needed to make the products, calculate the quantities of parts, and determine when the parts are required: the time that parts are required will be calculated by working backwards from the time that the product is required, using the manufacturing and supplier lead-times. When all this information is calculated a check will be made against the stock file in order to determine what parts are already available, or what parts will be available on the basis of orders already placed. This information is used to refine the materials requirements. The output of this processing may be in the form of a suggested orders list. Suggested purchase orders list those parts that have to be bought and, on the basis of the supplier lead-times, will indicate when orders should be placed. Suggested works orders relate to work that has to be done on the shop-floor and again will indicate dates that work is required and when it should be started, allowing for the defined works' lead-times.

Thus, the MRP provides information which can be used to schedule purchases and allows detailed scheduling of shop-floor processes on the basis of parts' requirements. It is not a shop-floor scheduling system, as more detailed capacity planning and scheduling may still need to be done, but it is widely used and there are many computer packages available on the market which run on even the most basic of personal computers.

Just-in-Time

Just-in-Time (JIT) is a production philosophy developed in Japan. The aim of a JIT system is to ensure that goods are available just in time, whether these be parts, subassemblies or products — the consequence of this being increased productivity and flexibility. Increased productivity means that product can be made in the shortest possible time, with minimum resources. Flexibility means the company's ability to react to changing circumstances, whether this be a change in customer order, or a modification to the product design. JIT manufacture is typified by a make-to-order, single-unit flow-line production system, where essentially product is made in batches of one.

The JIT aim is achieved by a disciplined approach which involves three principles applied to the organisation:

1. The elimination of waste.
2. Total quality control.
3. Total employee involvement.

In the JIT environment any non-value-added activity is deemed to be waste. These activities are costly to the company but do not increase the value of the product, i.e. the amount the customer will be prepared to pay. There are many non-value-added activities to be found in a manufacturing organisation using more traditional production methods, including inspection of parts and holding stock of parts and finished goods. Elimination of the waste produced by these activities can be achieved by eliminating defects and by not over-producing (making to order).

Total quality control leads to the elimination of waste by eliminating defects. However, within the JIT environment the aim is not to detect defects but to prevent them occurring in the first place by tracing any problems back to their source. This involves the whole organisation from product development, ensuring that new products can be manufactured to specification, through to purchasing who must ensure that bought-in parts achieve the specifications required for manufacture. The emphasis within the manufacturing area is on statistical process control and in-process testing rather than on inspection after processing. This ensures that processes can achieve specification instead of finding out afterwards if they have achieved it.

The third principle of JIT is total employee involvement, which means that management must provide leadership which results in employees wanting to be involved in what is happening. Management must also provide opportunity and encouragement which includes providing education and training, and using teams at work.

A number of specific techniques are used in JIT to achieve increased productivity and flexibility. All of these follow from the three JIT principles. They include work-in-process reduction, the use of group technology, reducing set-up times for machines, training workers for multi-skills, and supplier integration.

Work-in-process reduction is achieved by reducing batch size, the target being a batch size of one. As batch size reduces the amount of work waiting to be processed is reduced, as is the time to process any batch. This leads to advantages for the company because less space is required, and less stock, while lead-time for a product is reduced, thus providing improved customer service. In addition the non-value-added activities of handling and administration are reduced. Flexibility is obviously increased. If there is a small amount of work in process then changes do not have to wait a long time before being introduced.

Multi-skilling the workforce is of great importance in the JIT system. It is also a requirement for group technology. Multi-skilling provides increased production flexibility by making it possible to redeploy workers, when required by changes in product or demand, with minimum disruption to the production line and with minimum losses in productivity. It also has a great advantage for the workforce. Workers can contribute more effectively to problem-solving because they understand the wider implications of their suggested solutions; it also leads to increased motivation.

The role of suppliers in the JIT system has major significance. If suppliers do not supply on time or meet the specifications set, then the system cannot work. Therefore a company using JIT must work closely with its suppliers. The aims of building a good relationship are as follows:

1. To improve the design and quality of goods and services.
2. To improve logistics, i.e. the movement of material from the supplier to the customer.
3. To reduce the requirement for high levels of stock.

The strategy used for achieving these aims involves reducing the supplier base to a minimum, and using local suppliers where possible. It requires the building of a relationship which benefits both supplier and customer, and thus the companies work together, the concept of teamworking extending beyond the internal production system.

JIT has many benefits: companies that employ the JIT techniques are happy to talk of the savings in time, costs and space, that have resulted from the use of the system. These companies have also found that worker motivation has improved. The difficulty that some companies experience with JIT is that in order to work effectively it requires a *culture* change, particularly in the way that management and workers work together.

3.3.4 Management activities in the operations area

As well as the production planning and control activity there are many other management tasks that have to be undertaken within the operations area. A few of these are described below.

Management of people

In a manufacturing company a large proportion of the workforce will be directly employed within the operations function. There will be people from many different backgrounds within the same department. Whatever the number of people involved and their skill level and position in the organisation, they all have to be managed in a way that ensures that the objectives of the operations function are met. In this section we will look at some of the issues that relate specifically to the operations function; however, more general personnel management is dealt with in detail in Chapter 4.

The factors that will affect the way in which an Operations Manager will manage the operations workforce include the number of people in the operations function, the way in which the workforce is organised, the level of automation in the operations area, and the personnel policy of the company. The activities involved in managing the personnel of the operations function include general personnel management, manpower planning, and training.

Manpower planning in the operations function is strategically important to the company as it will have a direct effect on output. It involves considering the skill levels of people available and those required to achieve the company's output targets, and then aims to match the available capacity to requirements. For example, if a worker can produce two units of output per hour then available capacity can be calculated by considering how many hours that worker is available. When this is done for the whole workforce it is possible to calculate any differences between what is available and what is needed. This allows decisions to be made about recruitment, redundancy, working hours including shift-work and overtime, and allows planning for sickness, holidays and training.

Training is a particularly important activity in a manufacturing environment because of the variety of skills that is needed and the fact that people need to be retrained in new technologies. Decisions regarding training requirements may be made following employee appraisal interviews, or manpower planning. Training can then take place on- or off-site as appropriate. The effects of training have to be carefully considered by the Operations Manager because of the changes it may make to output, particularly if using experienced people to provide on-the-job training to less experienced people.

The other aspect of training that has to be considered in this environment is that of external controls that may be applicable. For example, if you are providing a graduate training scheme that has been developed in conjunction with one of the engineering institutions, or if you are operating a nationally recognised apprentice-training scheme, there will be specific training and review requirements. There may also be requirements for attendance at external training courses, or for external progress monitoring.

Management of machines and processes

Manufacturing requires the use of many different machines and processes and there are many aspects of machine management that have to be considered. The major factors influencing the management of machines and processes are the type of production being used, the level of automation and the complexity of the processes. The actual processes used will depend upon the products being made and the way in which they have been designed, the numbers of product being produced and the skills available in-house. For example, if a product is to be made in large volumes then it should be designed to be produced using processes that allow high-volume manufacture. The management activities relating to the processes will include making decisions about the types of production used, which processes are used, and monitoring and control of processes.

Monitoring and controlling processes is particularly important. Standards of workmanship and product specifications must be set and then the process controlled to ensure that design requirements can be met. Many monitoring techniques have been developed, statistical process control being an enduring favourite. All such techniques aim to inform of impending difficulties before the process has drifted out of specification.

One of the main activities in the management of machines will be selection of equipment to be used. This is not a continuous task as such capital investment decisions will not usually be undertaken on a regular basis; however, it is important to make the right decisions. Factors that will have to be considered in selection will relate to the processes used in the company, the desired capacity of the company, the return that the company would expect from any machinery purchased, as well as any plans that the company has for expansion or development and any technological changes that affect the company's processes.

3.4 Marketing

Marketing ensures that a match is found between the customers' needs and the attributes of the product, at a profit. It examines how future markets and products may be brought together to the benefit of the company.

Some of the tools of management have been with us for a long time. Human nature has always been the same and the concept of payment for goods has always been with us. A conversation between two Romans bargaining over price would, no doubt, follow almost identical lines to two modern Europeans doing the same. Marketing, however, has not always been with us and the obvious question is: why has the need for it come about? Marketing was born with mass production. Trade before mass production was characterised by individuals meeting each other to conduct the business. Supply met demand face to face. Manufacturers did not sell to people they had not met in some way and they could not produce large numbers of goods without pouring in proportionate quantities of labour. The improved communicaitons and massive increase in productivity of the Industrial Revolution and then the advent of twentieth-century mass-production methods changed this situation for ever. It is now possible to produce almost unlimited amounts of product, and the larger the numbers of units of production, the cheaper production becomes. For example, a manufacturer of electronic microchips could produce billions of units from a manufacturing site the size of an ordinary warehouse. The manufacturer is very unlikely to meet the people who end up using the microchips — it is much more likely that they will be sold to an organisation that will incorporate them into a product of their own. This product may, in turn, be incorporated into the product of another manufacturer or trading organisation. The original manufacturer may only have a partial view of who actually ends up using the product and the purpose to which it is put. For manufacturers in such a position, the question is not whether the product can be made, but rather whether it can be sold.

It is the task of marketing to ensure that sales are made from such a situation. The selling process is described in Section 3.4.5, and is clearly important in this process and consequently sales and marketing are often thought of together. However, marketing involves much more than just selling. The purpose of marketing is to ensure that customers are satisfied with the products on offer and thus to further the success of the organisation.

In order to achieve this, marketing must play a role in the product development process. It should develop a detailed knowledge of the markets into which the product may be sold. Understanding how the different markets for the organisation's products behave is of paramount importance to the marketing function: only by understanding its markets can an organisation be reasonably sure of exploiting them effectively. It is this preoccupation with understanding and exploiting the markets of the organisation that has given the function its name.

In this section we shall examine some of the key tasks facing the marketing department. To start with, in Section 3.4.1, we shall see how this task is organised. We will then, in Section 3.4.2, briefly examine the activity which provides the data upon which marketing activities are based — market research. In the next section, we will consider some of the techniques that are used to assist in the marketing process from a strategic point of view, and then we shall introduce some of the ways marketing personnel divide the market up in their quest to exploit it. In Section 3.4.5 we investigate some of the tasks in which marketing personnel must be involved. Lastly, in Section 3.4.6, we examine the role that lead-time plays in marketing and introduce strategies that may be adopted in order to shorten the product development process.

3.4.1 Organisation of the marketing department

In many companies, particularly small ones, there is no physical distinction between the marketing function and any other, although it is likely that the various aspects of marketing are being addressed since businesses can only sustain themselves if they can produce what customers want. In addition it is likely that they will only be able to attract funds from external sources, such as banks, if they can indicate how the company will develop over the next few years.

Some companies will organise themselves such that the functions of both sales and marketing are covered by one department; it tends only to be the large companies that have a separate marketing department. For those companies who operate without a separate marketing department there are several organisations who specialise in carrying out market research on a contract basis.

In many organisations, however, particularly those formed during the 1960s and 1970s, it is common to find a separate marketing department. In such organisations the task of understanding, examining and suggesting ways to exploit markets was given over to a separate and distinct function. The advantages of being able to specialise and of fitting in with the existing structure were seen to outweigh the disadvantages of separation and communication weaknesses. The marketing department interacted with the other functions as necessary and by having a director of marketing on the board the important issues raised by the function found their way into company strategy.

More recently, organisations have come to the conclusion that the communication disadvantage that a separate function places on the product development process is too great to ignore and the separate function has been replaced with a change in attitude within the product development process itself. In such an organisation it is common to find the engineers involved in the product development process also directly involved in marketing issues. They meet customers, they analyse the needs of the different markets and, as part of the product development process itself, the engineers undertake marketing activities.

It does not particularly matter which method of meeting the organisation's marketing needs is employed — the important thing is that the organisation is provided with acceptable market intelligence in order to meet its strategic goals.

3.4.2 Market research

Market research is the name given to any activity undertaken by the organisation with the express purpose of gaining information about potential customers. There are four key elements to market research:

1. Analysis of current activities.
2. Market intelligence.

3. Market analysis.
4. Product evaluation.

The analysis of current activities will include monitoring current sales, and trends in sales, and so will give a picture of who is buying what, and where. Market intelligence involves knowing what is happening in the market and would include analysing new departments, in particular knowing what the competition are planning. Market analysis is an examination of the consumer reaction to new pricing policies and new advertising campaigns and an assessment of the market potential of new products by contacting and gathering information from customers and potential customers. Evaluation of product after launch is done to determine the effectiveness of the strategies adopted for that product and makes apparent any changes that may be required for the future.

Market research provides the base for the decisions affecting marketing which then focus on the product and on how it is promoted, distributed and priced. We will now look at how market research provides a marketing input to the product.

3.4.3 Strategic marketing

Marketing is by its nature forward looking and so will be concerned with determining the needs of customers for future products or modifications. For this reason marketing should have a formal input into the product development process. It will also be able to comment on the ideas of the designers in relation to their commercial viability. It is able to do this because a significant factor of market research is determining what the basis is for customers' buying decisions, thus it can build a picture of features that are essential to sell the product, and those that are desirable and might improve sales. These features may not be restricted to particular product features but may relate to such things as availability of spare parts, geographical location of distributors or speed of delivery.

Marketing should also make significant input to business planning and corporate strategy by providing forecasts of demand in terms of products and quality which are then used to produce budgets and prepare cash-flow forecasts, as well as allowing decisions on company expansion — or contraction — to be made. It should provide the data on which the company can make decisions about how it is to develop. For example, any product will have an expected life during which it will generate different levels of profit depending upon the stage that it has reached — this is called the *product life-cycle* and is shown in Figure 3.10.

At the peak of its life a product will generate a steady profit, with

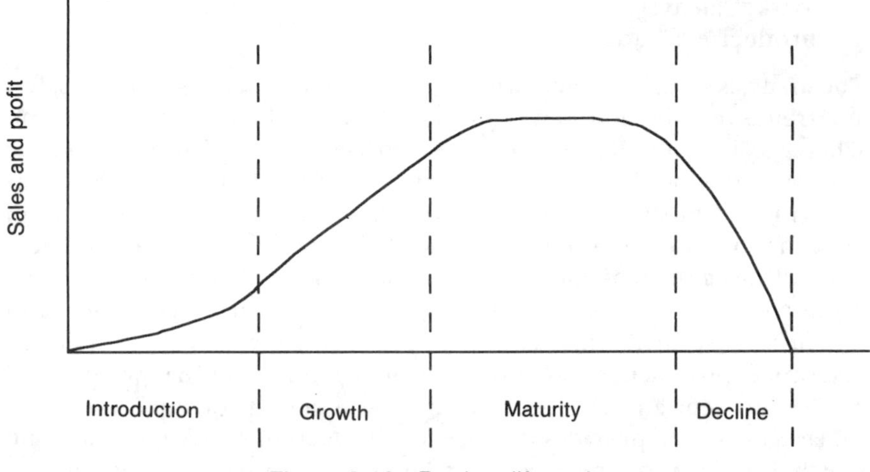

Figure 3.10 Product life-cycle

demand being fairly constant; after a while, however, there will be another product that does a better job, another company that produces the same product more cheaply or it may be that new technology renders the product obsolete. Whatever the reason, sales of the original product will start to decline. It is the responsibility of marketing to predict when this decline will occur and to put forward proposals for the company to maintain its profitability. There are four options that a company can consider: these are shown in Figure 3.11.

Market penetration involves maintaining sales by attracting to your product more customers than the competition, or by getting the customers of your competitors to buy your product. You might achieve increased market penetration through lower pricing or by increasing awareness of your product through advertising.

Product development involves targeting new or modified products to previous customers. This is where marketing will make a direct input to the design and development process. In some businesses, particularly those greatly affected by changing technology, such as personal computer manufacture, a certain rate of product development is normal. Product development is also used to refresh the market by providing something that has a slightly better performance, better appearance or is easier to use, thus prompting customers to make a further purchase and upgrade their existing product.

Through *market development* a company aims to improve sales of its current product range by selling to groups of people who have not previously purchased, or who have but intended a different use for the product. An example of this is the way in which baby shampoo is sold to adults who wash their hair frequently and thus need a mild shampoo. A further example is the way in which yoghurt, previously a healthy

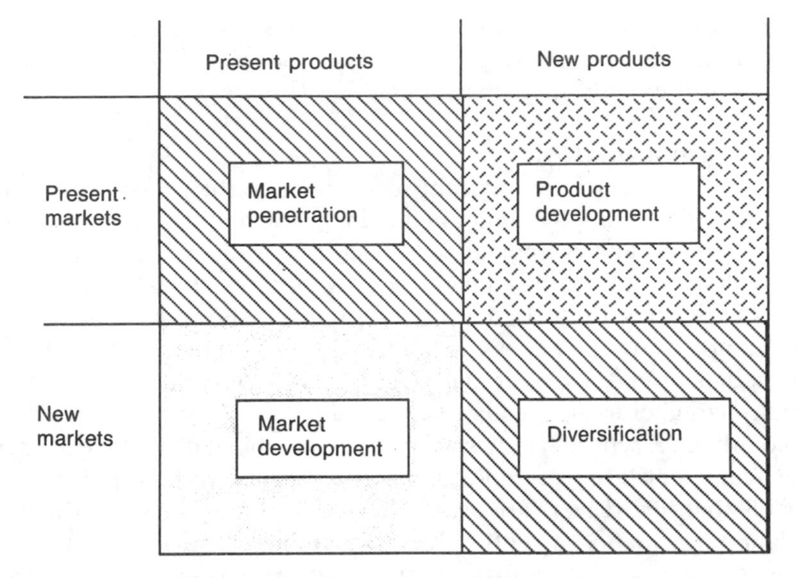

Figure 3.11 Strategic marketing options

dessert, is promoted as a low-fat healthy alternative to ice-cream and is marketed as a snack food.

Product *diversification* is the riskiest strategy for any company, as it means putting a new product into an unknown market. Again the marketing input to design is crucial and has to be supported by extensive market research.

In addition to these activities marketing will also have to develop strategies, for both current and new products, in terms of pricing, promotion and distribution. The pricing policies will require input from the engineers who will have to indicate costs, or give estimates of future costs, for manufacture of the product. Marketing will also have to consider the additional costs of distribution and promotion, as well as considering what is happening in the market-place. Promotion raises awareness of the product with potential customers and therefore includes the way it is packaged, the way it is advertised and any promotional literature that is used to describe the product. Distribution describes the way in which a product is transferred to the customer and so will involve an examination of the sorts of outlets that are required, their geographical location and the way in which the product is sold.

3.4.4 Markets

Marketing seeks to optimise the organisation's use of the markets available. Many terms are used to describe either activities that assist in this process, or particular types of markets themselves. In this section we shall introduce some of the the most important terms that relate to markets and explain their meaning.

Segmentation

Market segmentation is the name given to the process of dividing up a market into simpler components, each of which has its own distinct characteristics. For example, a manufacturer producing cars would not simply describe itself as selling into the car market. The market-place for cars is enormous and may usefully be subdivided into smaller markets, for example into hatchbacks, saloons, sports or luxury. A brief look at the models offered by a car manufacturer will soon show the extent to which manufacturers 'segment' these markets even further in order to exploit them. A typical family saloon car is available with a range of engine sizes from very basic and under-powered through to high-performance models that few drivers are ever likely to be able to drive to the limits of their performance. After this the variety of accessories offered is wide, as is the choice of trims, instrumentation, colour and warranty agreements. When these variants are considered together there are an alarming number of different combinations available. This can be borne out by investigating any large car-park: although there may be many examples of a given generic type of car present, the number of examples of identical specifications is extremely low. Why is this enormous range offered? The answer is that the organisation is segmenting the market. In their efforts to avoid any potential customer not purchasing a car from the organisation, because they do not provide one that is ideal, the marketing functions of the car manufacturers have led their manufacturing departments to provide enormous choice. Car production has changed a great deal under the influence of marketing since the days of Henry Ford, who maintained that his customers could 'have any colour you like as long as it's black'. Few would doubt that the developments are desirable.

Taken to the limit, this 'segmentation' becomes one-off, *customised* production. The ideal solution to the needs of a customer in the motor industry is unlikely to be met by one-off production since the initial expense and maintenance cost of an individually designed and produced motor car would be prohibitively expensive. It costs millions of pounds sterling to develop a new model. By segmenting the market and taking advantage of the engineering capacity to mass-produce and by introducing almost limitless number of combinations, customers are getting the advantage of a car that is almost individually tailored to their needs and yet are benefiting from the economics and serviceability of mass-produced goods.

Refreshment

After a market has been segmented and an organisation has sold extensively into the different segments a situation may arise when demand starts to fall because all the customers already have the best product available to meet their needs. It is clearly of interest to the organisation to identify ways of stimulating the customer base to buy products again. In order for these customers to buy again something new must be provided which satisfies a need not previously met. Market refreshment is the process by which this may be achieved.

The basis on which the process depends is that customers who buy a product do so not because of an absolute need for a given level of performance, but because of a relative need for performance. This concept can easily be illustrated. For example, someone who buys a camera will regard themselves as being of some particular level of competence in the use of cameras — if they regard themselves as a person of average photographic abilities they will conclude that a camera of average performance is appropriate for their needs. This will lead them to prefer to buy such a camera. This purchaser will do so with complete disregard for the absolute performance of the camera. One never hears such a person say that the average camera of forty years ago had less features and so that is the camera they require, yet one does hear them expound the benefits of their purchase as being suited to their needs. This can only be because such consumers arrive at their needs by considering their position in the market-place in relative terms. This observation is borne out by observing the features provided on top-of-the-range cameras. Such cameras have far more features and abilities than are actually required by many of their purchasers, yet the cameras sell, simply because of the customers' perception of themselves as being 'professional' photographers who need a top-of-the-range camera.

This creates a very interesting commercial possibility: namely that if a new product of superior specification can be found to move into top position, everybody in the market will have to buy the new product in order to stay in the same position. Those who regard themselves as being at the top will have to buy the new model, those who feel centrally placed will have to own the next model up and so on. When executed successfully, this market refreshment can stimulate everyone in the market to upgrade and buy again. Clearly a second-hand market may develop in such a situation but the organisation will certainly cash in on the new demand and can expect to sell many of the existing models throughout the range as people move up.

Naturally, there are limits on such a process — one cannot simply go on bringing out new products with higher and higher specifications and expect everybody to trade up every time. Eventually the gap between what is actually needed and what is on offer becomes too large for the customer to bear and the technique fails. When applied successfully the technique can have a powerful effect on the profitability of an organisation and it is most likely to work where the dimension of the product used to refresh the market is technologically limited. An example of this might be the power-to-weight ratio of a jet engine. As technology advances and it becomes possible for a manufacturer to launch a new engine offering a higher power-to-weight ratio, so those customers who wish always to own the state-of-the-art engine — such as the military — will be forced to replace some of their existing engines with the new models. Much of the market may be refreshed by each new advance and the amount of profit from such a development may therefore be substantially larger than the profit obtainable on the sales of the new product itself. It is for this reason that organisations are often prepared to undertake very expensive product development programmes that promise to extend significantly the technical performance of their products.

End-user

The 'end-user market' is the name given to those customers of the organisation that purchase a product and then actually use it for its extended purpose. In the end, of course, every product is intended for an end-user but it is rare for products to be sold directly to them. For example, consider an electronic chip manufacturer. The manufacturer will almost certainly sell the product to distributors and wholesale stockists who then sell the product on, either to smaller retail outlets or to manufacturers of other products that include some quantities of the original product. By one route or another the chips may end up in a multitude of different manufacturers' products. The chain does not end here, for the products of a manufacturer may go through several more stages before finding their way to the end-user. It is quite common for basic components to pass through up to ten different hands before reaching the end-user. From this observation we can see that, contrary to popular belief, the end-user may represent a very small portion of the organisation's markets, if indeed any sales are made to end-users. Organisations that do sell directly to end-users have the advantage that their normal trading brings them into contact with their customers. Such organisations do not have far to go when gathering information about their markets.

Original equipment manufacturer

The 'original equipment manufacturer' market, or OEM as it is usually called, is the name given to a particular type of customer. As the name implies, OEM customers are those who purchase products from the organisation and then incorporate them in products of their own which are then sold to end-users. For example, a company might be manufacturing magnetrons for generating low-power microwaves. Several other organisations may then purchase these magnetrons and incorporate them into microwave ovens. These are then sold directly to consumers. In such a situation the company manufacturing the magnetrons is said to have several OEM customers, or an OEM market.

Such markets are very desirable since they represent continuous business for the organisation. OEMs tend to want deliveries scheduled months, or even years, in advance. It is common for OEMs to make contracts of supply with the organisation and these may run for several years, stipulating maximum and minimum volumes of supply. Naturally such agreements involve much negotiation since each side is intent on getting the best deal. The co-operation between a supplier and an OEM customer can be very great, each passing information to the other in the interests of the continuing success of the collaboration. However, OEM customers have their dangers too, particularly if an organisation has only one such customer. In such a situation the customers may ultimately wish to take over the supply of the product themselves, in order to reduce costs (perhaps by acquisition of the supplier or by alternative supply), or they may try to eliminate the need for the product by changing the design. In either case, an organisation with only one OEM customer is vulnerable to commercial change. The most desirable situation is therefore to supply several OEM customers, all of whom are in competition, selling into the same market. In such a situation, the organisation is not tied to the fortunes of just one OEM, and can be sure that the collection of competing OEMs will exploit their market as comprehensively as they can and in so doing provide the biggest market for the organisation's product.

Retail

The 'retail market' simply means selling directly to individual customers. We are most familiar with it as the market in which we buy products such as hi-fi, DIY goods and food, as consumers. Retail is used to describe the final point of sale of products that require extensive distribution and it is this that distinguishes the 'retail market' from the 'end-user'. It is normal for a manufacturer to supply distributors who buy and sell in bulk at 'trade prices'. These suppliers then sell in bulk to a retailer who carries smaller quantities and sells individually and at a higher 'retail price'. In this way products requiring extensive distribution can find their way to the customers.

Displacement

In a 'displaced' market, the customer has to be separated from the purchaser. In such a case, although the product must meet the needs of the actual customer, it must also meet the needs of the purchaser; this may not always be possible if the opinions regarding the important product attributes are not the same. For example, it is common for scientific equipment purchased for research to be used by postgraduate and research staff. However, the people who sign the purchase orders for such equipment are, occasionally, the senior scientists in charge of the research and, most commonly, the purchasing committee of the institution. In such a situation the users of the equipment may regard certain labour-saving features as worthwhile whilst those who make the purchase do not. A situation can occur where the organisation cannot please both parties, the price and attributes that are acceptable to one group being unacceptable to the other. Competition and lack of understanding between the parties can make selling into such a situation difficult. Other examples of displaced markets include that of toys. Clearly, with many toys, children do not have the purchasing power to buy the ones they want and therefore selling of the toys is directed at the parents, who do. Advertisements show parents pleased with the happiness brought by the toy to their children, not children playing with the toy at the expense of other, rival toys.

3.4.5 Sales

Selling involves ensuring that buyers are found for the products that are available, and so it is concerned primarily with current activities. It may also include ensuring that customer needs are translated into production targets and providing data for marketing and product development. In this section we will examine the organisation of the sales function and consider some of the activities that have to be carried out.

Sales involves getting the product to the customer. In a sales department there will be three main types of activity: (i) selling; (ii) providing customer contact; and (iii) translating customer requirements into production orders.

There are several ways in which a sales department can be organised. Some companies organise their sales teams according to the particular products that they sell, some by geographical area, others by the type of selling that is used. The types of people in a sales department can be varied but in the engineering world, particularly in high-technology companies, you may meet the 'Sales Engineer'. This person may or may not be an engineer by training and qualification — however, the term usually implies that this person has a more in-depth technical knowledge of the product than a 'salesperson'.

Selling

The type of selling selected by an organisation will depend upon the products made and the market into which they are sold. A common method is selling by catalogue in which the company makes a number of standard products and then aims to sell them to customers off-the-shelf. In this case the specification for the product is known, the price is set and the delivery lead-time will be known. Sales personnel need only a superficial knowledge of the product but will require good sales skills as products sold in this way usually have a lot of competition.

Another selling method is that of selling by tender or quotation. This is commonly used when companies are providing a more specialist product or service, and is very common when goods are capital equipment purchases. In this case the sales personnel will need extensive knowledge of the product and will have to work closely with the other departments in the company. The procedure for selling in this way is as follows: the company wanting to buy the goods (the purchaser) sends out a tender or quotation request to potential suppliers. The suppliers may be chosen because they are known to make the goods required, or a more general tender request can be issued through an advertisement in an appropriate journal. If a company wishes to tender for the goods it then has to follow the procedure given in the tender request — this will usually mean that a bid is prepared and is sent to the customer for evaluation. The bid will provide full details of the goods to be supplied. This might mean that the customer's specification is accepted or that some modifications are made and the supplier includes a full specification of what they can supply. In addition to the product specification the bid will contain all the contractual terms to which the supplier is prepared to submit, for example cost, delivery, after-sales service and support, and warranty arrangements. All bids are sent to the customer and evaluated. If the customer accepts a bid then a legal contract is made; it is therefore imperative that in this type of selling the sales personnel understand the implications of any claims made in the bid.

Providing customer contact

The sales department will provide the front-line contact for customers in terms of answering queries and sorting out problems. This has two advantages: firstly, it means that the customer has a contact, often a personal one, and will find it easier to build a relationship with the company; secondly, it ensures that people in other departments are not interrupted by customers when they may not be in a position to give the attention a customer deserves.

Translating customer requirements into production orders

The sales procedures must ensure that when a customer places an order with the company, they receive what they want. In a company that sells off-the-shelf products this might be as simple as placing a request with the finished goods stores. However, if the product is not in stock, or is manufactured to order, then the task is to ensure that the customer requirements are fully understood by the manufacturing department. This may mean filling in a production order form, or may mean keying requirements into a sales order processing package that fronts an MRP package.

During this activity it may be necessary for the sales personnel to refer back to the customer, particularly if there are any problems with delivery or specification; at a minimum the order should be acknowledged.

3.4.6 The marketing process

The process of marketing naturally begins with the organisation's customers. In this section we shall examine some of the tasks that must be accomplished in order to market the organisation's products successfully.

Market information

Section 3.4.2 described four techniques for gaining information about the market and these are used in the initial phase of the marketing activity. Clearly this process of gathering information from customers is a communication issue and Section 9.3 deals specifically with this. The important issue is that the organisation gains a clear picture of what its customers need in order that it may provide products that meet these needs.

Product definition

Once gathered, the needs and 'wants' of the customers may be defined. This information is then used to develop a specification of the product. This specification is not in any way a technical description of the product — indeed, it may well contain points that are very uncertain from a technical point of view. A marketing specification tries to describe the attributes that a product will need in order to be successful in the market, in terms of the customer needs. For example, the marketing specification of a new product to compete in the domestic washing-machine market will include details of ideal maximum load, power consumption, size and all the other attributes upon which a decision to purchase may be based. The marketing specification will also define the numbers in which it is expected that the product will be sold, the consumer groups who will purchase it and all the other commercial information regarding the product that will affect the engineers responsible for converting this paper description of an imaginary product into a real, profitable product from which customer satisfaction and organisational wealth may flow. Once a marketing specification is prepared it is used by the product development process and any factors likely to compromise the specification must be examined and agreement reached. Any new information gained from further market analysis must be included.

Promotion

Product promotion can take many forms and usually involves marketing personnel. The style of promotion used depends very much on the product in question and often organisations have products that require different promotion strategies. Advertising a custom storage shelving system for warehouses would require different advertising and promotion to plastic bins that might be used for storing components in a small laboratory, yet both these products might be made by one organisation. Promotion means more than just advertisements on television. Advertisements can be placed in newspapers, journals and magazines. It can mean conferences, shows, trade fairs, hosting visits, assisting others, sponsorship, discount schemes, games, competitions, customer visits, exhibitions or publishing papers. Even lodging patents or fighting court cases can be seen as promotion. In marketing anything that affects the customer's perception of the organisation or the product can be manipulated to promote the product.

Image and identity

An important part of the marketing process is the act of establishing and maintaining image and identity. An organisation selling cameras, for example, may wish its involvement with film sales to be made public in order to promote its image as a comprehensive supplier of photographic equipment. In reality, being a retailer of film offers little or no advantage to the camera manufacturing process and the association is made simply to foster an image in the customer's eyes. Image and identity can be created through many means and almost any product attribute that is visible to the customer can be used in this way. Certain car manufacturers, for example, use a particular combination of lines in the bonnet or the radiator grill to make an image and give identity, the purpose being to reinforce in the customer's eyes that this is a product made to the particular manufacturer's specification.

Finally, marketing is a continuous process and the separate tasks described above do not happen in order. Market research is continually under way and is used to support and understand the markets even for mature products.

3.4.7 Simultaneous engineering

In this section we examine a marketing weapon called 'simultaneous engineering'. It has become clear in the last five sections that a critical function of marketing is to ensure that products reach the market-place in a form that will meet the customer's requirement for performance and benefits whilst also meeting the organisation's financial performance requirements. This process is inextricably entwined with the product development process and we have seen that in some organisations no formal distinction is made between the marketing function and the product development function.

The advantage given to an organisation that can get new products into the market-place quickly is enormous. It is not so much a question of getting into the market-place first — many organisations trade very successfully by choosing to be second or even being deliberately late into the market-place. Such organisations benefit from all the market development money poured in by the leaders and have the advantage of being able to choose products to develop after the market has had time to recover from the initial products. Products developed later are more likely to be accurate in meeting the customer's needs even if there is less market share available by this time. The issue is not being first or even last — the important issue is that, when an organisation makes a decision to search out a new product, it is able to move from initial market investigation to market-place sales in the minimum possible time. Simultaneous engineering (SE) is a means by which this may be achieved.

SE simply means that, as much as is possible, the tasks that have to be completed in order to launch a new product are performed in parallel. At its most basic, SE is simply good project management and the techniques described in Chapter 8 are of importance. At its most sophisticated, SE is a marketing weapon that will persuade customers to buy from the organisation because they believe that the organisation's product development process is so good and so quick that they will be receiving a better and more up-to-date product because of it. This customer perception can have a very powerful effect in market-places that change quickly. For example, engineers buying computer equipment will often prefer to buy from organisations whose product development is fast and proven since they are more likely to benefit from a coherent product development strategy, software subsequently released is likely to work on older systems and the benefits of tomorrow's research and development can be used, when it is available, on today's equipment.

The two tasks that are overlapped to the greatest degree are the preparation of the marketing specification and the technical aspects of the product development process itself. In an idealised product development process the marketing specification is completed before the technical work starts. There are clearly economies that can be made, however. For example, certain technical requirements may be identified very early in the marketing process. Once a technical aspect has been identified work may start on it and, even though it is not known exactly what form it will take in the finished product, work can progress. The advantage of such an approach is clearly that it makes the whole process shorter — however, the major disadvantage is that it is likely to be more expensive. This is because unnecessary technical work may take place before the technical requirements are accurately known. Another disadvantage is that the work is more difficult to manage and consequently it is more expensive. Difficulties in management may come from trying to plan future blocks of work on outcomes that are not yet known and from the de-motivational effects of personnel being involved in work that is not ultimately used.

In this chapter we have briefly introduced the most important aspects of marketing. This is a management textbook and within this chapter we are considering the tasks that organisations have to undertake in order to trade successfully. Clearly marketing is something that must be accomplished successfully; however, it is also clear now that marketing is not a management tool in the way that, for instance, personnel management, finance or teamworking are. For this reason we shall not investigate this subject in any further detail.

3.5 Finance

The finance function is concerned with money — bringing it into the company, looking at how it is used and paying it out. Only by effective control of these activities can a company stay in business. The aim of the finance function is to ensure that this control provides the organisation with the financial ability to meet its corporate goals.

In this section we will look at the finance function as it would be organised through a finance department and consider some of the activities that have to be carried out. Specific finance and accounting techniques will be dealt with separately in Chapter 7.

3.5.1 Organisation of the finance department

Because of its importance the responsibility for the finance function in medium-sized and large organisations will usually lie with a director of the company, and will then be operated through a finance department. In a small company the owner or an executive director will have the responsibility for the task but may use the services of an external adviser.

The finance department will be responsible for the management of all the company's financial activities. In addition to accountants, the finance department may employ ledger clerks, payroll clerks and others with specific responsibilities for certain tasks, such as dealing with taxation. It may even have responsibility for the purchasing function. The size of the department will depend upon the size of the company and the amount of financial management required. The finance personnel, with the exception of the finance director who will be responsible to the board, often have no line management responsibility outside the department but they may be responsible for setting and monitoring financial policy, such as defining spending limits and procedures for committing expenditure, which will then have to be used by managers in other areas. Because of this the finance department must have good communications with all the other departments in the company

In addition to internal communications personnel in the department may have to deal with external organisations such as the Inland Revenue, banks, customers and suppliers.

3.5.2 Activities of the finance department

We will now look at some of the specific activities carried out by the finance department and the requirement to report on these activities.

Bringing money into the company

The finance department is responsible for capital planning. This means that it must look at plans put forward by the directors and departments in order to determine the financial requirements that allow these plans to be achieved. It will then have to prepare proposals indicating the implications if the plans are to be followed and, if they are approved, will have to find and manage the funding. This may involve, for example, looking at taking loans, using reserves or increasing shareholding.

In addition to this strategic activity the finance department will be responsible for the accounting activities of the company, including invoicing for payment of goods and services supplied to customers.

Using money

In order to operate efficiently a company needs to ensure that its money is being used effectively. To this end it needs to know how much its products cost, which in turn allows sales prices to be set and cash-flow requirements to be known. The finance department will be responsible for determining this information on the basis of data supplied by the operations department. The analysis of the production costs will provide information about profit margin and estimates of turnover, and will allow budgets and spending plans to be prepared.

Cash-flow forecasting on the basis of production data and planning is an important task, which involves defining what the company wants to achieve over a period of time and then analysing what this means in terms of cash that has to be paid out. For example, a company may be planning to make a number of units of product A. The product takes three months to make and will sell for £30,000 per unit. However, before you can get any money in from the sale of the product you need to spend money to buy materials, pay staff, purchase services and so on. The finance department's job here will be to look at when the sums of money have to be paid out and when sums of money will come into the company. It will then have to make decisions about how to deal with any problem, such as there being a shortfall of cash-in in one month.

Paying money out

Money has to be paid out of the company for the goods and services that is uses in order to make a product. Paying for labour in the form of salaries is an important part of the finance department's function and allied to this will be the collection of taxes, national insurance and pensions. The department may also be directly involved in setting pay scales and determining bonuses.

Payment for goods and services purchased from other companies will be made on the basis of invoices received. As when sending out invoices for payment it is important to balance payment of goods in order to control cash flow. A problem many small companies have is that of their customers not paying invoices on time which causes serious problems if they have had to pay their invoices on time in order to secure supplies. This can put companies out of business because they are no longer able to finance their activities.

Reporting on financial activities

A company will require financial information for internal use. There will need to be information on budgets, sales figures, capital spend, cost of overheads and cash flow. This will be used to determine the financial constraints within which the company, and its departments, will have to operate. The finance department will be responsible for collecting the data needed to generate this information and for putting the information into a format that can be understood and used by managers throughout the company.

In addition to the internal information, all private and public companies are required by law to produce financial statements. For the majority of companies these take the form of balance sheets and profit and loss accounts. The provision of this information is a prime responsibility of any finance department. There may also be requirements for financial information to be provided for other external sources, for example if a company is applying for a loan.

The role of the finance department is very wide, as can be seen above.

ts activities it is imperative that the information that is collected sseminated has integrity, as the whole company relies on it to stay iness.

3.6 Product development

Product development is a function of an organisation and it is a task that all organisations have to accomplish, although there are many ways of achieving it. Product development, which includes design, is often thought of in conjunction with manufacturing organisations, although all creative organisations have to go through the process. Even abstract products such as computer programs or banking services have to be designed and developed.

Organisations seek to ensure that the products they sell meet the needs of their customers. It is, however, the product development function that fashions the product and so it is this function that must ultimately take responsibility for the physical arrangement of the product itself. Ensuring that the product does indeed meet the needs of the customers will involve many interactions with other areas of the organisation. Primarily, the product development function will require information from the customer about what features and attributes are required. This may be achieved by design engineers meeting the customers themselves, although often a marketing department facilitates this process. The designers also need to design products that meet the needs of the organisation. For example, no engineer will be rewarded for producing designs that involve manufacturing techniques beyond the scope of the organisation. The abilities and limitations of all the departments must be considered when developing a product. External contractors can be used but where new processes or techniques are to be used a careful relationship must be built up to ensure their operation.

In short, the product development function is ultimately responsible both for the product accurately meeting the needs of the customer and also for instructing the other departments of any developments that may be required to support the new product and realise its profitable sales. Responsibility for this onerous task begins when the first ideas are put forward and does not finish until the product is withdrawn from manufacture.

In this section we shall not be looking at the technical aspects of design, where engineers put their technical skills to work designing products, but rather at the way the overall process is being managed. Producing a new product is a risky venture, there are always uncertainties, and the process must be managed to prevent wasted money. Product development can easily degenerate into a bottomless pit of expenditure — there is no upper limit on the amount of money engineers can consume developing a product. Careful control is necessary to ensure that the engineers are confident of success whilst the venture as a whole remains profitable.

Design and development is the essence of creative engineering and the terms relate equally to products and to processes. In this section we will examine the process of product development. We will consider the way in which this function can be managed and look at some of the factors that affect it.

To look at the management of the process we will examine how a company can organise its design and development activity, the activities involved in product development, the operation of a drawing office and the role of research.

3.6.1 Organisation

The organisation of the design and development activity can take many forms. At one extreme the organisation is divided up into separate departments, each having very specific responsibility for a particular aspect of the development process. One department will research the market and suggest products, another will design and produce prototypes, another will continue the development to allow products to be manufactured. There will be a department with responsibility for planning the actual manufacture of the product, one for costing, and finally another will manufacture. In this functionalised approach the cross-department communication can be difficult, often there will be no particular engineer who stays with the product through its life, and the design process is characteristically slow but is very thorough. Staff are employed in those areas in which they are most competent and there may be a high level of specialisation. However, care has to be taken at the points where responsibility is transferred from one department to another to ensure continuity and to prevent duplication of effort; following transfer of the product a department is free to start on the next project. In this approach you will probably find an 'engineering department', although this term is used very broadly and may cover something very specific such as CAD/CAM or may be rather general including production engineering, design and production control.

The opposite approach is to treat the design of a new product as a discrete project in its own right. One person is put in overall charge and they are responsible for all the necessary tasks. The continuity is good — often individuals see the product right through from conception to manufacture. The process is characterised by speedy product development and a good match between what the customers want and what they receive. Accountability is good and control is easier to achieve than in the functional organisation. The main disadvantage of this type of organisation is that the engineers involved need to have many skills and this requires that they work on tasks other than those at which they are best. With a project-based approach it can be difficult to ensure that useful work of a suitable level is always available for everyone, and it is not always possible to regulate the work to flow evenly, causing periods of frantic activity and periods of under-loading. To operate effectively with this type of organisation requires much more flexibility from both the organisation and the individual engineers.

3.6.2 The product development process

The process of taking an idea and turning it into a product is the process of product development. It involves developing an understanding of the needs of the potential customer, knowing the requirements of the company, and being able to apply engineering knowledge in order to meet those needs. It may also refer to the process of taking an existing product and modifying or extending it in some way, in order to react to changes in customer need or changes in technology, or due to problems in manufacture.

The process requires effective management because of the need to address many different requirements and to prevent resources such as time and money from being wasted. It is possible to strive for perfection in designing a new product but if the process takes too long or the final product is too expensive, then the opportunity for the product to reach the market is lost.

Organisations seek to ensure that the products they sell meet the needs of their customers. This will involve many interactions with other areas of the organisation. Primarily, the design function will require information from the customer about what features and attributes are required of the product. This may be achieved by design engineers meeting the customers themselves, or it may be developed from data and information provided by the marketing department. As well as the customer's technical requirements, marketing data may also dictate the maximum cost of the final product, the way in which it should be packaged, the amount of time available for the design process, and other commercial requirements.

Figure 3.12 shows the idealised product development process — it is obviously iterative.

Design specification

The first step in the process is the development of the design specification which takes all the inputs and sets the limits on the design.

The market possibilities limit the design specification insofar as there

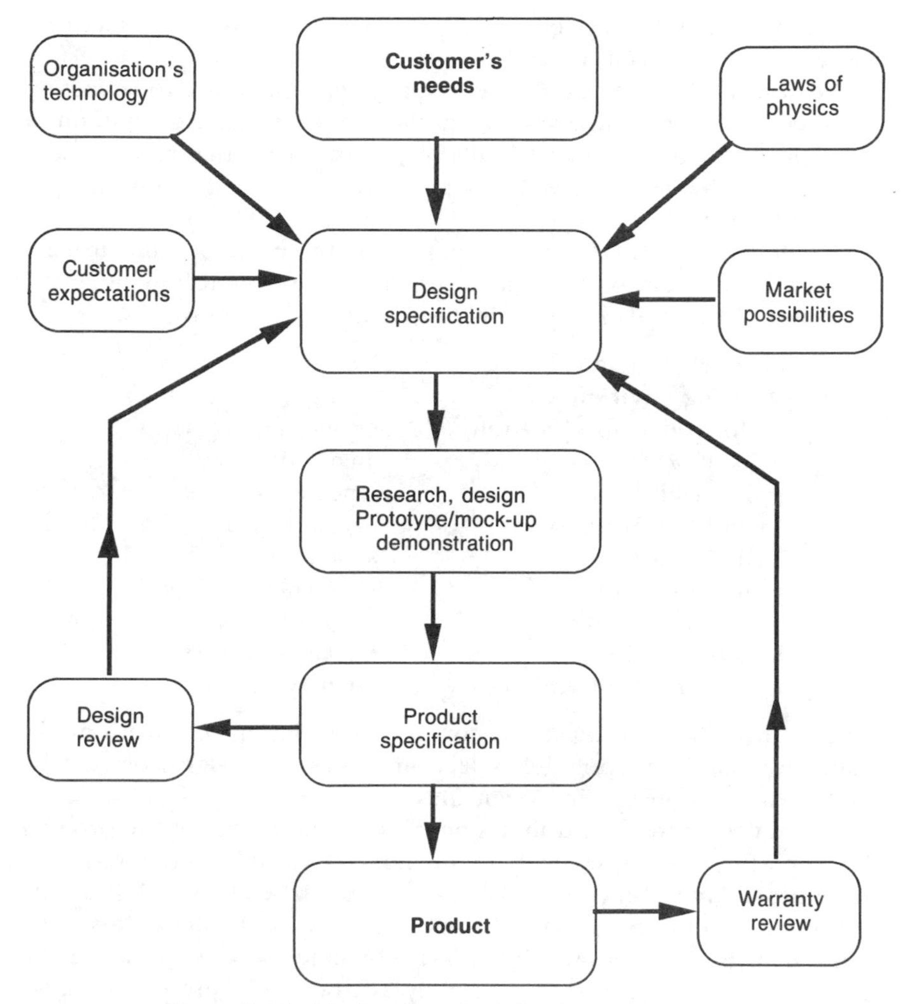

Figure 3.12 Idealised product development process

is no point in having elements of the specification that cannot be accepted by the market. These limitations may come from the size of the market: for instance, if the market for a new product is only a few units this may rule out the exploration of certain expensive technical possibilities, the costs of which cannot be recovered with few sales.

The designers also need to design products that meet the needs of the organisation; for example, no engineer will be rewarded for producing designs that involve manufacturing techniques beyond the scope of the organisation. The abilities and limitations of all the departments must be considered when designing a new product, and if external contractors are to be used, where new processes or techniques are required, then a relationship must be developed which will allow best use of the specialist knowledge of the contractor.

The organisation may have other requirements that have to be met by designers. There may be a requirement to use parts that are common to other products, there may be limits on the materials that can be used, or there may be certain regulatory safety or equipment standards that have to be met.

The design specification considers all of these factors and usually consists of a series of headings with a description following each. Examples of some of the usual inclusions are given below.

Physical principle	Size
Working environment	Weight
Performance specification	Service requirements
Reliability	Chemical effects
Product life	Appearance
Vibration sensitivity	Surface finish
Delivery rate	Noise emissions
Failure modes	Shipment technique
Sales price	Patent protection
Raw material supply	Special processes
Compliance to standards	Instrumentation

Notice that the specification contains no design calculations or justification of design principle selection — all such design work will be contained subsequently in design files.

Once a design specification has been prepared, a plan of the product development process through to manufacture can be prepared, and deadlines and milestones for various activities can be identified. Planning techniques that can be used for this are described in Chapter 8. It is likely that the major deadlines will be imposed by external factors such as the need to meet a specific customer delivery date or to have product available for an exhibition.

The design process

Designing is the process by which engineers creatively apply their technical knowledge of maths, science and manufacturing, and combine this with their commercial knowledge of the organisation's customers and markets to meet the strategic needs of the organisation for new products.

The way in which tasks that constitute the design process are carried out will vary depending upon the company and the product; however, the smaller the company the more likely it is that the designers will be involved with all aspects of design. In addition to the overall product design the sorts of tasks that may have to be addressed include: sourcing of new parts; preparation of purchase specifications, drawings, user manuals and documentation for manufacture; costing; building prototypes; testing. All of these tasks must be identified at the planning stage and responsibility assigned to people for carrying them out.

At various stages in the design process there should be clearly defined design reviews. Design reviews should be carried out formally and by people who are not intimately involved in the design. The aims of design review are both organisational and technical. The organisational aims are to check that the design is going to plan, and to confirm that the design is correctly defined. The technical aims are to ensure that the specification is being met, that appropriate design options have been considered, that the design is feasible for manufacture, inspection and test, and that critical calculations and decisions are correct. When changes are made to the design following review they should be documented so that the reasons for the changes are clear to everyone, even long after the decision point has passed. This ensures that the rationality of the changes is open to scrutiny and that work is not retraced.

Completion of the design results in a product specification which defines the product, its operating characteristics and physical attributes. It must also include a very clear identification of the aspects of the product that are crucial to both the function of the product and its safe operation. Documentation that defines the product for manufacture should also be produced (this has been discussed in Section 3.3).

Design modification

Completion of the design does not necessarily mean that the designer's involvement with the product is over. Particularly in small companies, a designer will be closely involved with manufacture of at least the first batch of product and may be called on after this if there are particular technical difficulties. There may also be a need to carry out some modification to a product or part and this should be referenced back to the original designer if there is a chance that the suggested modification may affect the original design parameters.

Unless it is likely that a proposed design modification or change will affect the basis of the original design, it is likely that it will be carried out by the drawing office. Any modification should be checked and approved and a drawing updated before it is carried out on the shop-floor. Obviously if there is an urgent production problem which requires a speedy modification, this procedure may put a significant delay into the system which is unacceptable. Some companies have an emergency procedure to deal with this situation that involves use of 'marked-up prints', when the modification is shown on a copy of the drawing and is approved by a designer but is not fully documented and checked before being actioned. This procedure obviously has risks associated with it and use of such a procedure must be a management decision. An engineer in such a situation must judge whether the severity of the marking-up is too great and it cannot be employed or whether the marked-up prints are unambiguous and practicable, and so may be used without danger.

In addition to the implications for the design, modifying products and parts will have implications for the business, its suppliers and customers, and these should be considered before a modification is made. If a product specification is changed by a modification then future customers need to be aware of the new specification. It must always be the case that a customer receives a product of the specification ordered, regardless of how difficult complexities of the manufacturing system and numbers of units make it to achieve. Particular problems can arise if a customer is expecting a number of products all to the same specification and the modification would mean a change part-way through the order. Similarly if the company has placed orders with suppliers and then needs to change the specification or number of parts, then it should ascertain whether this is contractually viable and whether the supplier is prepared to agree to the change in requirements. For the company itself a design change may result in stock becoming obsolete or processes being changed, both of which can result in significant cost.

3.6.3 Drawings — the language of engineers

Engineers imaginatively create new things. They develop technology to produce new solutions to all manner of problems. Describing such solutions is not easy — even language cannot always cope. Imagine the confusion and ambiguity that would result if you tried to explain in precise detail something as simple as an electrical plug to someone who had never seen one before. Few people would depend on such a description to ensure accurate manufacture. The problem will become greatly exaggerated when one tries to describe something as complex as a car.

A car contains thousands of individual parts. Some will be bought directly from suppliers as standard components, others will be produced by other manufacturers to specific requirements, yet others will be produced in-house. Initially the components are assembled together into small sections, each with its own function or integrity, called *subassemblies*. A gearbox, for example, or a dashboard, are both subassemblies of a car. Such subassemblies may require extensive testing and calibrating before they can be built into the final product. One would not manufacture a car, for instance, by putting all the pieces together and then testing it, only to find that a minor fault in a small subassembly caused it to fail the final test. Subassemblies are joined to other subassemblies and components to form progressively larger sections of the product. A car engine, for example, is a subassembly of the product but is itself composed of many smaller subassemblies. In the final stage of the production process the last level of subassemblies is brought together, and the product is assembled in its entirety. Occasionally single components are also brought together in the final stage of assembly. For example, the engine of a car, containing thousands of components, is fitted on the production line, yet so is the windscreen which is clearly a single component.

So how do you describe a product as complex as a whole car — where do you start? The description must be sufficiently accurate to allow repeated manufacture of identical cars. It must be utterly unambiguous and must deal with the very different ways in which components may be made. The answer that engineers have developed is a language of their own, which copes with this specialist communication problem. It is the language of drawings.

Drawings in practice

In this section we shall not consider the technical aspects of drawing preparation. Draughting is a skill in its own right and engineers study it separately. There are many standards that govern the preparation of drawings and each organisation has its own preferences and chooses particular national standards. We will, however, consider drawings from the point of view of information control. We will do this by looking at some examples.

Figure 3.13 shows a drawing of a component. The drawing shows not only the engineering aspects of the component but also much management information. The important features of the drawing are described below.

The details: The majority of the area is taken up with the details of the component itself. The dimensions, tolerances, manufacturing techniques and all the other important aspects of manufacture are described in such a way that repeated manufacture of acceptably similar components can be guaranteed. Of course, no two components can ever be absolutely identical and so the choice of tolerance bands is an important part of the drawing. The drawing does not show many properties of the component which may be important in its operation — its second moment of area, perhaps, or its thermal conductivity. This is because such properties are themselves a function of the dimensions, materials and processes specified, they are 'implied' attributes and therefore stating them is a repetition.

The bill of materials: After the details themselves, the bill of materials is the most important item on the drawing. The bill for the

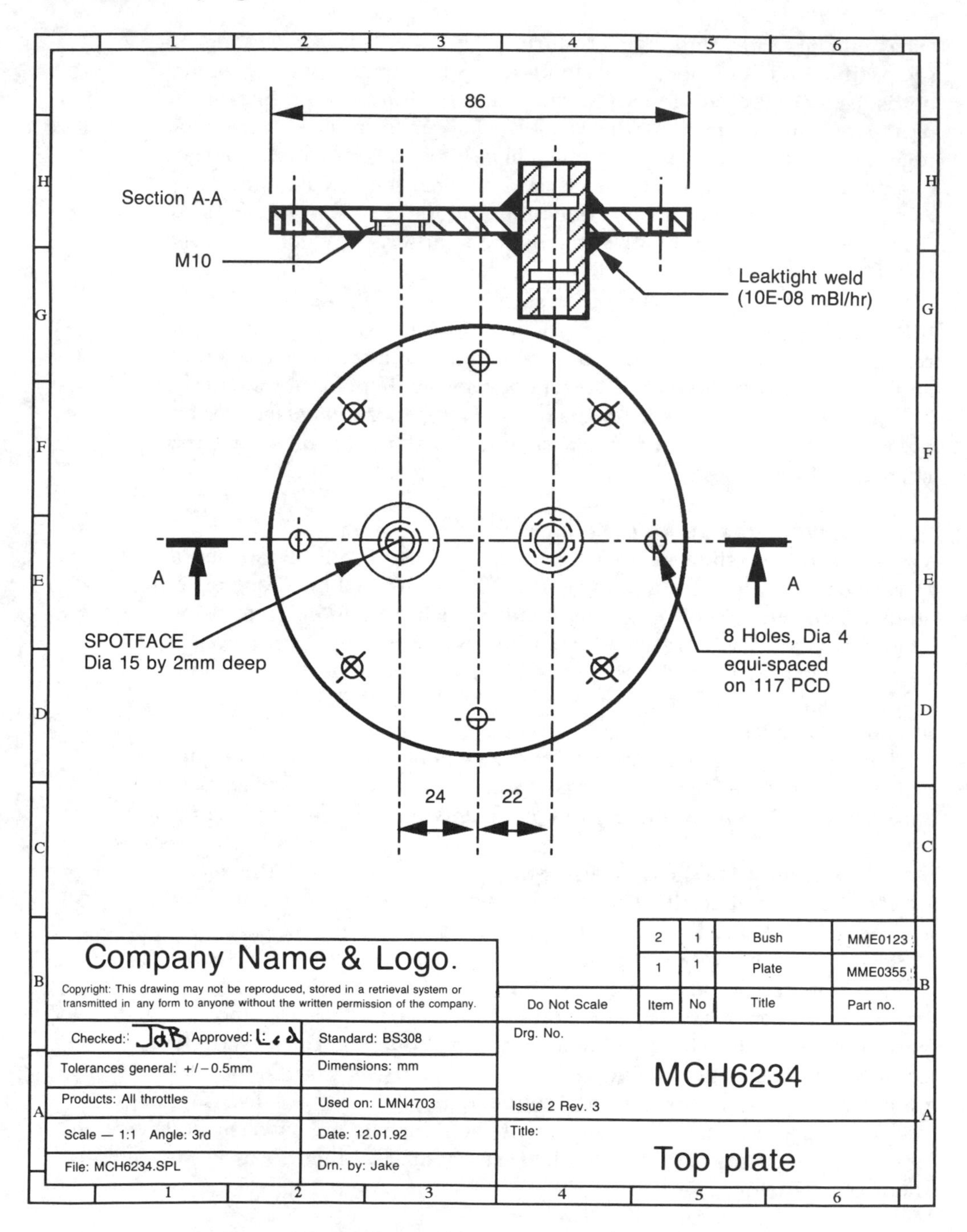

Figure 3.13 A drawing of a component

drawing in Figure 3.13 can be seen above the title block, and contains two items, a plate and a bush. From the drawing it can be seen that these two items are welded together to form the top plate. Of course each of these components will in turn need a drawing to describe them and elsewhere in the organisation the drawings MME0123 and MME0355 will be stored. These items are said to be 'called up' by the bill. Each drawing calls up the components necessary to manufacture the object it depicts. In the end, drawings will be called up that show items that are very simple and are not composed of more than one thing. Such drawings are called 'component' or 'detail' drawings.

Border: The drawing is surrounded by a border which not only allows the edge of the drawing to become damaged without making the details unreadable, but also contains a grid reference system. Occasionally this can be useful when engineers are, for instance, talking about a drawing over the telephone and need to specify the area they mean. It is of particular help when a drawing is very complex.

Company name and logo: This is usually displayed on a drawing together with a copyright reminder. By virtue of being prepared on company paper the company own the copyright of the drawing. If the organisation is involved in restricted work, such as defence, the drawing may carry other warnings about reproduction.

Drawing number: All drawings have to be uniquely identified. There is obviously no point in having two drawings with the same number. The simplest way of doing this is to number them sequentially and many organisations do this. As the number of drawings increases it helps both the filing and interpretation of the drawing if a number that carries meaning is assigned to each drawing. Often the number will indicate something about the item — if it is machined, bought in, welded, or what class of component it is, for example. It is common for the engineers to become so familiar with these numbers that items are referred to by number rather than descriptions.

Title: A title, as specific and descriptive as possible, is assigned to each drawing to help in understanding the overall function of the component.

Reference data: At the bottom of the drawing, on the left, are ten boxes that contain more information. Two of them, 'Products' and 'Used on', are described in the next paragraph. The 'Scale' box states what scale has been used and in this case the projection system (third angle).

'Date' of drawing records the date of first issue, the 'File' box records the file name under which the drawing is stored on the computer system and the 'Drn' box records the draughtsperson who drew it. It is common to include such matter for assistance in controlling drawing production and to assist in rapidly locating drawings within complex filing systems. The remaining boxes are self-explanatory and they record the tolerances used throughout the details, the dimensions that have been used, and any drawing standard that applies.

Bill of material references: As well as the bill of materials which shows those items necessary to make the component, it is normal to reference those items on which the component is used. Two boxes in the drawing refer to this, 'Used on' and 'Products'. The first box indicates the next item 'upwards' in the bill, i.e. that item whose bill of materials calls this drawing. The second indicates the ultimate destination of the component, usually a product. Of course, a component may be used on more than one item and there may be a list in the 'Used on' box. These two pieces of information are of particular use when engineers are discussing possible modifications to drawings since they can see at a glance how many other components or products are affected. There are clearly problems with the number of entries in this box if a given part is used on tens, or even hundreds, of products. For this reason some companies ignore the reference and use the bill of materials directly to trace the usage of a part. In this way the bill of materials relates every single component that forms a product to all the others. To find out more detail about components one moves down the bill to progressively more detailed drawings and, to gain more of an overview or see how components relate to each other, one moves up the bill to drawings that progressively show more and more of the product but that have less detail.

Figure 3.14 shows a complete assembly in which the component shown in Figure 3.13 is used. Such a drawing is known as a *general assembly drawing*. Item 6 on the bill of materials is the top plate, drawing No. MCH6234. In the details of the drawing the top plate can be seen surrounded by the other components that form the phase change and sensor assembly. However, in this drawing none of the top plate details are shown.

Management of drawings

The system of describing physical objects outlined above is very effective and is adopted in some form by practically every engineering organisation in the world. It has clear advantages of economy, thoroughness of description and ease of interpretation. However, these advantages can soon be lost if the drawings are not managed in an intelligent way. In the following sections we will mention the most

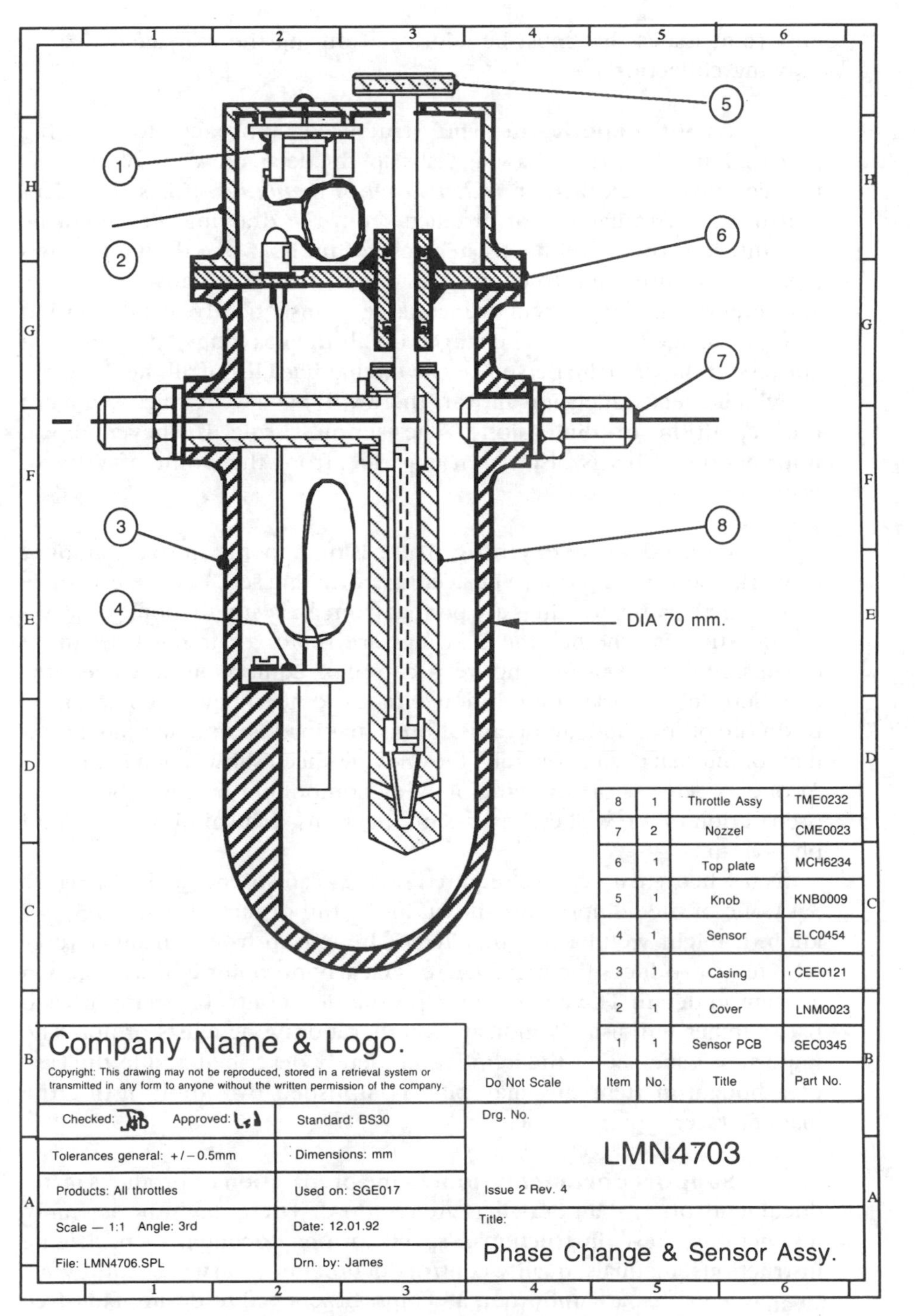

Figure 3.14 A general assembly drawing

important issues that must be addressed during the management of a drawing collection.

Avoid repetition: The structured approach to drawing preparation, where each drawing 'calls up' the detail drawings of the parts that compose it is called an *indented set of drawings*. It is self-evident that quality manufacture of products from the drawing set cannot be guaranteed if there are inconsistencies within it. If one drawing shows a particular dimension to be one value whilst another shows it to have a different value, errors will be inevitable. Consequently, if a dimension is to be changed, it must be changed on all the drawings upon which it appears. If the draughtsperson is to avoid having a list of all the drawings on which every dimension appears there will have to be rules to govern the repetition of dimensions. The simplest rule is 'never repeat information'. This is also a sensible rule from the point of view of economy.

Don't draw everything: There is no upper limit to the amount of work that can be put into perfecting a drawing set. The issue at stake is not whether the drawings are perfect from an aesthetic point of view, the question is whether the drawings are fit for their purpose. In an organisation mass-producing very complex equipment the need for complete and accurate description extends to nearly every component. If, on the other hand, an organisation is making just one demonstration unit for internal evaluation only, the drawings need show much less detail. At one extreme is the drawing of every component, at the other is the preparation of a few sketches, circuit diagrams, critical dimensions and photographs.

Even when great detail is required savings can still be made. In Figure 3.14 one of the components shown in the bill of materials, item 5, is a knob. It might well be that this item is bought in from a manufacturer who mass-produces the item. There is clearly no point in drawing such an item in detail. The entry on the production control system for each part number will usually indicate whether a drawing exists or not. The important issue is that the item is completely described and in the case of a bought-in item this may be accomplished by simply listing the manufacturer's part number.

Support documentation: In the preparation of products many documents other than drawings are required. These might be assembly instructions, test instructions, specification documents, packaging instructions, manuals, quality control documents, safety procedures. All these documents are important and must be obeyed to ensure fault-free production. With an intended drawing system it is easy to relate each of the documents to its assembly by the part number. A database of all the part numbers, often held on computer, states which documents apply to the part in question. This technique allows the organisation to have as much or as little detail as is appropriate yet still know for each part what documentation applies. In some industries there are specifications that apply to the preparation of such documentation. For instance, some military specifications state that every component must be drawn and whenever two components are joined together a procedure, referenced in the drawing, must be followed to assemble them. This process may extend throughout assemblies as complex as complete jet fighter planes. It is clearly onerous and involves copious amounts of documentation but it does bring excellent product control. It is estimated, for example, that the documentation associated with a jet fighter weighs far more than the plane itself. The drawing structure therefore provides a framework upon which other documentation may be referenced.

Drawing preparation and release: In an engineering organisation using an indented system of drawings it is clear that the production of drawings can become a major management issue. It is common for a separate department to handle the drawings, although many organisations also operate a distributed approach with draughtspeople assigned to the individual departments that need them. Many organisations have their own standards or have adopted existing standards, and draughtspeople often have a lot of in-house standards to learn when they join an organisation. Computer sytems greatly ease the labour associated with maintaining a drawing set. Modern computer draughting systems can be purchased at practically any level of complexity, from the electronic pad through to integrated draughting, production control and manufacturing systems. They bring with them the capacity to manipulate the drawings with previously unachievable speed and versatility, but their management is not straightforward and costs can be prohibitive.

The aim of the drawing production system is always to be able to provide accurate drawings to meet the needs of the other areas of the company. In addition to the preparation of the drawings there are two particular areas that need special attention in order to achieve this aim. The first is checking and the second is drawing release.

The first issue which is vital to drawing production is *checking*. It is clearly important that the drawings that are used throughout the organisation are correct. As the design of a product improves it may well be the case that modifications to the components within a product are required. For this reason the drawings that define a product are often being changed and so it is not a question of simply getting the drawings right once and for all, but rather of developing a system of knowing if the drawing is up to date. There are several points of view from which a drawing must be checked — from its draughting integrity, its design, its ease of manufacture and its suitability for use with other drawings. Responsibility for checking drawings from a draughting point of view is usually given to a draughtsperson other than the one who drew them and a box is provided for an approval signature, as shown in Figures 3.13 and 3.14. Responsibility for checking the technical content of the drawing is usually given to one engineer.

Drawing release means sending copies of the relevant drawings to the manufacturing areas that require them. In an organisation running a formal production control system there will be an organised process to respond to incoming orders and ensure that all the relevant departments receive the drawings they require. Exactly how this procedure operates will depend on the nature and size of the organisation concerned; however, it is easy to see that several departments will require the same drawing. Sometimes it happens that a drawing that needs to be released has some modifications required that have not yet been implemented. It takes time to modify drawings and if the release of the drawing is urgently required something must be done to get round the problem. The solution is to issue 'marked-up prints'. This describes the process of taking a print and then making the modifications in long hand on the drawing, on every copy that is required. The modifications must be initialled by a draughtsperson and an engineer in order that the modification is authorised and to prevent the organisation working from unofficial marked-up prints. The modifications may then be made later. Clearly there is room for error in this technique and it is therefore normally only used in exceptional circumstances.

The purchasing department may need to refer either to the drawing itself or to the bill of materials in order to purchase raw materials. The manufacturing function will clearly require the drawings in order to machine and produce the component. There may well be several different production versions of a drawing showing the different stages of manufacture. After this the inspection department will require drawings with which to inspect the components; often these are separate 'inspection' drawings showing the procedures and important issues. The test department may require drawings in order to test and optimise a product and finally there will often be a packaging drawing showing the way in which packaging is to be carried out and what items are to be included for shipment.

3.6.4 The drawing office

The primary role of a drawing office is to control the drawings used by the company. Depending upon the way in which the design activity is organised, the drawing office may have responsibility for producing all or some of the company's drawings and it will usually have responsibility for carrying out any modifications to drawings. Some companies keep design and drawing as separate areas, in which case the drawing office provides a service to the designer by producing engineering drawings based on sketches provided by the designer. It is becoming more common for drawing offices to be seen as extensions of 'design offices', with designers using a computer-aided design package and then producing their own drawings.

Irrespective of the organisation of the drawing office, the control of drawings will have to be carried out at three stages: during preparation, prior to issue, and when finalised.

During the design process and the preparation of drawings there must be a well-defined system for revision, particularly following design review. Because the product is evolving during design the drawings must also evolve, reflecting the current state of the product, which can make control at this stage quite onerous. It is essential that the system used to carry out this control is appropriate to the level of control required, the company and the product, and it must not inhibit the design process. The control method should ensure that the reasons for design changes are adequately documented. There must also be effective controls to ensure that drawings in preparation are not inadvertently released to the shop-floor for manufacture.

During drawing preparation any accepted or required standards for drawings must be adhered to, although if using a separate design and drawing office these standards may only be applied to drawings prepared for use by manufacturing.

When a product is finally developed the drawings will form part of the product definition package and it is therefore important that before the drawing is deemed 'final' it is checked and approved. There should be a system for checking the drawing for dimensional accuracy, that drawing standards have been followed, that the drawing is complete and that the design specification has been met. Following this check there must be a procedure for approving the drawing for manufacture.

Parts lists and bills of material may be produced by the drawing office when the final drawing set is complete, or they may be produced and updated throughout the design process, with the drawing office having responsibility for checking part numbers and format before they are entered into the production control system.

Following completion of the drawings the drawing office's responsibility is then to release, update, store and retrieve the drawings as appropriate. The release procedure is the way in which copies of the master drawing are made available to others for the purposes of manufacture, review, preparation of quotation and so on. The procedure used must ensure that the people utilising the drawing always have the current drawing available, even though this may mean that a new set of drawings is released every time a new batch of product is to be made.

Effective storage and retrieval of drawings is important not only because of the requirement to release drawings to the operations department for manufacture, but also because the drawings form a record of the product at each stage in its life. The importance of this is illustrated when a customer who bought a product some time ago returns and either wants more products built to the same specification, or wants spare parts for the existing product. The manufacturer must know what was used in that product.

Drawings will also have to be accessed and revised as the product is modified.

3.6.5 Research

Companies often carry out development work under the general title 'Research and Development' (R&D). However, the term 'research' really relates to work that is speculative in nature and consequently risky. It is risky because, although the company knows what it is working towards, it does not know whether or not it will achieve its goal. Another term that you may hear is 'blue sky' — this is research that is conducted with few constraints. R&D is important because it provides both new and next-generation products. Many companies also use the fact that they carry out extensive R&D as a marketing tool, and to attract new recruits into the company.

In Figure 3.12 we saw the idealised product development process. The customer's needs start the process off and lead directly to the preparation of the design specification. However, various things impinge upon the design specification and limit what may be included in it. The diagram identified four such factors: (i) customer expectations; (ii) market possibilities; (iii) the laws of physics; and (iv) the organisation's technology. The effect of the last item is of special interest to engineers since they are people who are in a position to change the technological abilities of the organisation. When they investigate such a possibility, engineers are performing research.

In its scientific sense, the word 'research' is used to describe enquiry into the fundamental workings of nature and is often a very limited intellectual process. In commercial engineering the word is used in a much broader sense and describes extending the technologies of the organisation. This may involve fundamental research, as is the case with, for example, the microprocessor manufacturers who constantly research the fundamental operation of their miniature circuits. Alternatively, commercial research may involve bringing established technologies into an organisation, for example a manufacturer of small metal equipment cases may wish to expand into supplying plastic products too. The use of injection moulding for this purpose is not new in principle but it may well be new to the organisation. Research will have to be carried out to see what effects the change might have, how it will be managed and how the new products and manufacturing techniques will fit into the organisation.

The process of research is often imagined as the stereotype of mad professors in isolated laboratories proposing impractical gadgets that have no real customers. In practice, research could hardly be less like the stereotype. Successful organisations have well-managed research laboratories peopled with intelligent scientists and engineers who maintain the technical position of the organisation within their budgets. Research is, by its very nature, an investigation into the unknown. The characteristics that bring success at it are creativity, fertile imagination, technical excellence, good personal management and an ability to work with others — all characteristics that attract engineers. It is clear that managing a research function will have some problems that are peculiar to this unique discipline; however, it is also clear that the function must be managed, as must all the functions, to ensure that the overall objectives of the organisation are achieved. Research is usually a very expensive pastime and if it is not well managed it can rapidly consume an unacceptable amount of profit. There is usually no upper limit on how much money can be employed to solve a research problem, and good research managers are therefore those that produce the results required but only consume the resources they originally asked for.

Communication issues and links with other departments are very important in research. Good communication is required because of the importance of establishing customer needs and determining manufacturing constraints. It is also of vital importance that the research personnel are able to communicate the importance of what they have designed and researched, and why. They will have to sell their ideas to others in the company and this may not be an easy task. Typically the work involves a vast number of choices of routes to follow and ideas to evaluate. Keeping track of the work and chronicling the decision process becomes a considerable task in its own right; for this reason documentation quality is always under scrutiny in the research department. In addition, due to the nature of the work and the long time-scales that can be involved, the company needs to ensure that it is able to carry on with a project if a member of the team leaves; new members will have to understand why certain decisions were taken or why something was done in a particular way.

In some ways the strategic management of research is similar to that of other functions. The organisation has its own agreed mission statement and from this the executive management produces objectives for each department and successive layers of the organisation until everybody has their own personal or departmental objectives and budgets. Clearly, objectives for research activities are much more difficult to set than for other, more predictable, activities but plans can still be prepared and Section 5.4 gives details of some techniques that can help. There are many issues involved in the successful management of research such as how much profit should it consume, how radical may the organisation's new products be, and how is the link between research and the customer to be made? These are complex questions and complete textbooks have been written about each one.

Research can take a great deal of money to finance, with a risk on return — therefore the financing can often be difficult. Some companies carry out very little research, preferring to develop markets rather than products, others license products (which means that they pay for the right to manufacture a product that has been developed and tested by someone else); other companies fund research from profits or use external funding provided by government agencies or customers. Some spend as much as they can on research and others spend as little as possible.

Because of its very technical and innovative nature the people involved in research tend to be more highly qualified and specialised than in other areas. Also, in order to suit the nature of the work, the organisation in research tends to be much less formal and more relaxed than elsewhere. Personnel, project and financial management are all very important issues in research, as are policies on publishing results, intellectual property rights and patenting.

In summary, research is the process of extending the organisation's technological abilities. Excellence in technological provision gives an organisation an advantage over its competitors and so is of special interest to organisations. The process of research is a creative one and, by its very nature, research involves investigation into the unknown. For this reason it is intrinsically difficult to plan; however, it is clearly important to the success of the organisation that research is controlled and that the results that are required by the organisation are identified with an acceptably small level of expenditure. The strategy employed by the organisation to provide its research needs can take many forms, ranging from a complete and separate department through to a subcontracted project placed with another organisation altogether. How research is achieved is not particularly important; what is important is that the technological needs of the organisation are provided for.

3.7 Quality

Quality means many things to many people; however, in the engineering industry there is a clear understanding of the term provided by the definition from the international standard ISO 8402 Quality Vocabulary which gives:

> **Quality: The totality of features and characteristics of a product or service that bear on its ability to satisfy stated or implied needs.**

The ISO 8402 definition of quality is quite clear but many other definitions are also used, usually incorporating ideas relating to meeting customer requirements and cost. However, quality should not be confused with expense. For instance, high quality doesn't have to mean high cost — a cheap pen that does the job for which it was intended and is then thrown away can still be a quality pen if that was what the customer expected. There is also confusion sometimes about the terms 'quality assurance' and 'quality control'. Quality assurance is a management tool consisting of a number of activities that provide confidence that quality requirements can be met, whereas quality control is one of those activities and is concerned with determining whether or not a product, or service, meets the specification set. It follows from these definitions that any engineer will therefore have some involvement with 'quality'.

In this section we will examine what a quality system is and consider the implications of operating a formal quality system. We will then consider how the management of the quality function can be achieved, the specific requirements for a quality system, and we will examine the cost of establishing and operating such a system.

3.7.1 Quality systems

The definition of a quality system, from ISO 8402, is:

> **The organisational structure, responsibilities, procedures, processes and resources for implementing quality management.**

Most companies have some form of quality system which ensures that the company's policies relating to quality assurance are implemented, used and monitored, many of these following the requirements for quality systems given in the international standard ISO 9000 Quality Systems (equivalent to BS5750). Total Quality Management (TQM) is an extension of this standard, although there is no written standard for TQM. The philosophy behind TQM is that quality should pervade all aspects of the business of the company, whereas the standard is limited to certain sections of the business. In addition there is much more emphasis in the TQM philosophy on employee and customer involvement.

The aim of any quality system is to provide a systematic approach to design and manufacture in order to prevent failures, and to provide objective evidence, usually documented, that an agreed quality level has been achieved. If the quality system meets the requirements of ISO 9000 it can be registered as such by an independent third party approved by the National Accreditation Council for Certification Bodies (in the UK). This reduces the requirement for assessment by individual customers, and also the need for customers to define their own quality system requirements. Registration may also mean that a company can address new markets and thus increase sales. Many markets require very high standards of reliability and traceability, for example the nuclear market, the defence industry and the medical market. The customers in these markets require manufacturers to operate proven and effective quality systems; registration provides assurance that this is the case. In addition, as more companies are achieving either a total quality system or registration to a quality system standard, they are requiring their suppliers to have similarly rigorous procedures to assure the quality standards of their goods and services.

There are many other advantages of having an effective quality system, some of which relate to the aspects of the system aimed at preventing defects. Any company needs to make profits to survive and broadly speaking the profits are determined by the difference between the cost of making a product and the price at which it is sold; therefore any activity which cannot be charged as part of the manufacture cost reduces the profit, as does any activity which adds no value to the product. Two activities that have these effects are rework and warranty work. Rework is that work carried out in order to rectify faults or problems — these could be due to faulty equipment or poor workmanship, or even damage due to the part being dropped. The term generally refers to work that is done in the factory prior to shipment of the product, and the extra costs associated with rework will include parts costs and labour costs. Warranty is the work that is done under guarantee, after the product has been shipped to the customer, although it may be carried out either back at the factory or on the customer's premises. The costs associated with carrying out warranty work will include shipment and packaging, service engineer's time and travel, parts and labour. The use of an effective quality assurance system will reduce the need for both rework and warranty work, thus reducing the costs that would be set against the profits.

Any parts requiring rework or warranty work will have to be dealt with in addition to the normal production activities. This has two implications: if the amount of extra work is considerable, either extra capacity has to be maintained in order to avoid normal production schedules being disrupted or disruption has to occur, causing delays to customers and reducing the efficiency of the manufacturing organisation. Consequently, the reduction of rework and warranty work will reduce the potential disruption, or alleviate the need for extra capacity.

It follows that if a company has a reputation for producing high-quality and reliable goods, then the company's sales potential will improve. If they are also able to demonstrate that their quality is better than the competition, this will improve their sales — thus registration can be a powerful marketing tool. Advantages relating to more rigorous documentation and assessment of product, as well as more rigorous design and manufacture controls, include improved safety and reduced risk of facing product liability charges.

Quality systems not only provide these very tangible benefits but can also improve worker morale and motivation by increasing interest in the job, giving a feeling of pride in the work.

There are potential problems with quality systems, such as them being overly bureaucratic, inappropriate to the company, ineffective and very costly. These areas should be considered carefully and addressed when a quality system is being implemented. We will look in more detail at these areas as we consider the requirements and operation of a quality system.

3.7.2 Organisation of the quality function

Quality control is all the activities that are carried out to ensure conformity of parts or products to specification, and so will include both inspection and test activities. The way in which a company deals with quality control is usually based on the way in which 'quality' was introduced into the company. In many companies there are inspection and test departments which may be a part of the operations department, part of the quality department, or a department in its own right reporting to a chief inspector who in turn reports to a senior manager or director. However, increasingly more inspection and test is done at source by operators, thus giving them 'ownership' of their work. It is important that authority to stop defective material and parts is also given to the operators working in this way, this authority always being part of an inspector's job.

Due to the modern views of quality management, particularly because of the concept of 'Total Quality', quality as a function is now more commonly found spread throughout the organisation. However, many companies do retain a quality department which has responsibility for implementing, co-ordinating and monitoring procedures to ensure that agreed quality levels are maintained. In this case the quality department is similar to the personnel department in that it does not have direct authority but it will need to integrate with all aspects of the business. It will also have to liaise with outside organisations such as certification bodies, customers and suppliers.

Management of the quality function, as with the personnel department, is complicated by this lack of direct authority but many companies alleviate some of the problems by giving a director, or a senior manager, responsibility for the function. Indeed if the company operates in accord with ISO 9000 the standard requires this to be the case. It is of equal importance that the responsibility and authority of everyone working within the quality system is clearly defined so that they too can carry out the tasks assigned to them.

Operation of the quality system relies on four major factors: the quality manager, the quality manual, the procedures and audits.

The quality manager

The quality manager is the person with defined responsibility and authority for ensuring that the requirements of the quality system standard are implemented and maintained. This person should have senior manager status, and should report directly to the Board of Directors in all matters concerning quality, irrespective of other duties.

The quality manual

The quality manual is the operating document for the system. It may contain all the procedures that are operated in order to meet the requirements of the quality system, but is intended to give an overview of the complete system.

Procedures

These are the actual documents that define the tasks to be carried out in order to meet the requirements of the system. Any procedure must give the job title of the person responsible for carrying out the task, and the job title of the person that has authority for making decisions about the work being undertaken. Procedures clearly define the way in which the task is to be undertaken and the records that are to be kept in order to verify that the task has been carried out correctly.

Procedure writing requires much care and practice to ensure that ambiguities are avoided and that the steps involved are complete and appropriate to the intended user.

Audit and review

Auditing and review of the quality system should be undertaken by the senior managers of the company as well as by an accreditation body, if appropriate, in order to examine evidence to:

1. Verify the existence of the system.
2. Check that the system is being followed.
3. Check the effectiveness of the system.

In order to maintain objectivity audits should always be carried out by people that do not have direct responsibility in the area being audited.

3.7.3 Main activities in the quality system

The main activities that form a quality assurance system fall into the following areas: contract review, design control, purchasing, manufacturing control, inspection and test, calibration and maintenance, nonconforming product disposition, product identification and traceability, and handling, storage and despatch. In addition there are a number of activities that relate to all of these areas, these being the system procedures. The system procedures are document control, records, corrective action, training, and audit and review. It must be stressed that the way in which companies address these areas must be appropriate to the company and the way in which it operates. There are no universally effective procedures.

We will now describe each of the quality system areas of activity.

Contract review

Contract review is the process by which you ensure that you can meet your customers' requirements. It may require some sort of review meeting before tenders or quotations are sent out, or it may just be an approval of delivery time from the operations manager for a catalogue item. It will involve design, manufactuirng and marketing people at the very least.

Design control

Design control ensures that when design is carried out the requirements of the design are fully specified, and that when the design is complete it is reviewed to ensure that the specification has been met. A major aspect of design control is the design review, which should be undertaken by someone other than the person responsible for the design. The design review should ensure that all aspects of the design have been considered. It should be an ongoing process, and not just take place at the completion of the design. It is also important that a recording system is operated for recording assumptions made in design, recording criteria for component selection, and noting any critical calculations.

It is in design that quality levels are set, as it is the design that dictates the parts and processes that are to be used in manufacture. Care must be taken when defining a design control procedure to ensure that it does not add any unnecessary delay to the system.

Purchasing

Purchasing is concerned with obtaining goods and services. In purchasing it will be necessary to define clearly what is required in terms of technical specifications and quality level, as well as contractual specifications such as delivery time. However, there is no point defining a tight specification for a product and then placing the order with a company whose abilities and probability of successfully meeting the requirements are unknown. There must be within the purchasing activity a method for assessing the suppliers and determining what specifications they can achieve, in order to allow you to decide whether or not these are appropriate to your company's requirements. Supplier appraisal involves an initial assessment of potential suppliers to ensure that they can meet your requirements and then becomes an ongoing activity of monitoring the quality of supplier services and goods. Supplier appraisal may involve visits to suppliers and audits of their systems or you might accept suppliers on the basis of them having a quality system registered by an independent third party.

Manufacturing control

When an organisation has a well-defined product and parts that meet the required specifications, it needs to ensure that the manufacturing activities that are used are adequate to produce the right quality of product. Manufacturing control ensures provision of the instructions that people need to carry out particular processes. It is normally achieved using work instructions or procedures, which not only define the processes but will define the skill levels required and the level of workmanship that is acceptable for any given process on a specific product. In many engineering companies work instructions take the form of drawings.

Process control

Process control involves control of the actual processes used, such as machining and moulding, whereas manufacturing control is more concerned with the overall manufacturing function. In order to control processes you need to identify the actual processes used, determine the process specifications and institute a procedure for monitoring to ensure that processes are carried out within that specification.

Some processes are defined as special processes and these are those that cannot be monitored whilst they are being done: it is only possible to carry out monitoring of quality after completion. Welding is a typical special process. For special processes the way in which you achieve control is by ensuring that equipment and operators can meet the required standards — this may require special training and qualification for operators.

Inspection and test

Inspection and test procedures define the quality control that is to be carried out on the manufactured product in order to ensure that it meets the design specification. They have to define the actual form of the inspection or test, the results expected including the tolerance bands, and what should happen to inspected product. They should also define the level of qualification required for the inspector/tester.

Calibration and maintenance of equipment

There is no point in inspecting and testing unless you have confidence that the equipment being used to carry out the inspection or test is capable of achieving the required level of accuracy, and that at the time it is used it will achieve that level of accuracy. It is therefore essential that in any quality system where there is a requirement to carry out either inspection or test there is a procedure that, if followed, ensures that equipment is appropriately calibrated and maintained.

Nonconforming product disposition

A nonconforming product disposition procedure acknowledges that sometimes things go wrong. Nonconforming products are those parts, or products, that do not meet specification, for whatever reason. The procedure must provide a method for identifying the parts, taking them out of the manufacture process, and then dealing with them in the most cost-effective way. Some options are to reject and scrap parts, to reject and rework parts, or to ask for a concession from the customer. The concession is an agreement from the customer to take the product even though it does not meet the original specification. Each of these options has a cost and when determining the option to be used this cost and the effect on the business have to be carefully weighed.

Product identification and traceability

Product identification ensures that at any time you know the inspection status of material and can therefore prevent non-specification material from entering the manufacture process or being sold. Traceability was discussed in Section 3.2, and involves being able to trace parts and products to their source.

Handling, storage and despatch

Methods for assuring the preservation of product quality are required at all stages of manufacture and shipment, until the product is no longer the liability of the manufacturer. This might involve anything from ensuring that delicate goods are not stacked on top of one another, to defining special packaging that will protect the equipment as it is being loaded into an aircraft.

Document control

Document control is the way in which documentation is managed to ensure that it is appropriate, accurate and available where it is needed. It is particularly important in the manufacturing area where it is essential that up-to-date drawings and part specifications are used. Document control is fundamental to the effective running of a business and is therefore dealt with in more detail in Section 3.7.4.

Records

Records will be required throughout the quality system: they provide documented proof that procedures have been followed. You need to ascertain what records are required, what information is needed, the most appropriate method for keeping the records, and there should be a system for maintaining them.

Corrective action

Corrective action is another procedure that acknowledges that things go wrong. In this case it is used to review areas of the company systematically and objectively and identify any problems with the operation of the quality system in order that they can be resolved. It also defines the way in which problems will be followed up to check the effectiveness of the solution.

Training

Everyone should be trained to do the job that is expected of them. In order to ensure that this happens, a company should have a training policy and a procedure that ensures that training is reviewed at regular intervals. The subject of training is dealt with in detail in Section 4.5.

Audit and review

When a company has procedures in place to cover all these areas it needs to know whether they are effective. It will therefore require some form of internal audit and review to take place on a regular basis. An audit is an objective analysis of the way in which procedures are operated, concentrating on looking for evidence of how things are being done. Audits should always be done by someone who does not work in the area being audited and by people who are trained in auditing techniques.

3.7.4 Document control

Document control is the means by which companies ensure that the correct documents are available where they are needed and that obsolete documents are removed from use. It also allows changes in documentation to be monitored and controlled.

In companies operating a formal quality system, such as that defined by ISO 9000 Quality Systems, there is a specific requirement to have a document control procedure. However, it is also necessary business practice in manufacturing.

A document control procedure should cover all the documents that define the product, although it may be extended beyond this if a company feels it appropriate. The types of documents that should be included are as follows:

bills of material
part and product specifications
drawings
assembly, test and work instructions
process instructions and specifications
instruction manuals
operating procedures
inspection instructions

The way in which these documents are controlled is first of all by putting control information on the document and then by regulating modifications and the distribution of the document. Figure 3.15 shows a document with control information.

The unique identifier for each document usually follows a particular house format, as in Figure 3.15 using QP to indicate Quality (system) Procedure. It ensures that there can be no ambiguity when referring to documents. This is particularly important when using part numbers to identify documents and you need to differentiate between the drawing for a part and its bill of material.

The issue number shows how many times the document has been issued after being amended in full. Initially the issue number will be 1 (the original document). At this stage the revision number of the document will be 0. When an amendment is made the document will be redistributed with an increased revision number and the amendment clearly shown. However, it is usual after a number of small revisions or after one or two large revisions to re-issue the document so that the issue number increases and the individual modifications are incorporated fully into the document. After this process the revision number reverts to zero.

The number of pages should always be shown on a document so that the recipient can be sure that they are in receipt of the complete document.

The approval signature and date shows that one person has authority for that document and has taken responsibility for its accuracy and

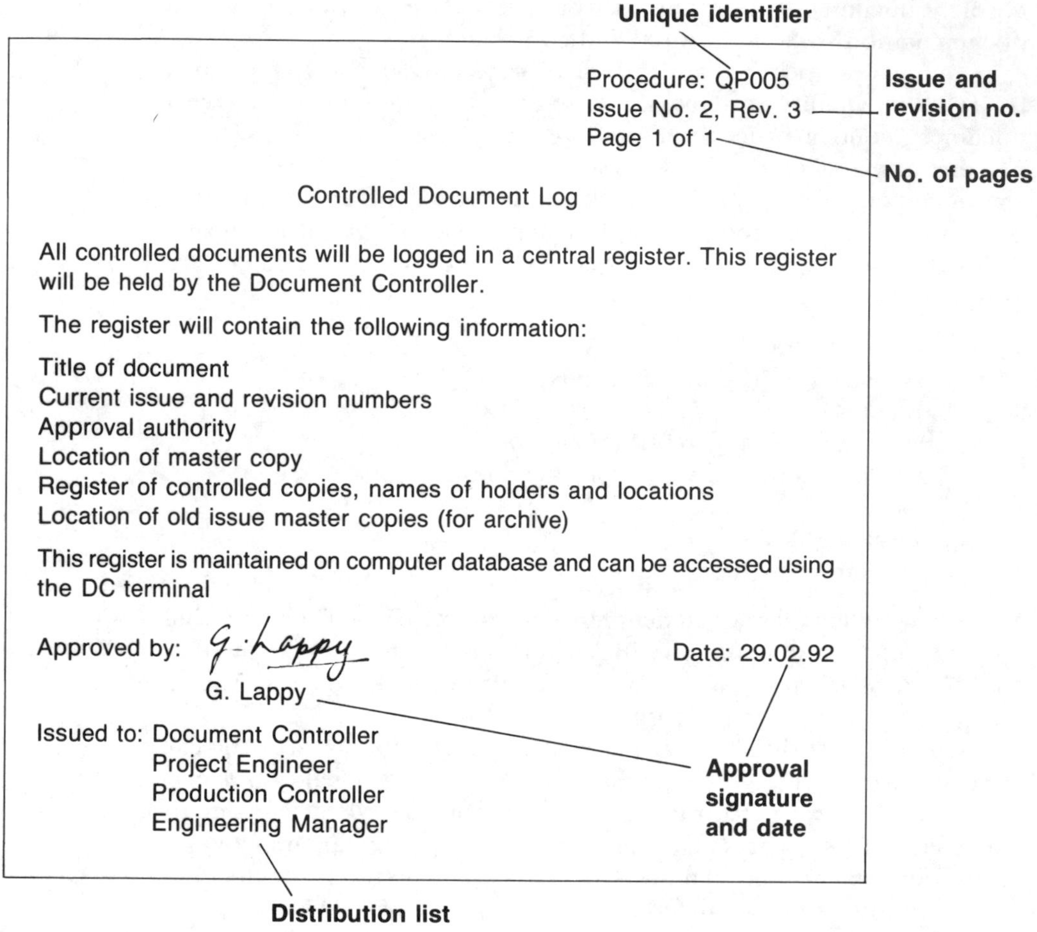

Unique identifier

Procedure: QP005
Issue No. 2, Rev. 3 — **Issue and revision no.**
Page 1 of 1 — **No. of pages**

Controlled Document Log

All controlled documents will be logged in a central register. This register will be held by the Document Controller.

The register will contain the following information:

Title of document
Current issue and revision numbers
Approval authority
Location of master copy
Register of controlled copies, names of holders and locations
Location of old issue master copies (for archive)

This register is maintained on computer database and can be accessed using the DC terminal

Approved by: G. Lappy   Date: 29.02.92
G. Lappy

Issued to: Document Controller
Project Engineer
Production Controller
Engineering Manager

Approval signature and date

Distribution list

Figure 3.15 Document with control information

completeness. Subsequent modifications to the document should not be made without further approval from this authority. This is particularly important when changes are being made to design or specification because the approval authority will have access to information about why a design was made in a certain way and will be able to evaluate the effects of the proposed change.

The distribution list indicates for whom the document has been copied.

In order to control the document a master list of all documents has to be kept; this may be called a *log* or a *register* but should contain the information, defined below, relating to all controlled documents.

The titles and identifiers
The current issue and revision numbers

The approval authority
The location of the master copy
The distribution list for holders of controlled copies of the document
The location of old issue master copies

The document log ensures that at any time it is possible to identify the current version of a document and who has copies of it. When a document is to be redistributed the people on the distribution list can be contacted and the old copies of the document withdrawn. This is particularly important if one is to avoid old versions being used inadvertently. New copies can then be issued.

In some circumstances a company might not wish to have all copies of all documents controlled, for example if a drawing has been sent to a workshop for quotation, or if a product manual has been supplied as a sample. In these cases the copies must be clearly labelled as 'uncontrolled' and the recipient must understand that these will not be updated and should not be used for manufacture.

Document control procedures should be as simple to use as possible and avoid excessive bureaucracy. Only in this way will they be operated effectively.

3.7.5 The cost of quality

When considering quality systems, particularly in smaller companies, there is a cost of implementation in the short term and in order to justify this cost you need to have a means of calculating quality costs and potential long-term savings. Unfortunately very few companies have such a system.

Quality costs are the costs that are associated with not getting things right first time and are generally defined by three categories: appraisal, prevention and failure.

Appraisal costs are those that come from trying to find out if goods meet the specifications set, and are therefore fairly easy to calculate. In a manufacturing company appraisal costs will include the cost of inspection personnel, and the cost of equipment and overheads to maintain an inspection department.

Prevention costs are more difficult to quantify because they arise from the tasks performed to try and stop substandard goods being produced. Prevention costs would therefore include the cost of training people in order that they can achieve defined levels of workmanship.

Failure costs fall into two categories. The first category is internal failure cost and this should be reasonably easy to calculate: it is the cost associated with scrapping, reworking and warranty. The second category, external failure cost, is the most difficult to quantify of all the quality costs. This is the cost associated with failure outside the company and would include the effects of customers not buying your product again, or telling others that your product is not worth buying.

So in calculating the cost of having a quality system many companies are faced with a dilemma because they do not have accounting procedures that allow accurate calculation of the cost of poor quality. Against this are the costs of implementing and operating a quality system which are all too easy to quantify and which can be excessive if care and judgement are not exercised at the planning stage.

The costs of implementing a quality system will include personnel costs — someone must be available to implement the system and co-ordinate the various activities. The training bill will be very high as employees will have to be trained in the use of the quality system, and there may also be a need for specific technical training. Training costs are often high because not only must the direct costs of paying for training and expenses be allowed for, but also time that is not being used for manufacture. Documentation costs can be high as they involve the time to produce the documentation, as well as the costs of stationery, printing and secretarial services. In addition to all this there may be costs because of the need to purchase equipment for inspection or test, or to have current equipment calibrated.

If a company decides to have its quality system assessed with the aim of accreditation, then there will be costs for registration and audit.

3.8 Personnel

Aspects of the personnel function will be carried out in all departments throughout the company. Because of the need for every manager to manage personnel this topic is covered in detail in Chapter 4. However, this chapter is examining all the functions within organisations and so in this section we will look at how the personnel function can be organised and consider some of the activities that have to be carried out.

3.8.1 Organisation of personnel

Anyone carrying out a management role will be involved in the management of personnel. However, a personnel department will normally act as central co-ordinator for this activity and will both prepare and monitor personnel policy. In many small companies the personnel department will be one person, perhaps in a part-time post.

3.8.2 Activities of a personnel department

An important activity of a personnel department is a strategic one, defining the personnel policy for the company. The personnel policy will define the procedures and rules that will apply when managers in different departments deal with personnel issues. The personnel department may deal with some issues but generally personnel staff will have no line management responsibility outside the department. Their prime tasks will be to establish personnel policy, provide advice and training to people who will carry out the policy, and monitor the effectiveness of the policy to ensure fairness for all employees.

Some of the specific personnel issues that will have to be dealt with are described below.

Manpower planning

Manpower planning matches the company's requirements for people and skills with those that are available. It should highlight deficiencies and surpluses in advance of them making an impact on the business. Manpower planning will be particularly important in the manufacturing area but will have to be addressed throughout the company. The personnel policy should indicate how manpower planning is to be performed, and possibly when; it should also indicate any constraints that managers will have to work within.

There are two levels of manpower planning, the highest level being the overall company requirements to meet corporate objectives; these would be identified in the business plan and would not normally be very detailed. However, at a lower level the line managers in each department would have to plan in detail their manpower requirements in order to meet department objectives. For example, the business plan might say that in order to meet expansion plans the manufacturing area will have to grow by three people over the year. However, the manufacturing manager will have to specify what those people are to do and exactly when they are required. The manager will also have to plan in more detail to allow for holidays and sickness amongst personnel in the department.

Employee appraisal

Employee appraisal is used to assess performance of individual employees, following discussion, and then allows decisions to be made about the employee's future progress in the company. If an employee appraisal system is used the personnel policy should define the way in which the appraisal is carried out, what records are made, and how often appraisal should take place. The personnel department would be responsible for ensuring that the managers carrying out appraisal are trained to do the job, and it would be responsible for monitoring the application of the appraisal system to ensure that all employees were being treated fairly and that consistency is maintained throughout the organisation.

Recruitment and selection

This activity involves identifying requirements, following manpower planning, and then preparing job descriptions, advertising and carrying out selection interviewing in conjunction with managers. It should also include induction of new staff. The personnel policy should define the recruitment and selection procedures to be used. The personnel department would normally provide appropriate training and advice to managers actually involved in recruitment. In addition it will play an administrative support role in distributing application forms, arranging interviews, sending out contracts and advertising the post. The personnel department would also be required to provide advice on the legal aspects of recruitment and selection.

Training

This activity involves the provision of training following from the needs identified during appraisal. It should also include monitoring and control of training provision, which is also a requirement of the quality function.

Health and safety

There are specific legal requirements for health and safety policies in companies and many will administer and monitor these policies through their personnel departments. There is also a requirement for first-aid facilities, first-aiders and an accident log, and these may be administered by the personnel department.

Welfare

The attention an employer pays to the welfare of its employees will affect levels of morale and motivation. The personnel department will be concerned with all policies to maintain and improve employee welfare. A counselling service for employees is often provided through the personnel department.

Consultation and negotiation

This may involve negotiating with unions, staff associations or individuals, on areas of company remuneration and working conditions, as well as in cases of dispute.

The personnel department will usually be responsible for preparing contracts of employment and for defining payment systems, both of which may require extensive negotiation with employee representatives.

Dismissal

There are legal regulations governing dismissal and all companies should have a policy regarding dismissal. The personnel department has to formulate the policies, monitor them, and may be required actually to undertake dismissal proceedings.

Records

A company needs to keep records for each of its employees, and these will be maintained by the personnel department. The records would include application forms, references, appraisal forms, pay and contract details, and any records of grievance or disciplinary proceedings. All of this information will have to be held securely and access limited, these issues being addressed in the personnel policy.

3.9 Company operation and the role of engineers

Companies have to carry out many different functions in order to conduct business. We have looked in detail at the major business functions in the last seven sections. However, it would be wrong to give the impression that each of these functions exists in isolation. For any company to transact business successfully there must not only be good communication between the various functions but also some meshing of the functions. Similarly, people cannot work in isolation and the smaller the company you work for the greater the need to work closely with others in different functional areas.

To illustrate how companies really do function and to show the roles that can be taken by engineers we have prepared three case studies. The first case is about Matthew Hobbs who works as a development engineer in a company employing sixty people. The second case is about Martin Twycross who, although an electrical engineer, is employed as an area sales manager in a company employing nearly 600 people. Finally we have the case of Simon Tu who is a quality engineer for a company in Singapore employing more than 1000 people.

Case study: Matthew Hobbs

Matthew Hobbs works as a development engineer at Warwick Power Washers Ltd, Oxford, UK. It is a small company employing approximately sixty people, and it has been trading for twenty-six years. The main business of the company is the design and manufacture of power washers, which are used by both domestic and industrial customers; they also sell the consumables, such as detergent, for the washers. Some of the larger industrial machines are made to individual customer specification; these will often take the form of making modifications to a product made previously.

The company has a technical department which is responsible for the design and development of all products; this is where Matthew works, reporting to the Chief Engineer. He graduated in mechanical engineering and has worked in the company for a year, since completing his studies.

The technical department consists of four people including the Chief Engineer, and Matthew is able to work in a semi-autonomous way, being able to organise his own time and priorities within the constraints imposed by the department's objectives. In designing new products Matthew is involved with the complete design from customer specification through to getting the product into manufacture.

The on-site sales team of the company consists of four or five people, but selling is also done through a network of distributors and by service engineers. This sales team has prime responsibility for customer contact; however, because of the nature of the business the technical department also liaises with customers. Matthew, like the other members of the technical department, is sometimes contacted by customers or dealers who make a direct input to the design process; he also answers their technical enquiries. Sales make the major input to the design process by passing on customer requirements in the form of a design specification, giving general customer feedback on product performance, indicating general market trends, and advising on new EC safety guide-lines. They also take part in the design review which takes place after a prototype has been built by the department.

The company has a buyer who is responsible for buying all the parts for manufacture. In the design process Matthew is responsible for ensuring that wherever possible standard parts are used; however, if this is not feasible then he has to define a purchase specification and, sometimes, source a supplier; the buyer has responsibility for negotiating contract terms after the sourcing has been done. In addition, when new subcontractors for machining processes are sourced Matthew may visit them to consider their manufacturing capability for possible future requirements; he will also gain technical information which can be used to improve product design. Very occasionally he may visit to check on the quality of goods and service, although this is usually undertaken by the company's quality engineer.

Following the design process Matthew is responsible for ensuring that the new product moves smoothly into manufacture. This involves preparing drawings, designing jigs and fixtures when required, and producing assembly instructions; bills of material are prepared by the department's technical administrator. Matthew will show the manufacturing supervisor how to build the new product. When the product is in manufacture he may be called on whenever there are production problems; he retains technical responsibility for the product. ■

Case study: Martin Twycross

Martin Twycross graduated with a BSc in Engineering, specialising in electrical engineering. For the first four years after graduation he was in the British

Army; for the last four years he has been working in the Sales and Marketing Department of Research Machines.

Research Machines is based in Oxford, UK. It was started 21 years ago and now employs 550 people in Oxfordshire. Initially the company supplied electronic components, but it now produces a range of industry-standard personal computers (PCs) and local area networks (LANs). It is a main supplier of PCs to the education sector in the UK.

Martin joined the company as a salesperson and following a series of promotions is now an Area Sales Manager; he has been in this post for three months. It has never been a company requirement for people in these jobs to be engineers — however, working in a technical company he does find the technical background helps in his work. Martin has three main responsibilities: meeting sales targets; managing his team of four people; and ensuring customer satisfaction.

In order to meet sales targets Martin is in touch with potential customers, discussing their requirements and proposing solutions. This may involve surveying buildings for LANs and specifying particular systems to meet the customer need. Because of his technical education Martin can do some of this work on his own, but in complex situations he will draft an initial specification and then discuss it with the technical department.

When the customer requires a system to be installed this is carried out by the customer support team. In these cases Martin will liaise with them and with the customer to ensure that all the equipment is available and that the requirements are fully understood.

Research Machines has a quality ethos, operating on the principle of 'Quality Improvement'. As part of this every group is required to have an ongoing quality improvement project. For the sales department this may be something like reducing the time to answer customer enquiries. The projects are presented to the Sales and Marketing Director. At this time Martin hasn't set a project for his group — this will be one of his management tasks when he has settled into the job.

Martin and his team provide customer feedback information to engineering, via marketing, as part of the product development programme. He feels that this way of distancing sales from engineering is important in the fast-moving environment of computer development. There is an inherent danger with direct links because of the need for sales to deal with the latest product whilst engineering are likely to be dealing with the next product. However, Martin may have some direct contact when there is a need for engineering to talk directly to the customer and in this he is the link. ■

Case study: Simon Tu

Simon Tu Yeou Mou has been working as a quality engineer for Seagate Technology International in Singapore for one year. He returned to Singapore after graduating from a British polytechnic with a degree in Mechanical Engineering.

Seagate Technology employs more than 1,000 people in its Tuas, Singapore, factory.

This site is responsible for manufacture of high memory capacity computer disk drives which are made to Seagate designs from the United States. There are eight functional departments within the Singapore operation. Simon is employed in the Quality and Reliability Department. This department is responsible for defining the requirements and responsibilities within the manufacturing operations to assure conformity with specified requirements. This involves appraisal of all parts and finished goods, auditing of suppliers, in-process inspection, tools and equipment calibration control, and clean-room environmental auditing.

Simon's main responsibilities include implementation and monitoring of statistical process control techniques. To do this he has to work on the shop-floor occasionally; however, he has a supervisor and two technicians reporting to him who do most of the shop-floor work. He is also responsible for carrying out quality system audits and establishing networks to disseminate quality system information. Obviously these activities require him to liaise with all departments in the company.

Specific tasks include certification of all products from the United States to Singapore. This requires the inspection of all incoming parts and involves qualification of the plant's capability by process auditing. Simon carries out these tasks and then presents his audit results to a meeting in which all departments are represented. It is Simon's responsibility to identify any corrective action required and he will also explain what the particular process requirements are. In addition he may give guidance on how the requirements can be met. However, the responsibility for taking the action will rest with the department concerned.

A further aspect of Simon's job is running quality improvement meetings. These meetings involve representatives from test, engineering, manufacturing and quality. On all issues relating to product or process quality Simon has authority, including the ability to shut down the whole product line if he feels that is what is required.

As can be seen, Simon's job is very much concerned with people in other departments and then presenting information which requires action from them. He obviously has authority in the company but must exercise judgement before wielding that power if he is to maintain the respect of the managers to whom he is presenting information. ■

In business none of the functional areas described can operate in isolation and in the three case studies we have seen some examples of how the functions interact in three successful companies by looking at the role of one person in each company. As an example of these interactions we can focus on the operations function and look at some of the issues that it must address while working with other functional areas. The operations function in any company will have to be able to communicate effectively and interact with every other function. The types of interactions required in a manufacturing company are as follows:

1. *Operations/Purchasing*

These functional areas must interact in order to determine what parts, services or equipment are required, and what are available, for manufacture.

2. *Operations/Sales and Marketing*
The interaction here will be concerned with what products can be made, and when, and what are required for sale.

3. *Operations/Finance*
These areas must work together to produce product costing and budgets; in addition there will be payment of wages, determination of spending restrictions, stock-taking and so on.

4. *Operations/Product Development*
This is concerned with the process and skill constraints imposed on product development by manufacturing. It also involves development for manufacture and feedback on actual process and testing.

5. *Operations/Quality*
This is concerned with quality assurance procedures to ensure product quality meets the correct standards, and with feedback on actual performance of product and processes.

6. *Operations/Personnel*
This interaction will involve many things including defining manpower requirements, recruitment, pay and welfare.

The case studies show that there are also many other necessary interactions.

3.10 Summary of Chapter 3

In the introduction to this chapter we identified the seven main functions that are required to operate a manufacturing business — purchasing and stores, operations, sales and marketing, finance, product development, quality, and personnel. We have examined each in turn and looked at some of the activities that form the function and at some of the factors that affect the way in which they are managed. We have also seen, by considering the role of three people working in engineering companies, that none of these functions can operate in isolation. In order for a company to transact business there needs to be many interactions between the different business functions.

3.11 Revision questions

1. What are the main functions to be found in a manufacturing company? What information transfers are necessary between each function and operations?

2. What is the role of a Store? Describe the main activities undertaken by a Storekeeper.

3. What benefits can a company gain from operating a central purchasing and stores system, and what are the disadvantages of this method of operation?

4. What data are needed to operate a Materials Requirements Planning system? How can these be generated?

5. What are the main activities carried out by a marketing department?

6. What is the role of marketing in the product development process?

7. In a project team charged with developing a new product what would be the role of a quality engineer?

8. What are the implications for companies moving away from the traditional engineering department plus drawing office to the use of a 'design office' where all engineers work on CAD stations?

9. Why is the ability to communicate well, both orally and written, so important for a development engineer like Matthew Hobbs?

10. What benefits can a manufacturing company expect to achieve by operating a quality management system?

11. How could an engineering design consultancy (five consultants) justify the expense of implementing a quality management system?

12. Determine what information flows are necessary between the personnel and quality functions.

13. How does the quality function impact on the product development function?

14. What is the primary role of a drawing office and how do its personnel interact with the operations function?

15. Why is it necessary for an engineer, working in product development, to have an understanding of finance?

16. (i) Define the terms 'quality assurance' and 'quality control'.
(ii) Describe the operation of a quality assurance department in a manufacturing company, indicating the main tasks that you would expect to be carried out by the department.
(iii) In what way is the operation of the quality department analogous to the operation of the personnel department?

17. A company manufactures standard products as required by customer order. The products are manufactured using parts and materials that are kept in stock as well as some parts that are ordered specially. All products are machined, assembled and then inspected prior to despatch.

Draw a flow chart of the way in which a customer order is processed until the product is despatched, indicating the production system documentation used at each stage.

18. (i) List six functions, other than manufacturing, that you would expect to find in a manufacturing company.
    (ii) Briefly explain some of the information transfers that are required between each of the six functions and manufacturing.
    (iii) Explain the term 'bill of material' and describe, with the aid of a flow chart, how this is used in operations scheduling using Materials Requirements Planning (MRP).

19. Why is document control important within an organisation? Make a list of points. For each point describe an example problem that would occur if document control was not used. What sorts of documents need to be controlled?

20. What benefits does document control bring to people outside the organisation?

21. Describe a method of document control.

22. A customer takes an obsolete camera into a dealer for repair. Describe the role that document control plays from the dealer first receiving the camera to its being returned to the customer in good working order.

23. What disadvantages might a document control system bring to an organisation? Do these outweigh the advantages listed in question 19 above? Discuss.

24. Describe how drawings are used to assist the engineering description of complex products such as a car. What advantages do drawings bring?

25. Describe the management information normally contained on a drawing.

26. What is a bill of material?

27. Explain in your own words the following terms:

    Marked-up print
    Drawing release
    General assembly
    Detail drawing

28. Describe good practice in the management of drawings. For each point you make give an example of the sort of error that might occur if the practice is not followed.

29. Who normally checks a drawing and why?

3.12 Futher reading

General

Batty, Dr J. (1982), *Business Administration and Management*, Hindhead, Surrey: Learnex (Publishers) Ltd (ISBN 0905847-27-X).

Bennett, R. (1987), *CIMA Study & Revision Pack Stage 2 Management*, London: Pitman Polytech (ISBN 0-273-02812-X).

Lewis, D. (1988), *Basics of Business*, London: M&E Handbooks (ISBN 0-7121-0794-0).

Needle, D. (1989), *Business in Context*, London: Van Nostrand Reinhold (International) (ISBN 0-278-00005-3).

Purchasing

Baily, P. and Farmer, D. (1990), *Purchasing Principles and Management*, 6th ed., London: Pitman (ISBN 0-273-03124-4).

Operations

Harrison, M. (1990), *Advanced Manufacturing Technology Management*, London: Pitman (ISBN 0-273-03085-X).

Lockyer, K., Muhlemann, A. and Oakland, J. (1992), *Production and Operations Management*, 6th ed., London: Pitman (ISBN 0-273-03235-6).

Sales & marketing

Elvy, H.B. (1977), *Marketing Made Simple*, 2nd ed., London: William Heinemann Ltd (ISBN 0-434-98561-9).

Kenny, B. with Dyson, K. (1989), *Marketing in Small Businesses*, London: Routledge (ISBN 0-415-00921-9).

Product development

Ray, M.S. (1985), *Elements of Engineering Design*, Englewood Cliffs, NJ: Prentice Hall (ISBN 0-13-264185-2).

Quality

Sinha, M.N. and Willborn, W.O. (1985), *The Management of Quality Assurance*, New York: Wiley.

Personnel

Farnham, D. (1990), *Personnel in Context*, 3rd ed., London: Institute of Personnel Management (ISBN 0-85292-451-8).

Part 2

The Tools of Engineering Management

Chapter 4 Personnel management

Overview

Every engineer is responsible to someone and so personnel is an issue for all. In addition there are few engineers with no responsibility for personnel. Even in the early stages of their careers engineers may be called on to supervise technicians or fitters involved in the manufacture or test of equipment designed by them, and so the management aspects of personnel may be important. In this chapter we will look at the issues that affect the management of personnel and consider their practical application.

4.1 Introduction

Case study: Hilary Briggs

Hilary Briggs is an excellent example of how an engineer can use her engineering talent and expertise, combined with effective management skills, to progress rapidly through a company.

Hilary joined Rover Group after graduating in Manufacturing Engineering. Seven years on she is the Logistics Director for the Large Cars Business Unit at Cowley in the United Kingdom. The company is Britain's largest motor manufacturer and operates on several sites in the UK; it also services a number of international operations that manufacture under licence.

The Logistics Department covers production programming and material control for Cowley, KD operations and logistics development. Production programming involves taking information from the commercial department, in the form of sales projections and orders and translating this to production programmes and shop-floor schedules. Material control includes determining material requirements in order to meet the production programmes, transferring this information to suppliers, and then ensuring that the materials are available for the production line as required. This activity also includes quality assessment of suppliers and supply, in conjunction with the purchasing and manufacturing departments.

KD operations organises the export of kits of parts to companies that carry out final assembly under licence, and provides associated technical support. Logistics development has a long-term focus and is involved with looking at the way in which the different operations in the department are carried out, seeking ways to improve in the light of new technologies or improved understanding of the processes involved.

Hilary has been Logistics Director for eight months, since the department was formed; she previously worked as a Manufacturing Manager. The Logistics Department is obviously of great strategic importance to the company — it has a total of 365 employees and Hilary has five managers reporting directly to her. Hilary must provide the direction for the department and ensure that it meets company objectives.

Her main requirement is for effective personal and personnel management. Although she only has five people reporting directly to her, Hilary has to establish good communication links and ensure that the whole department is working to meet the company goals. She also needs to control the myriad of activities that she is now called on to perform. Because this is a new venture for Hilary much of the information that she needs to do the job is held by the five managers; they also have considerable expertise in the logistics area and she needs to draw on this resource. As a first step Hilary has established a fortnightly meeting with all the managers, in which they can discuss the role, objectives and performance of the department. This meeting helps to ensure that everyone knows that they are part of a team, and is followed up by a weekly meeting with individuals where targets for particular activities are determined and agreed.

One of the major ideas that Hilary is currently developing is a training package to ensure that all the people in the department understand the processes used in the operation of the department, in particular the information flows in and out. This is particularly important in a large engineering plant where 'process' is often thought to refer to the production process only and there are in fact considerable benefits to be gained from applying a production process-type approach to information flows, which are critical to the functioning of a company. This will probably form part of a much wider discussion with the managers when Hilary carries out their appraisals at the end of the year.

Hilary is also part of another team, that of the other directors who are responsible for developing strategy for the business unit at Cowley. In this role she needs to be able to step back from the detail that she is involved with in the department and must be able to look in overview at the operation of the company. It is imperative that the Logistics Department fits in with the overall company objectives, that logistics implications are taken into account when planning strategy for other departments, and that, as a whole, the Large Cars Business Unit meets the Rover Group objectives.

At this stage in her career Hilary doesn't use the specific engineering knowledge that she gained from her degree course, although she does use the problem-solving and more general aspects of her engineering training. However, the engineering background does allow her to talk the same language as the design department and her manufacturing experience ensures that she can empathise with, and understand the problems of, the Manufacturing Department which is serviced by the Logistics Department.

You can see from the above case study that personnel management is very important to Hilary's job. She needs to be able to motivate her staff to meet her departmental objectives, to have some method of determining their performance and then has to consider ways of developing and training people to modify or improve that performance. We saw from the case study that she will be involved in appraisal interviews and is developing training.

In the event that any of Hilary's staff leave the company or the department is planned to expand she would also be involved in the recruitment of new people, or she may delegate this.

Because Hilary has reached a high position in the company her personal management load is considerably more onerous than would be the case for a more junior engineer. Nevertheless the majority of engineers will be involved in personnel management at some level during their career and therefore in this chapter we will look at some aspects of personnel management and consider some techniques that can be used by the manager.

We will examine the methods by which the people that are needed can be identified and how they are recruited. We will then consider some of the management issues relating to stimulating people to work effectively. We will examine methods of developing people through training. Finally we will consider job design and reward of staff.

The learning objectives for the chapter are as follows:

1. To understand the process of employing people and to consider some of the legislative and financial factors that affect the process.
2. To consider the effects of motivation on productivity and to look at how motivation factors can be determined.
3. To consider methods for developing people in organisations, including appraisal and training.
4. To look at the role of job design and its effect on employee motivation.

4.2 Employing people

In this section we will examine the legal and contractual issues associated with employing people. We will start by considering the recruitment and selection processes used to attract potential employees to an organisation and to select the person most likely to meet the needs of the organisation. In order to recruit likely candidates there needs to be a clear definition of the job to be filled. We will therefore examine the role of job specifications and the preparation of job descriptions.

Following selection a formal job offer is made. We will look at the way in which this is done, including an examination of a contract of employment. In addition to contract law, employment practice is regulated by a considerable amount of legislation in all countries. In this section we will look specifically at some of the legislation that applies to recruitment and selection in the United Kingdom.

Following the offer of a job a new employee will join the organisation. Alternatively, an existing employee will take on a new role within the organisation through an internal appointment. The process of introducing the employee to their new job is called *induction*. We will examine this process and the effects that it can have on both employer and employee. We will also look at the costs associated with employing new staff.

Occasionally it may be necessary to terminate someone's employment and so we will also look at the different ways in which this can take place and what must be done if it does.

4.2.1 Recruitment

Recruitment is the process of attracting potential employees to the company in order that the selection process can be carried out. It follows from a need being identified through planned expansion or loss of staff. Prior to recruitment it is always advisable to carry out a job analysis to ensure that the vacancy exists, that the vacancy really does need to be filled, and that the requirements for the job-holder have not changed. It is very easy to replace a person who is leaving without assessing what job they were really doing and determining what really needs to be done. A job analysis should result in a clear definition of the job in the form of a job description. In addition the manager responsible for the job will have to consider the type of person that is needed to do the job well. A personnel specification should be developed to provide this definition.

Recruitment policies

Formal recruitment policies should ensure that recruitment is carried out in a similar way across the organisation and that everyone is treated equally. Recruitment policies may include rules relating to where jobs are advertised, the geographical location of employees, the rules relating to internal recruitment, expenses paid for interview, and the way in which interviews are carried out. The policy should define the way in which the job descriptions and personnel specifications are prepared, the way in which information about jobs is communicated, and the way in which information about potential candidates is gathered.

We will now look at the preparation of job descriptions and personnel specifications, advertising and the use of application forms.

Job descriptions

A job description documents the main tasks and objectives associated with a job. Many employers now provide job descriptions for their employees although there is no legal requirement for them to do this in the United Kingdom. However, some employers positively dislike job descriptions because they are felt to be symbolic of a bureaucratic system, and because they are seen as removing management flexibility by defining responsibilities. This can be a problem both in new and well-established companies and it is therefore essential that any system of implementing and using job descriptions should not place a large administrative burden on the company. Similarly there needs to be an appropriate level of flexibility built in to the job description that does not detract from the purpose of the job description, which is to define a job.

The job description is equally useful for both the employer and the employee. For the employer it forces important issues to be considered, addressing the areas relating to why the job exists and what is to be achieved by the job-holder. In addition it forces analysis of the needs of the job-holder, in terms of resources to do the job described.

The job description should give a broad indication of the responsibilities of the job and indicate how it links with other jobs in the company — in particular, how it contributes to meeting the company's goals. In preparing this statement the employer is forced to think about the reason for the job and if it is really necessary for the prosperity of the company. It also allows the employer the opportunity to consider whether or not filling this job is the most effective way of contributing to the company goals.

Objectives should be included in the job description, allowing the employer to state what is most important for the job-holder to achieve. These are especially important for reviewing performance and progress (see Section 4.4).

The job description indicates tasks that are to be carried out by the job-holder. In defining these tasks the employer is able to confirm that they need to be done and it provides an opportunity to review whether or not there are other tasks that also have to be done. If a job description is being prepared following someone leaving their post, then consideration should be given to all the tasks that were carried out by that person against what was stated in the job description. Job descriptions define *jobs* and should be separate from *personalities*.

The employer needs to consider what level of authority the job-holder should carry in relation to the resources that can be committed, such as the authority to commit financial resources, the authority to recruit staff, and the authority to dismiss or suspend staff. This is an important issue for it also ensures that when considering someone for the post the employer considers their ability to carry this authority effectively. Similarly, consideration needs to be given to the level of responsibility required to do the job effectively. The employer should define the resources that are available to the job-holder, such as the budget for the job, number of staff, and the equipment, plant and buildings.

For employees the job description is important because it allows them to know what is required, what authority they have, and what standards are expected.

The form of a job description: There is no defined form for a job description but a hypothetical one would normally include the following:

1. Job title and grade.
2. To whom you report.
3. What authority you have.
4. Definition of those for whom you are responsible.
5. Your main objectives.
6. Key responsibilities and tasks.
7. Reporting methods and requirements.

A clause may also be incorporated which provides the employer with a degree of flexibility when asking the post-holder to carry out specific tasks.

An example of a job description is given in Figure 4.1.

Personnel specifications

The preparation of personnel specifications follows from consideration of the job. Many small companies do not prepare formal personnel specifications, but if carrying out recruitment they must have views on the type of person that is being sought. The formalisation of these views is essential if a number of people are involved in the selection process, or if the person doing the selection will not be in direct contact with the job-holder.

When preparing a personnel specification there are a number of things that can be considered. Four of these are described below.

Qualifications: The level of qualification essential for the job, and that which is desirable, should be defined. In order to do this there will have to be a decision made about what is expected from the job-holder and how the job is likely to develop. It is important that neither

Job description: Technician engineer

The technician engineer in the Engineering Department will provide engineering support to other members of the department. The key tasks of the person being:

To provide technician support to members of the Engineering Department.

To test, assemble and develop components and products, as instructed, in accordance with the requirements of the quality management system.

To assist in the production of documentation for components and products including work instructions, bills of material and operating manuals.

Any other appropriate tasks required of the Engineering Department.

Job title:	Technician engineer
Grade:	3
Responsible to:	Head of Engineering Department
Responsible for:	Engineering support technician
Purchase authority:	None
Contract authority:	None
Reporting:	You will be required to produce a written progress report for the Engineering Department progress meetings.

Figure 4.1 Job description for a technician engineer

too many nor too few qualifications are required — both of these can lead to future problems. An under-qualified person will have difficulty with the tasks and will not be able to do them as well as required and may become de-motivated by lack of achievement. An over-qualified person will become bored and either not do the job satisfactorily or in a rather perfunctory manner, or will use the job as a fill-in operation until a better job can be found.

Skills: The type of personal skills required might include defining someone who is a good teamworker, someone with leadership qualities, someone who is self-motivated or who will be led, or someone who is a good communicator. In addition the technical skills required for the job will have to be considered.

Experience: Consideration should be given to requirements for previous experience, whether appropriate experience is required and if so whether it should have been in a similar company, or in a similar job, or at a similar level. It is important to balance this carefully in light of what is currently offered and what the prospects are. For example, if the job requires some specialised knowledge and there is little training provision it will be necessary to bring in someone that already has that knowledge. A balance of skills and experience should also be achieved by a consideration of how the experience of the job-holder will complement that which is already available in-house.

Attributes: If there are any physical attributes that are essential in order to be able to do the job then these must be clarified. This will require care to make sure that there is no inadvertent discrimination against any particular group by the imposition of unnecessary restrictions.

Consideration should also be given to other requirements such as an ability for geographical mobility, either nationally or internationally, and possession of a driving licence. A person's ideological beliefs may even be important.

Advertising

The job description and personnel specification provide the blueprint for the post to be filled and the person required. Advertising is the process of communicating the vacancy to potential applicants.

When advertising, the job and working conditions should be expressed in a way that makes them attractive to the right sort of people. It is usual to include some detail of the job and the terms and conditions of employment. At the same time there is a need to limit the number of potential applicants to those that are most likely to be able to do the job. It is, however, important to avoid the opposite extreme. An example of appropriate advertising for the post of an engineering manager is given in Figure 4.2. The advertisement includes details about the job, limitations on applicants, the application procedure, something about the company, and the closing date for applications. It should attract the attention of potential applicants. The placement of an advertisement should be appropriate to the audience that is to be attracted. This might indicate use of national rather than local press, or the use of certain trade or professional journals. In view of these many issues many companies resort to the use of advertising agencies.

In large organisations the chances of being involved in preparing and placing job advertisements is slight. This is more likely to be an activity undertaken by the personnel department. However, this is not the case in small companies and often line managers will be involved.

Application forms

Application forms are an effective way of ensuring that you get the right information from applicants and that it is in an easily analysed

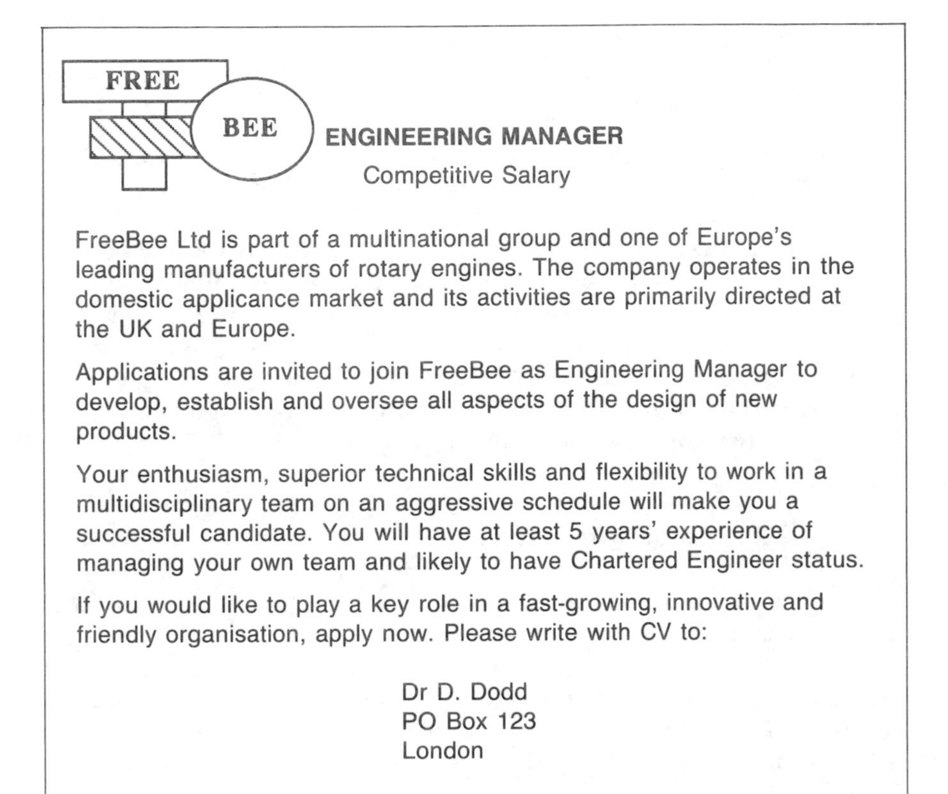
FREE BEE

ENGINEERING MANAGER

Competitive Salary

FreeBee Ltd is part of a multinational group and one of Europe's leading manufacturers of rotary engines. The company operates in the domestic applicance market and its activities are primarily directed at the UK and Europe.

Applications are invited to join FreeBee as Engineering Manager to develop, establish and oversee all aspects of the design of new products.

Your enthusiasm, superior technical skills and flexibility to work in a multidisciplinary team on an aggressive schedule will make you a successful candidate. You will have at least 5 years' experience of managing your own team and likely to have Chartered Engineer status.

If you would like to play a key role in a fast-growing, innovative and friendly organisation, apply now. Please write with CV to:

Dr D. Dodd
PO Box 123
London

Closing date for applications: 23 June 1993

Figure 4.2 Advertisement for an engineering manager

form. Depending upon the type of person to be employed a very simple form that asks for facts might be used. Alternatively there could be a requirement for a longer, more complex form that requires the applicant to argue their case for employment. Some companies ask for a *curriculum vitae* rather than send an application form but this can result in vital information being omitted and can result in inconsistencies in the way in which the information is presented.

Typically the organisation will want to know some basic information, such as:

1. What job the person is applying for.
2. Where did the person find out about the job (this gives feedback on the effectiveness of different recruitment methods).
3. Name and contact address.
4. Qualifications and experience.

5. Health and any disabilities.
6. Age, if appropriate.
7. Interests outside work.

This gives some idea about the type of person you are dealing with as well as providing information on which to base informal conversation at the interview.

In the more detailed application form you might also ask open-ended questions such as 'What can you bring to the job that others may not be able to?', 'Where do you see yourself in five years' time?', 'Why have you applied for this job?'. The answers to these types of questions can tell you a lot about the candidates and will help when preparing a short-list. However, as with all forms they can be used by candidates to put in only information that shows them in a good light.

4.2.2 Selection processes

Having attracted applicants for the job the next stage is to select the candidate that is most suitable for the job and the company.

The first round of the selection process is short-listing. This is the process of reviewing the application forms against your job specification and selecting the candidates that are likely to meet your specification. When using a recruitment agency they will normally do the short-listing for you.

Following short-listing the company should have a list of the people that are likely to meet their requirements and are worthy of further investigation — this is usually done by interview (some companies also use tests). References are used to provide assurance that someone meets the job requirements. Interviews, tests and references are discussed below.

Interviews

Selection includes at least one interview, and many companies use two, one of which may involve the candidate making a presentation. The aim of the interview is to determine which candidate from the short-list best meets the requirements of the job. It is also to present the job and the company to the candidates, and as such is very much a two-way process because the organisation will require good candidates and will be competing with other companies for their services.

The interview should allow the company to determine what the candidate is like, whether or not they could do the job, and whether they will fit in with the existing team. This can only be achieved if candidates talk frankly about themselves and their abilities. Candour is important since the company will want both to employ and to retain the best candidate. It is important that promises are not made if they cannot or will not be kept. Effective interviewing is an essential part of all managers' jobs and is discussed in detail in Chapter 9.

Tests

Tests are becoming increasingly common in the selection of graduates and others at junior and middle-manager level. There exist many types of test and fashion seems to play as large a part in the selection of tests as does scientific basis. Tests range from simple intelligence quotient (IQ) and reasoning tests to residential outward-bound courses which test candidates' ability to work with others, identify leadership qualities, examine behaviour under stressful conditions, and test physical stamina. Handwriting tests (graphology) are common in some countries, such as France, and are becoming more widely used in the United Kingdom.

References

References from previous employers, college lecturers or people that have known the candidate for a significant period of time are usually required. The value of references is not high since it is unlikely that anyone would name a referee if they thought something bad would be said. However, references do give the opportunity to get a different perspective of a candidate and are usually taken up after a job offer is made 'subject to satisfactory references'. Anyone who cannot find someone to say something positive about them should be treated with caution — for instance, they may have left their previous employment under dubious circumstances. For this reason the presence of references is often more significant than the content.

4.2.3 Making a job offer

After the process of selection, interviewing and testing, the company must make a decision about which — if any — of the candidates it wants to employ and make that person an offer. Many companies allow a period of time between carrying out interviews and making an offer. This can be useful as it allows time to consider each of the candidates carefully. However, it can also be disadvantageous in times of high employment or if the company is recruiting in a skill shortage area because while the decision is being made a good candidate may have received other offers. In addition it should not be forgotten that selection is a two-way process and the candidate that is chosen may not choose the company, so it is always worth considering the fall-back position. If an offer is made on the same day as the interview it allows an offer to be made to a different candidate if the first refuses, thus avoiding a whole new round of selection.

When a person is employed they are contracted to work in return for remuneration and the terms and conditions under which they work will be defined in a contract of employment. The contract of employment as used in UK companies is described below, as are the legislative requirements defined in English law.

Contract of employment

A contract of employment exists from the moment the two parties reach agreement on the basic terms, the parties to the contract being the employer and the employee. There is no requirement for the contract to be in writing, although this is usual. However, under English law a written statement of the main terms and conditions of work must be made available to each employee within thirteen weeks of them starting their employment.

Contract law: Under contract law there are four main points that have to be satisfied for the contract to be legally binding. This applies as equally to the contract of employment as to a contract to purchase a house. The four points are as follows:

1. *The parties must have legal capacity to contract.* Those who do not have legal capacity to contract include minors, the mentally ill, and people who are obviously drunk at the time of making the contract.
2. *The parties must have intended it to be legally binding.* This is automatically assumed by English courts.
3. *There must be a clear offer that has been accepted.* The offer does not, in law, have to be made in writing; however, it is sensible to ask for a written offer, particularly if promises have been made at interview. Acceptance of the offer must be positive, that is the acceptor must do something to confirm acceptance, and it must be unqualified. Any qualification is a counter-offer and would have to be accepted by the other party. For example, if you receive a job offer with four weeks' holiday per year and your answer is an unqualified 'yes' then this is a positive acceptance. If, however, you reply 'yes, but only if the holiday is increased to five weeks', this is a qualified acceptance and requires further acceptance by the company. In practice, going to and fro like this is quite common, particularly when salary is being negotiated.
4. *There must be an element of consideration.* Consideration is what you give in return for what is offered. In the employment contract this is the work that you promise to do in return for the remuneration package offered.

In addition to these four main points the objects of the contract must be legal. A contract to do something illegal — committing a crime — is, not surprisingly, illegal.

The existence of a contract gives rise to both explicit and implied terms of employment. The explicit terms should be stated in the contract as described below. The implied terms are that an employer is expected to pay wages, to take reasonable care of the employee, to indemnify the employee for expenses and liabilities incurred in the course of their employment, to treat the employee with courtesy, and to provide a safe place to work. In return the employee is expected to do the job agreed, to take reasonable care in the performance of duties, to obey reasonable instructions, to act in good faith towards the employer, and to refrain from impeding the employer's business.

Main terms and conditions

An employer is required to notify the employee in writing of the main terms and conditions of work within thirteen weeks of the start of the employment. The terms and conditions that must be addressed are the following:

1. *Names of the employer and employee.* These are the parties to the contract.
2. *The date of starting employment.*
3. *Whether any previous employment counts towards the employee's period of continuous employment.* This may be important as the period of continuous service is used to calculate sickness, maternity and paternity leave, and pension payments. It is also used by some companies as a way of allocating bonuses or determining eligibility for fringe benefits such as share options.
4. *The date at which the contract expires if it is for a fixed term.*
5. *The rate of pay.* This should state the normal rate of pay and is usually given as either a weekly or annual sum. It should also be clear when overtime payments will be made, if at all, and at what rate overtime will be paid. The majority of 'professional' jobs are salaried and overtime is not normally paid. However, a salaried person is often expected to work beyond 'normal' hours.
6. *The intervals of pay.* This defines how often payment will be made and will indicate whether it is made in advance or in arrears.
7. *The normal hours to be worked.* These will be the hours that will be paid at the standard rate. It is usual to indicate any fixed breaks such as lunch-time.
8. *Holidays, sick pay and pension arrangements.* This will define the number of days that can be taken as holiday and should specify if there is a requirement to take any holiday at particular times, for example during a factory shut-down. The rate of holiday pay should

also be given and the method of payment. The terms and conditions for sick pay entitlement, methods of notifying sick leave, method of payment, should all be defined. The pension arrangements should be included. Full details of the employee's contributions to the pension scheme and any made by the employer should be given; however, it is usual to refer to a separate document containing all the other pension details. Some companies will contribute to a personal pension. If this is the case it should be indicated.

9. *Terms of notice.* Notice is required to bring the contract to an end. If notice periods are not included in the contract then the statutory minima apply as laid down in the Employment Protection (Consolidation) Act (1978).
10. *Grievance and disciplinary procedures.* A statement about the disciplinary and grievance procedures operating in the company, or information about where these details may be found, must be included.

There may be other rules relating to the job, such as those concerning safety. These should be referred to in the terms and conditions indicating where full details can be found.

An example of a contract of employment is given in Figure 4.3.

Remuneration

Remuneration is the term used to describe the package of pay and other benefits that you receive from your employer in return for work. Usually the greatest part of any remuneration package will be the salary. However, the value of other benefits such as pension, company car and medical insurance, can be considerable. As part of the remuneration these benefits are generally taxable. Remuneration is discussed in detail in Section 4.6.2.

4.2.4 Legal aspects of recruitment and selection

Employment practice is regulated by legislation in all countries. In this section we will consider English law as it applies to recruitment and selection; however, it should provide a guide to the types of legislation that will exist in countries that do not use English law.

During the recruitment and selection processes care must be taken to avoid discrimination on the basis of gender, colour, disability or ethnic minority. Discrimination is not only held to be socially unacceptable it is also illegal in many countries. In the United Kingdom the Equal Opportunities Commission and the Commission for Racial Equality oversee anti-discrimination legislation. There are three main pieces of

Project Engineer

Terms and Conditions of Employment

1. Employer: . . . .
2. Employee: . . . .
3. Employment period
The contract is a fixed term of two years commencing . . . It is not renewable. However, in the last six months of the period . . . (the Company) will indicate whether or not a permanent position is to be offered.

4. Job details
The job description is attached. The job title is Project Engineer (PE). The PE will report to . . . .
Project monitoring will be carried out on a fortnightly basis with minuted meetings between the PE and the Project Manager, with progress being reported at the quarterly Engineering Projects Management Committee Meetings. In addition the PE's personal, professional and career development will be assessed at six-monthly performance reviews with supervisors.

5. Hours of employment
Normal hours are 37.5 hours per week excluding lunch breaks. The hours of work are 8.45–5.15 Monday to Friday.

6. Notification of intention to cease employment
a) *Employee:* One month's notice in writing.
b) *Employer:* One month's notice in writing.

7. Holidays
a) In addition to Bank Holidays, 20 working days by prior arrangement with the immediate supervisor, with each holiday limited to two weeks unless agreed by immediate supervisor with the approval of two directors.
b) In the event of termination of employment you are entitled to accrued holidays or, with management approval, pay in lieu at the current rate.

8. In case of absence from work
If you are unable to come to work for any reason you must notify your immediate supervisor as soon as possible or phone in to pass on a message.

9. Pay and pension
You will be paid £. . . . p.a. This will be paid in arrears, in twelve monthly instalments on the last Thursday of each month.
The position is non-pensionable.

10. Disciplinary procedure
a) A written notification will be given if disciplinary action is necessary for any reason.
b) If two written notifications for disciplinary action are issued then dismissal may result.

11. Grievance procedure
a) If there is dissatisfaction with any disciplinary decision or a wish to seek redress of any grievance relating to the employment the employee may apply in writing to the immediate supervisor.
b) The employee will be advised of any action taken as a result of the above within ten working days.

Figure 4.3 Contract of employment

legislation: the Race Relations Act (1976), the Sex Discrimination Act (1975) and the Equal Pay Act (1970). These are described briefly below.

Race Relations Act (1976): This Act is enforced by the Commission for Racial Equality which points out breaches and can apply penalties for future breaches. A breach of the Act is not a criminal offence; however, anyone who feels they have been discriminated against can claim damages at an industrial tribunal. Alternatively the Commission can issue a non-discrimination notice ordering a business to stop discrimination during the five-year period of the notice. An employer is liable if its employees discriminate unlawfully.

Employers can discriminate on the basis of race where there is genuine occupational qualification. This covers the following circumstances:

1. *Authenticity of entertainment.* For example, you can advertise for a black man to play Othello.
2. *Authenticity of modelling.* You can select people of a certain race if you are trying to, for example, illustrate certain cultural distinctions.
3. *Authenticity of food and drink establishments.* This only applies to premises where the food and drink is consumed, normally, on the premises. For example, you can advertise for Thai people to work in a Thai restaurant.
4. *For work with a particular ethnic group.* A particular ethnic group may be defined if someone is required to work with people in such a group. For example, an employer may advertise for a Bangladeshi person to work as a counsellor to Bangladeshi groups in the community.

Sex Discrimination Act (1975): This Act is overseen by the Equal Opportunities Commission. In a similar way to the Race Relations Act you can lawfully discriminate in such circumstances as the following:

1. To achieve authenticity of entertainment or modelling.
2. For jobs in single-sex establishments.
3. For jobs providing personal welfare and education.
4. For jobs where there are legal restrictions on the employment of one sex.

5. For employment in a private household.
6. For employment carried out wholly, or mainly, outside the UK.

With both sex and race relations indirect discrimination is unlawful, i.e. applying a condition or requirement which, although applied to both sexes or all racial groups is such that a considerably smaller proportion of one sex or racial group can comply — for example, stating a requirement for a senior engineer to grow a beard.

Positive discrimination can be unlawful, but is sometimes used in the form of special training schemes and initiatives for under-represented members of one sex or race in order to encourage more people from these groups into the workplace. In the United Kingdom it is also used to promote the employment of disabled people.

Equal Pay Act (1970): This Act provides for an implied term in the contract of every woman employee, known as the 'equality clause', which entitles her to similar rates of pay, holiday and benefits as those for male employees in the same organisation, where she can prove one of the following applies:

1. She and the man are employed to do like or broadly similar work.
2. She and the man are doing work rated as equivalent by a job evaluation scheme.
3. She and the man are doing work of equal value to the organisation. Whether or not this is the case will ultimately be determined by an independent expert appointed by an industrial tribunal.

The third clause was introduced in 1984 as a result of rulings in the European Court of Justice. However, although this is a potentially very useful clause it is limited because it only applies where comparisons can be made between men and women in the same employment. If the workforce is all one sex no claim can be made.

Many companies now implement equal opportunities monitoring in the hope of identifying problem areas and resolving them.

4.2.5 The induction process

Induction is the process of introducing new employees to the company and to their jobs. It should not be restricted to people who have just joined the company. It may be just as appropriate for people who have transferred from one department to another.

The alternative to having an induction programme is to 'throw people in at the deep end'. This idea, which still has its devotees, is rather outdated and is clearly impolite and arrogant, giving the impression that the new employee is not worth the effort of common courtesy. In addition, many people are very good at their jobs when given the opportunity to settle in but are not good at coping with the stress of being left alone in an alien environment. This method discriminates against them and can alienate them further.

The induction programme, which must be a planned activity, should allow people the time to settle in whilst enabling them to gather the information that will make them most effective in their jobs. Thus, any induction programme should include the following as a minimum:

1. Introductions and meetings with people with whom the new employee will have to interact.
2. A tour of the facilities where the employee will be working.
3. An introduction to the company, including some history and an overview of its operating systems including organisation charts.
4. An introduction to the department in which the employee will be working, particularly introducing the systems with which the employee will have to become familiar, such as the quality system and the design system, as well as any processes that the employee will have to use.
5. An introduction to the company's products
6. An explanation of the safety requirements and procedures operating within the company, including the issue of safety equipment and a tour of emergency exits and places where first-aid and safety equipment is kept.
7. A discussion of the objectives of the job and the setting of initial tasks.

Obviously an induction programme will take time and will cost money but its benefits will accrue from increasing the speed with which the employee will be able to settle into the job, thus reducing the time when their effectiveness will be limited. It can also be used to encourage teamworking if responsibilities for different parts of the induction are shared amongst the team, and it should encourage a transfer of loyalty from the previous employer by making the new employee feel 'at home'.

4.2.6 The cost of employing new staff

The cost of recruitment, selection and employment can be very high and should always be borne in mind when making decisions about employing new people. Even internal recruitment can have a significant cost.

The costs of recruitment and selection, as with any type of costing, can be broken down into labour, materials and expense. If we consider the labour costs first of all, these will include the time of people from the personnel department who will be involved with the adminstration of the process, including sending out application forms and arranging interviews. They will also include the time devoted to preparing job specifications, short-listing, and preparing and carrying out interviews. Consider that many interviews for graduate appointments take at least an hour, will require at least an extra twenty minutes for preparation and review, and will be conducted by more than one person. If two people are interviewing it will cost the company of the order of two days of time to interview five people and this is likely to be middle-manager-level time.

The materials cost is not likely to be significant and would probably be limited to stationery. However, the total expense could be very significant if you are paying a recruitment agency — they are generally paid on the basis of a percentage of the new employee's salary. If this is not the case there will still be the costs of interviewees' expenses, postage, telephone and advertisements.

Costs will be enhanced considerably if the company uses any form of testing requiring the time of a consultant, use of premises outside the company, or in recruitment overseas.

When the employee is in post the cost of employment is not limited to the salary paid but includes an amount for national insurance and what can be quite a significant amount for benefits and perks such as pension, car, health insurance and share allocations. In addition there will be costs for induction and training.

Because of the significance of these costs it is important to make sure that a job needs doing before you recruit for it, that you select the most appropriate person for the job, and that you retain that person by providing a good working environment and room for personal development. In Sections 4.3 and 4.4 we will look at some aspects of employing people that relate to their well-being whilst working. Unfortunately, however, it is sometimes necessary to terminate a person's employment and we will now look at this aspect of personnel management.

4.2.7 Termination of employment

An employer can terminate an employee's employment by giving them notice and then dismissing them. An employee can also give notice in order to terminate the contract of employment legally. In addition, employment may be terminated because an employee has reached the agreed pensionable age, or because the term of a fixed-term contract has come to an end.

We will now look briefly at notice and then consider some of the reasons that can be used for dismissing someone.

Notice

Terms of notice should be defined in the contract of employment; however, if they are not then there are statutory minima that apply, defined in the Employment Protection (Consolidation) Act (1978). If notice is given by an employer to an employee and that employee has been working for more than four weeks but less than two years the statutory minimum notice period is one week. Each additional year worked adds one additional week to a maximum of twelve weeks after twelve years' service.

The minimum notice from employee to employer, defined in the Act, is one week regardless of length of service. However, the typical notice period for salaried staff defined in a contract of employment is a month from either employee or employer, although the higher your position in the organisation the more likely it is that this period will be increased.

Notice can be written or oral and pay in lieu of notice can be agreed.

Dismissal

An employee can be dismissed for a fair reason. The reasons laid down in the Employment Protection (Consolidation) Act (1978) are conduct, capability, redundancy, legal prohibition and any other substantial reason. These are each discussed below.

Conduct: Conduct relates to the way in which employees go about their duties and to what is accepted with regard to custom and practice. It covers acts of gross misconduct where the action of the employee is such that the contractual relationship is destroyed. There is no legal definition of gross misconduct but it is usually seen to include acts such as theft, violence, drunkenness or other 'unacceptable behaviour'. Gross misconduct can lead to dismissal without notice.

Capability, including health: This would arise if an employee was no longer able to perform the duties required at an appropriate standard. For example, if a warehouseman with a bad back was unable to lift boxes, he could be fairly dismissed if this had previously been one of the tasks that he was required to do.

Redundancy: Redundancy occurs when the whole or main reason for dismissal is that the employer's need for an employee to do work of a particular kind in the place of employment has either diminished or ceased. Examples of when redundancy can arise include a reduction in the workforce due to a fall in sales, a move to a new site some distance away, or a change in work requirements due to implementation of new technology.

Rerdundancy is deemed to be fair dismissal but selection of candidates for redundancy must also be seen to be fair. This is normally accepted as 'last in, first out', i.e. the newest recruits to the company are made redundant first. This is not always practical because the company may be looking at particular skills or areas of operation. Nevertheless the onus is always on the company to be seen to be fair. It is unlikely that people being made redundant will think it fair and therefore they must always be dealt with carefully and sympathetically.

When a company decides that redundancy must take place any recognised trade union in the company must be informed, at least ninety days beforehand if more than a hundred employees are involved and thirty days beforehand for employees numbering more than ten but less than one hundred. At the same time the Secretary of State for Employment must be notified in writing. In addition employees who are being made redundant must be given reasonable time off to look for other work.

Many companies will negotiate redundancy terms with either a trade union or staff organisation; however, any terms must be equal to or better than those laid down by law. Minimum redundancy pay is only paid to those below the age of retirement, and in 1991 was limited to those people who had worked for more than two years, for at least sixteen hours per week, or who had worked for more than five years for at least eight hours per week. The minimum rates are calculated on the basis of age, the number of years of service with that employer, and the wage received just prior to redundancy. There are maxima set on the number of years of service that can be counted, and on the weekly wage that will be used for the purpose of the calculation. There is a state redundancy fund which pays minimum redundancy pay if a company becomes insolvent.

Rather than making people redundant a company may opt to offer alternative employment, in which case an employee can only claim redundancy if one of the following applies:

1. The alternative offer is inferior, such as being offered a lower level of pay or benefits.
2. The new job is unsuitable to the employee by way of skill or training. It is therefore not possible to offer someone alternative work for which they are unqualified and then avoid giving redundancy pay if they decline to accept it.
3. The new job is an unreasonable distance from home.

Legal prohibition: An employee may be dismissed if they are no longer able to carry out their duties because they are not legally allowed to undertake certain tasks. For example, a lorry driver could be fairly dismissed if an offence has been committed which resulted in the driver's driving licence being revoked.

Any other substantial reason: This is likely to arise if there are prohibited activities in the workplace that have been previously enforced, but have been carried out regardless. For example, if there was a company rule that decreed no one should consume alcohol at lunch-times and you had been doing so then you could be fairly dismissed, as long as this rule had previously been upheld or warnings that it was now going to be enforced had been given, and you had been warned that this behaviour was likely to lead to dismissal.

In almost all cases, for dismissal to be fair a series of warnings should be given — the exception would be an act of gross misconduct. An employer is required to give the reason for dismissal in writing, if requested, if the employee has worked for more than six months.

Cases where an employee believes that s/he has been unfairly dismissed must be taken to an industrial tribunal within three months of the dismissal. To claim unfair dismissal the employee must have worked for at least two years continuously, at least sixteen hours per week, for the same employer. The employee has to show that the dismissal was normal, which means that either the contract was terminated by giving notice or a fixed-term contract was not renewed, or that the dismissal was *constructive*. Constructive dismissal is where the employee is put into such a situation that they no longer feel they are able to carry on with their employment and so resign. In all cases the employer is required to show dismissal was for a fair reason. Remedies for unfair dismissal are reinstatement, re-engagement or monetary compensation.

4.3 Motivation and leadership

There are two main areas to consider when managing personnel if they are to work effectively: the first is motivation and we will look at how motivation affects people and how motivating factors can be assessed. The second aspect is leadership and the way in which people can be managed in order to achieve objectives. We will look at the area of leadership and consider some of the styles of leadership that can be adopted by managers. We will also review the factors that affect the style of leadership adopted.

4.3.1 Motivation

When you start to plan your career and your first job you will be concerned with many things: what type of work, salary, training are being offered. Many factors affect why people take jobs and the weightings applied to each factor will vary with each individual. When you take a job the way in which you approach your work, whether you are enthusiastic or indifferent, or whether you take on extra responsibility, will also depend on a number of factors which relate to you, to the job itself, the organisation and so on. It is important to know which factors affect you and the relative importance of each so that you can make the best decisions and end up with a happy and satisfying career.

Everyone understands the problems of a lack of motivation, when people do not feel like doing a certain job so keep putting it to the bottom of the pile (in the hope that one day it might just go away), or it gets done but not very well but at least it is out of the way. Apart from this, if your motivation to work is weak you may not worry if you get to work late, or if you miss deadlines. Motivation can therefore have a profound effect on productivity and on the quality of work. The implications for a company of having a poorly motivated workforce are therefore great and so managers have to understand what motivates their subordinates and provide an environment in which they will be motivated and will produce good-quality work at an acceptable rate.

Psychology has much to offer in the way of motivational theories and fortunately there are plenty of people who have done a lot of work in the field of motivation that an engineer/manager can draw upon in trying to determine what might motivate themselves and their subordinates. Some of the motivation theories that have been produced are summarised below.

By considering the motivation theories you will have some ideas that will help you to plan your initial strategy. They are not rules and they do not tell you how to do the job — people with experience may look at them and say that they are just common sense.

Maslow and Herzberg

The two motivation theories that are most commonly referred to are those of Professor A.H. Maslow and Professor Frederick Herzberg. There are many excellent texts that discuss these theories in detail and some are given in Section 4.10.

In summary, Maslow and Herzberg define specific things they found to motivate people based on experiments carried out in the 1950s and 1960s. Maslow's results are presented as a 'Hierarchy of Needs' in which needs have to be met at different levels in order that people can be motivated by the next set of needs at a higher level. The hierarchy, which is graphically represented as a triangle, starts with the most basic physical needs and works up to the pinnacle of self-realisation needs. The work of Herzberg led him to define two areas that affect people at work: he called these 'hygiene factors' and 'motivators'. The hygiene factors correspond to the more basic needs defined by Maslow and relate to working conditions. Herzberg concluded that if the hygiene factors were not adequate then morale and productivity would be low; however, improvement of these factors beyond a satisfactory level would not lead to an increase in productivity. This could only be achieved by improving the motivating factors such as the level of responsibility of the job and the type of work done.

A review of these two theories gives you a list of things that you should consider when trying to determine what gets people to work and what gets them to work better. Some of these are salary, job security, the tasks involved, the people and working relationships, and the working conditions.

Salary was found not to be a motivator by Herzberg but a hygiene factor. If salary was at a suitable level then it would enable someone to be motivated by other factors but would not in itself motivate. Whilst it is true that many people are not motivated by salary a large number do seem to be, particularly with pressures regarding material possessions in society and a higher standard of living.

Job security is very important to many people because people generally like to know what they might be doing next month and next year. The pressure of society encourages many people to live beyond their means or take out large loans, so the security of a job is of paramount importance. In times of high unemployment the need for job security will override considerations of salary, hours to be worked, working conditions and so on.

The job that you are required to do may be challenging and will stimulate you, it may challenge you a bit, or it may not challenge you at all. If you are not interested in the job you are not going to be motivated by it, but you may like the status or responsibility associated with it, or even find the perks invaluable.

The people that you work with can affect your motivation. If you have to work closely with someone who is well motivated and enthusiastic this will affect you and will increase your motivation. Conversely if you are working with someone who is dissatisfied with their work or depressed then this will de-motivate you. It is for this reason that some companies prefer to give pay in lieu of notice if someone resigns because their presence could spread bad feeling and reduce productivity.

The working conditions relate to the environment that you have to work in and the physical comfort associated with your place of work. In a foundry where the environment may be unpleasant people may find work uncomfortable and will minimise the amount of time spent in that atmosphere. Therefore thought should be given to trying to remove some of the pollution. Similarly office chairs that do not provide adequate back support can make someone feel sufficiently uncomfortable that they do not stay at their desk for very long.

John W. Hunt

An alternative to considering 'needs' has been put forward by John W. Hunt. In his book *Managing People at Work* Hunt proposes that the most effective way in which to predict people's behaviour is to consider their personal goals. From this you can motivate them by creating an environment in which their goals can be satisfied whilst the goals of the organisation are also met. He puts individuals' goals into the eight categories of comfort, structure, relationships, recognition and status, power, autonomy, creativity, and growth.

Whereas the theories of Herzberg and Maslow are presented as constant, in that the same needs have the same priority throughout life, Hunt puts forward the view that people's goals evolve and that they do so because the goals and the factors that influence them are based on people's backgrounds. Consider that at different stages in your career your motivation will be affected by not only the place in which you work but the external environment and, very importantly, your home life. For instance, engineers after graduation may give priority to getting into a company which offers a proven training scheme to management and engineering certification. When this is done they may be more interested in salary if they want to buy a house, or geographical location if they have a partner to consider. Later they may be more keen on a job which provides flexible working so that they can spend time with their children.

Motivation in practice

A problem with having to deal with motivation is that there are no correct answers. Many engineers, who have a very rationalistic approach to work, seem to have difficulty with this. There are no universal solutions. What motivates one person will not motivate another, therefore the engineer who manages has to be prepared to compromise and develop strategies that form the best overall solution for a number of people, including themselves.

Managers have to try to build up a picture of what their subordinates are really like and then establish what it is that drives them. If you are asked what motivates you it is unlikely that you will be able to find a satisfactory answer in a short time. Similarly, if a manager asks a subordinate this question it is unlikely that a reliable answer would be supplied unless they had been given plenty of time to think about it first. Because motivation is a complex issue, affected by many different things, it is difficult to quantify. Yet a manager needs to have this information. Whether the 'needs' of Maslow, the 'hygiene factors' and 'motivators' of Herzberg, or the 'goals' of Hunt are appropriate depends upon the manager's view, but they only provide pointers which allow the psychology of subordinates to be considered.

So, how can this information be collected? At the selection stage when employing people there is no reason why they should not be asked the appropriate questions; however, many people may give the answers they think the interviewer wants to hear. A more accurate method is to look closely at the way the candidate acts when not constrained by others — in particular what sort of outside interests the candidate has, what initiatives the candidate has instituted, and what the candidate feels is important in the job you are offering.

Whan a person is in post it is possible to look at what the person really does at work and their attitude towards it to indicate what they find most stimulating. The appraisal interview (see Section 4.4) provides an opportunity for pursuing this type of investigation.

Motivation can be improved, or the opportunity for improvement provided, only when a subordinate's motivating factors are known. Job descriptions and appraisal interviews can help in improving motivation as they can be used to set goals, clarify what is expected, and explore any expectations the employee might reasonably have. It is widely felt that in manufacturing particularly, for operatives that have rather tedious jobs, anything that can improve the job will improve motivation. A technique which is becoming more widely used is that of *quality circles* where people are encouraged to take an interest in the wider aspects of their jobs, in particular considering ways of improving working conditions and techniques to improve product quality. These are described in David Hutchins' book *Quality Circles Handbook*.

4.3.2 Leadership

Leadership is the way in which managers cause people to meet the objectives of the organisation. Managers have the authority to require their subordinates to do what they are told, but this does not always prove to be the most reliable method. The aim should be to achieve authority through respect so that people do what is needed because they themselves appreciate the need. Establishing a leadership style is difficult, particularly as many people have no experience of leading when they first start work. However, it is a management skill which can be developed like any other.

Leadership is intrinsically linked with motivation; people have to be motivated through sound leadership in order to meet the company's objectives. Unfortunately, as with motivation, there are no correct answers defining how to lead a particular group. As with all management skills the effective solution will be one that takes into consideration the situation being dealt with, the way the manager feels and likes to operate, and the subordinates. This means that as a leader the manager cannot formulate a plan of action to cover all circumstances. Every new situation must be assessed and appropriate action taken 'thinking on your feet'.

There is, however, guidance as to the types of leadership styles that managers use. In the same way that the motivation theories provide a framework for developing your own ideas the leadership styles broaden your view as to what is possible without laying down what is right.

Leadership styles

It is important that whatever leadership style is adopted it should be consistently used when dealing with specific people or situations, or it should be made clear that the style is being changed for a specific reason. Continuous changes in style will mean that subordinates never know what to expect and this can be very unsettling, and consequently de-motivating, for them.

The Continuum of Leadership Behaviour, developed by Tannenbaum and Schmidt, is useful for considering the different styles of leadership that can be adopted and how these relate to the style expected by subordinates. The Continuum defines a natural progression from extremes of the authoritative, allowing the subordinates no role in the decision-making, to a democratic style where the decision-making is delegated to the subordinates.

Two other contrasting styles are those of the task-centred leader and the employee-centred leader. The task-centred leader is concerned predominantly with the task to be accomplished and sees subordinates as tools that are used to get the job done. The employee-centred leader takes the converse view and is mainly concerned with the welfare and well-being of subordinates, with the view that if the subordinates are cared for then the task will be acheived through their commitment.

A view put forward by Douglas McGregor is that there are two ways of viewing man's approach to work, theory X and theory Y. According to McGregor many 'old-fashioned' managers are strong believers of theory X. This proposes that people dislike work and must therefore be manipulated and forced into it, and that people prefer to be told what to do. The opposite theory Y proposes that most managers have underestimated the worker and that work is as natural as play if the conditions are favourable. Control imposed from above and the threat of punishment are not the only ways for getting workers to work satisfactorily. People will exercise self-direction and self-control in the service of objectives to which they are committed. Theory Y also promotes the view that people will actively look for responsibility and that a large proportion of people have the ability to use their imagination, ingenuity and creativity to solve problems.

Factors that affect leadership style

The ideas described above — the authoritarian, the democrat, the task-centred leader, the employee-centred leader, theory X and theory

Y — all define potential extremes of leadership behaviour. We will now look at what determines the behaviour adopted by a manager. As mentioned earlier there is no correct way to lead: the most effective method will take into account you, your subordinates and the situation which obtains.

Your behaviour as a manager will be influenced greatly by your personality. You will also perceive any leadership problems in a unique way based on your background knowledge and experience. Among the factors which will affect you as a leader are your value system and the confidence you have in your subordinates. Your value system relates to what is important to you, such as whether you feel that individuals should share in making the decisions that affect them.

Your values might mean that you are convinced that if someone is paid to do a job then they should be able to make all the decisions associated with that job. Your value system will affect where you would be seen by your subordinates to be operating — between the extremes of autocrat and democrat. However, your behaviour will be further influenced by the relative importance that you attach to organisational efficiency, personal growth of subordinates and company profits.

Your confidence in your subordinates will be affected by your knowledge of their previous experience and performance. As well as evidence from previous records people do tend to vary in the amount of trust they concede to others; therefore there will inevitably be some people you could trust with one job but not with another. When given a task, as a manager, you should ask yourself 'Who is the best qualified to deal with this problem?' Unfortunately, many managers have more faith in their own abilitites than those of their subordinates and consequently are not good at delegating.

Before deciding how to lead a group of people you must consider the factors that affect the behaviour of the people in the group and the goals that are motivating them as individuals and as team members. You need to consider whether or not the individuals work well together and whether they are all interested in the task. You have to identify whether the team has the necessary knowledge and experience to deal with the task, and whether they are able to share knowledge and make decisions. Finally, a leader must consider whether the team members feel a need for independence or if they still want to be guided.

The situation will affect the leadership styles that you can choose because of the way in which the company expects you to perform and the culture that permeates the company. You will also be affected by the constraints that are imposed on you by the task to be undertaken, such as time constraints, whether or not the task will be repeated, whether there is a requirement for secrecy or confidentiality.

As you can see there are no rules about leadership style that must be used, or to indicate what style will be most effective. All situations must be assessed separately and you must consider your own feelings and those of your subordinates. The ability to apply an effective leadership takes practice and thought particularly in evaluating the effects after the event.

4.4 Appraisal of employees

Appraisal takes place all the time, with managers making assessments of their subordinates' work and considering their progress in the company. Informal appraisal like this cannot be stopped because judgements have to be made before a piece of work is allocated to someone or a decision is made about who to promote. However, one problem with informal appraisal is that it does not actively encourage the participation of the person being appraised, and it is left to the manager to decide when, and if, the judgements that are made are relayed to that person. Additionally the process is very dependent upon the objectivity of the manager's judgement.

Formal appraisal is used by many companies to overcome these problems, although it should be used as a complement to, not instead of, informal appraisal. In this section we will look at formal appraisal schemes, considering their aims, operation, and the costs involved. We will also look at the advantages and disadvantages of such a system for both employer and employee.

4.4.1 The aims of an appraisal scheme

Appraisal is the feedback loop used to assess the performance of employees. It usually takes the form of a documented discussion between manager and subordinate with the manager carrying out an assessment of the individual. An important factor of appraisal is that both parties should agree on the record of the discussion. Many companies use appraisal systems for some, if not all, of their employees. Appraisal is a judgement of an employee's performance in their job and therefore it should be done in a formal way at a time that is arranged by both parties. It should also allow a comparison of performance over time that is meaningful and therefore the time intervals should be appropriate to the employee's objectives. It is also a forum for discussion of the appraiser's approach to the subordinate and should allow a free debate without fear of repercussions.

Appraisal can be defined as an objective assessment of employee performance against objectives and it often involves the following:

1. A review of actual performance over the period covered by the

appraisal, and a discussion of improvements required in the performance, if appropriate.
2. A statement of the present skills and abilities of the appraisee.
3. A statement of the objectives and goals for the next period, including criteria by which performance will be measured.
4. Identification of the best way to develop from the existing level of skill and experience to one likely to prove more desirable in view of the objectives set.

From the point of view of the person being appraised there are four questions that should be answered during appraisal. Open discussion of these questions can help people be more effective in their work and their level of motivation should improve. The questions are listed below.

1. *What should I be doing?* Everyone needs to know what they are supposed to be doing, and what responsibility and authority they have. A job description will provide an answer to this question but a job description will change as the company's circumstances and needs change, and as the job-holder develops.
2. *How am I getting on?* The balance between what the company expects and what is being provided has to be clarified. An honest answer to this question allows assessment of achievements and will enable the appraisee to direct their efforts effectively and modify career plans if necessary.
3. *What will my next career step be?* The appraisee will want to know what the possibilities are for growth within the job, or for advancement into other jobs in the company. Again this allows individuals to manage their careers.
4. *How can I achieve my goals?* Finally, the appraiser should identify what help is available from the manager and the company so that goals can be achieved. This is the area of training and development which is dealt with in Section 4.5.

For people who don't have the opportunity to have appraisals at work, self-appraisal may be carried out. This is described in Chapter 6.

4.4.2 Formal appraisal schemes

There should be a clearly defined approach to appraisal, and in order to ensure fairness it must be monitored throughout the company. This consistency is generally achieved by everyone using the same appraisal form to record the discussion. Monitoring of the forms is then done centrally. The organisation of the appraisal system is usually done by the personnel department, although this requires careful management due to the problem of maintaining confidentiality. However, in order to be effective appraisals must be carried out locally between manager and subordinate, and anyone carrying out an appraisal should be trained in both counselling and appraisal interviewing.

Appraisals should be done on a regular basis: annual or six-monthly appraisals are common. Both appraiser and appraisee should have sufficient notice to prepare properly for the appraisal and to consider the issues that have to be raised. However, the appraisal system should never be used as an excuse to put off dealing with a problem at the time that it occurs. It will only develop and worsen the situation. What the appraisal can do is focus on the reasons why such problems occur and allow a strategy to be developed for long-term prevention.

The main part of the appraisal process is the appraisal interview which is documented using an appraisal form.

4.4.3 The appraisal form

An example of an appraisal form is shown in Figure 4.4; most companies develop their own forms so that they are appropriate to the

IN CONFIDENCE

PERFORMANCE APPRAISAL

SECTION 1 TO BE COMPLETED BY THE INDIVIDUAL BEING APPRAISED

NAME ....

JOB TITLE .... START DATE ....

PERIOD COVERED BY THIS APPRAISAL .... TO ....
TRAINING UNDERTAKEN DURING THE PERIOD OF THIS APPRAISAL

SECTION 2 TO BE COMPLETED BY THE APPRAISEE, PRIOR TO THE INTERVIEW LIST THE MAIN DUTIES OF YOUR PRESENT JOB AND ANY OBJECTIVES WHICH WERE AGREED FOR THIS PERIOD.

....

....

....

....

SECTION 3 TO BE COMPLETED BY THE APPRAISER
ASPECTS OF PERFORMANCE:
MARK EACH ASPECT ACCORDING TO THE FOLLOWING SCALE.
PLEASE ALSO COMMENT OR STATE 'NOT APPLICABLE'

Outstanding	A
Performance above requirements	B
Performance meets the requirements of the job	C
Performance not up to requirements	D
Unsatisfactory	E

a) Written communication — A B C D E
Comment

. . . .

b) Oral communication — A B C D E
Comment

. . . .

c) Technical knowledge and ability — A B C D E
Comment

. . . .

d) Application of latest technical knowledge — A B C D E
Comment

. . . .

e) Acceptance of responsibility — A B C D E
Comment

. . . .

f) Ability to get on with others — A B C D E
Comment

. . . .

g) Ability to produce constructive ideas — A B C D E
Comment

. . . .

h) Reliability under pressure — A B C D E
Comment

. . . .

i) Timekeeping and attendance — A B C D E
Comment

. . . .

j) Planning skills — A B C D E
Comment

. . . .

k) Ability to meet targets — A B C D E
Comment

. . . .

l) Analytical skills A B C D E
Comment

....

m) Ability to foresee and avoid problems A B C D E
Comment

....

SECTION 4 RECORD OF INTERVIEW TO BE COMPLETED BY APPRAISER
This should be an accurate record of the main points of the interview, drawing attention to strengths and weaknesses and including a statement of any disagreements between the interviewer and interviewee. State whether any of the identified weaknesses were discussed during the period of this appraisal (i.e. before the interview).

....

....

....

....

....

....

....

SECTION 5 RECOMMENDATIONS
Give details of any recommendations that have arisen from the interview

....

....

....

....

....

....

....

....

SECTION 6 AGREED RECORD OF THE INTERVIEW
a) Appraiser's Agreement

Signed .... Date

b) Appraisee's Agreement

Signed .... Date

Figure 4.4 Appraisal form

type of workforce being appraised. The purpose of the appraisal form is to provide an agenda for the subsequent interview. In particular it allows the prospective appraisee to consider the topics that will be raised and to plan what is to be achieved by the interview. The appraisal form also provides a record of the interview since it should be filled in during the interview and then signed by both appraisee and appraiser as a record of what took place, indicating what decisions or promises were made and what objectives were set. There are advantages to using standard appraisal forms because they ensure people are familiar with the topics addressed and it helps review and collation of data following appraisal. However, one has to counter this with some method to ensure that the appraisal form is not in itself seen as overly bureaucratic or indeed inappropriate to the situation. The form will ensure that the discussion covers all aspects of work but must be used in conjunction with good interview practice such as asking open-ended questions (see Chapter 9).

It is, of course, important that the appraisal form is not too complex or seen to detract from the most important part of the appraisal process which is the discussion between manager and subordinate.

4.4.4 The appraisal interview

The appraisal process will include at least one interview. This interview is carried out between the appraisee and their first line manager, the appraiser.

The appraisal interview should have a defined agenda, whether based on an appraisal form or not, and this should concentrate on reviewing actual performance, discussion of required improvements, and discussion of future prospects. It must be a two-way discussion if it is to be effective, with the appraisee contributing significantly to the discussion. As stated previously a record of the interview should be agreed and kept for reference throughout the period for review at the next interview.

The success of the appraisal interview relies upon the appraiser's interpersonal and interviewing skills to put the appraisee at ease, encourage discussion and ensure that all pertinent information is reviewed. Interview skills are discussed further in Chapter 9.

A basic agenda for an appraisal interview would include the following four items:

1. Review of performance.
2. Discussion of improvements.
3. Discussion of potential.
4. Objective setting.

The review of performance since the last appraisal interview should review the targets set previously and include a discussion of any problems, such as the reasons for targets not having been met.

There should be a discussion of any required improvements or changes in the appraisee's work which have been identified since the last appraisal.

Discussing the potential of an employee is an important part of the appraisal process; it is equally important to establish whether or not the appraisee is interested in further career development. It is usually achieved by assessing the appraisee against dimensions considered important by the organisation. In discussing potential the appraiser should examine what might be possible for the appraisee if given appropriate training or if allowed to develop certain themes. It is possible to use this discussion as a way of motivating people whose current performance appears unsatisfactory. The discussion should consider training and development. Training and development, as described in Section 4.5, are appropriate for those people who are willing to improve themselves and will be motivated by the training provision. However, substandard performance in those who are not motivated is unlikely to be rectified by training.

When carrying out this type of discussion it is important that the appraiser takes care not to promise what cannot be readily delivered, and also to be honest about the appraisee's potential. Everyone has their performance ceiling. Dishonesty and promises that are not kept can lead not only to demoralisation of the appraisee but also to a lack of confidence in the appraiser's ability to manage effectively. Any action that is planned for the future should be carefully checked with all the other parties involved before any commitment is made, even if this means deferring part of the appraisal interview.

Setting objectives and targets for the period to the next appraisal is the most important aspect of appraisal from the management point of view. Objectives allow performance to be measured as well as providing the input so that plans for work can be prepared. An objective is a prediction about the state of affairs that will exist at the end of an activity. The more precise this prediction can be, the clearer it is possible to be about what has to be done in order to make it happen. In the workplace objectives ensure that all resources are effectively employed in order to meet the company's goals. Objectives should be agreed by appraiser and appraisee for the next appraisal period. This can be very motivational since the objectives will provide goals for the individual and good objectives allow the opportunity for the employee to determine how the objectives will be met. Objectives can also be very de-motivating if they are unrealistic or adequate resources are not provided. Objective setting is discussed further in Chapter 6.

4.4.5 Two-interview appraisals

Some appraisal systems also use a second interview where the appraisee has a discussion with a person at a higher level in the organisation than their own manager, usually the manager's manager. This interview is not intended to duplicate the initial interview but to cover other aspects of the employee's job, in particular a discussion of the wider company objectives.

The use of the second appraiser helps to reduce favouritism or prejudice. This person will appraise many more people than an individual manager and will therefore be able to provide an overview of standards set and the way in which people are dealt with. It will also ensure that personality conflicts do not affect the objectives or standards set. This is important if an employee feels that they are being unfairly treated because they do not get on with their manager.

The second appraiser ensures that a uniform approach is taken across the organisation and allows the appraisee to air views more widely. The second appraiser should not overrule decisions made in the initial interview and should always avoid criticising the manager in order to show solidarity with the appraisee.

The operation of a two-tier appraisal system can absorb a significant expense and a company should be committed to the appraisal system and be aware of the costs that it will incur, before adopting such an approach.

4.4.6 The implications of an appraisal system

The implementation of an appraisal scheme needs financial as well as personnel planning. There are costs involved with training the appraisers and administering the scheme, but these can be small compared to the time that is lost to other business activities whilst appraisals are taking place. Although one should not limit the interview time, so that it becomes more a form-filling exercise than a discussion, it should be remembered that each hour of an appraisal interview is costing at least two man-hours of time. Therefore there must be some advantages to appraisal for the employer. These advantages are described below but as you will see they are largely intangible and it is difficult for a company planning to introduce an appraisal system to produce a numerical cost justification.

Advantages of appraisal for employers number the following:

1. It allows an objective assessment of an individual's performance to be made.
2. It provides a database of information concerning the personnel, their skills and abilities, on which the company can draw. From this, it can lead to better and more effective use of staff by means of transfers, training and planned projects.
3. It can identify difficulties and potential problems so that they can be dealt with before they become major problems.

4. It can improve the performance of personnel through the use of objective setting and increased motivation.
5. It is an ideal tool for use within a management system that relies on objective setting to control and motivate the workforce.

The disadvantages to employers, apart from cost, relate to the way in which the appraisal is carried out; if it is used as a means of criticising rather than developing personnel, it can have an adverse effect on performance.

The other problems that might arise are that the system may be seen by the employees as a control device rather than as a development procedure, and that any interview of this type will be affected by the manager/subordinate relationship. Appraisal schemes are less appropriate, and are less widely used, for employees whose work situation rarely changes, such as machine operators.

4.4.7 Linking appraisal to pay review

Many companies link the appraisal interview to pay review. This can be very cost-effective for the company as it means that only one process has to be administered. There are two other advantages for a company operating this system. It can provide a fair basis on which to divide the salary budget by awarding merit points for performance and giving salary increases on the basis of points achieved. In this way those who contribute the most get the best reward. In addition, it can increase employee motivation to reach company objectives.

However, there are disadvantages to linking appraisal and pay, and these are often deemed to outweigh the advantages. Firstly, it will increase the amount of defensive behaviour and reluctance to admit faults on the part of the appraisee, and some people will not listen until they hear news of their new salary level. In addition it can cause people to use the appraisal as an opportunity to justify why their salary should increase and this can lead to conflict between the appraiser and appraisee, damaging the personal relationship and undermining the purpose of the appraisal.

4.5 Training and development

Well-managed departments have clear objectives. In order to meet these objectives, a manager will require a battery of skills from the department's personnel. Since these objectives change with time, and because personnel come and go, there will frequently be times when the manager does not have the required skills and abilities within the department. The prudent manager has a constant eye for potential mismatches between the skills required and resources available, and aims to correct such a mismatch once it is identified.

There is a spectrum of response that the manager may use. At one end the manager may do nothing and leave the individuals to cope on their own. At the other, new personnel are recruited who have the required skills. In between these two extremes lies the middle ground of development and training. In this approach existing personnel are encouraged to develop and acquire the desired skills.

It is important to distinguish between development and training. Training presupposes that the desired skill is already within the capacity of the individual and that they only need to be shown how to do it. Development, however, involves preparation for tasks or behaviours that are currently beyond the individual's range of responses. A training objective might be 'to become competent in the operation of the production control computer system by the end of next week', whilst a development objective might be 'to acquire skills necessary to run the group of three sales engineers responsible for initial customer visits and after-sales commissioning within the next year'.

Landy and Trumbo have defined training as:

> Planned activities on the part of an organisation to increase job knowledge and skills, or to modify the attitudes and social behaviour of its members in ways consistent with the goals of the organisation and requirements of the job.

There are two prerequisites for successful training: these are intellectual capability and the desire to learn. One cannot train a mathematically inept individual to produce creative mathematical ideas, nor can one force knowledge into an unwilling pupil.

It should be realised that there are limits to how much a person may change as a result of training. Training cannot affect basic psychological attributes. An introvert will never be made extrovert simply through training. However, an individual might go through such a change in their own time and as a result of their own personality development, but such a process cannot fundamentally be altered by training. An individual's value system is equally deeply rooted and not amenable to modification by training. Attempted modifications to such attributes comprise brainwashing and the results are not only unethical but unreliable when ultimately tested. Brainwashing techniques include such things as 'reinforcement', where a particular idea is restated in different ways on hundreds of occasions, or the so-called 'love bombing' where acceptance of the idea being presented results in convincing affectionate behaviour from the 'trainers'. Sleep-deprivation is often used by filling the day and night with 'important' tasks that support the ideal being conveyed; also the absorption of material by repetitive regurgitation such as chants or the incantation of secret verses can be effective. After prolonged exposure to such techniques the individual becomes less able to apply rational criticisms to the material presented and even the most resilient individuals adopt it. Brainwashing works by replacing the normal supply of stimulus from the world with a biased and one-sided view. The more completely the biased presentation replaces the real world, the more complete the brainwashing becomes. Some extremist religious sects use brainwashing techniques to engender beliefs that are inconsistent with reality. Individuals who return to normal society soon recover their rationality and reject the brainwash. It is not therefore surprising to find such sects reluctant to expose their members to normal society which tests their brainwashing and usually overturns it. Whilst the isolation, and with it the illusion persist, loyalty, and usually money, are commanded by the sect.

4.5.1 Conducting a training programme

There are three phases to conducting a training programme:

1. Identify the needs: Using the organisation's objectives and a summary of the given individual's abilities as starting-points, generate a list of training needs for each individual in the form of objectives.

2. Select and apply the appropriate training method: There are many techniques with which people may be trained. Each is suited to particular circumstances. The most appropriate technique should be selected and applied. These are described in Section 4.5.2.

3. Monitor the effectiveness: At this stage the manager assesses the response of the group to the training. This should be measured both by how the individuals are performing and by discussing it with the trainers. The manager will wish to know whether any further training is required, whether the individuals are now able to produce the required skills, and whether value for money has been realised. There are implications for the monitoring system here. For instance, how will one measure whether value for money has been realised, and what criteria will the organisation use to assess the 'before' and 'after' case? Usually these questions should be answered in terms of the original problem which highlighted the need for training. Often the benefits of training take some time to be realised and are difficult to quantify. How does one quantify and measure an improvement in personal effectiveness, for example? The appraisal forms an excellent opportunity for the individual and the manager to examine the success of training and it is common to have a review of training received as a standing item on the agenda of an appraisal.

4.5.2 Methods of training and developing personnel

There are many ways in which individual performance may be enhanced and they all offer scope for either development or training to take place. The way in which the material is presented, the content of the material and the abilities of the individuals attending indicate whether development or training is in progress. The most effective methods for enhancing performance are listed below:

1. On-the-job experience.
2. Coaching.
3. Roleplay
4. Study.
5. Games, simulations and case studies.
6. Internal training courses.

On-the-job experience

This type of training occurs all through the working day. The trainer is anyone else involved in the daily running of the department but is usually more experienced or more senior. This type of training results from the daily human contact of colleagues at work but is a very effective teaching method. For example, an apprentice potter is assigned to assist a master craftsman, or an engineer has a desk in the project office in which many other engineers of different levels are at work: simply by being there much job knowledge is absorbed.

Training of this sort is extremely relevant and cheap. It affords ways for individuals to develop in their own environment. The training introduces the trainee to the whole of the job and hence offers a complete training system. The process does not disrupt the flow of work. Unfortunately, it is often necessary to choose trainers by position, rather than by their ability to teach. The need to perform the training in a working department makes providing a good learning experience difficult. If the trainers are coached to be effective teachers this very cheap training method is enormously successful.

As part of the training, individuals may be given particular projects within the normal tasks of the department. A collection of such projects is often a good way to introduce the individual to the operation of the department. Individual projects can also be used as training vehicles for larger, more risky projects that may follow.

Sometimes trainees need actually to visit and see a special process or situation. When dealing with the unfamiliar one can always be advised of how things are, but there is never a good substitute for seeing things yourself. A sales engineer joining a company may well be shown the various departments, special processes, or operations of the organisation. This is done even though the job does not directly involve such knowledge since benefits may be derived in terms of appreciation of what is involved and the ability to have knowledgeable conversations with customers. Simply telling the engineer is unlikely to provide the amount of appreciation and recall that even a brief visit gives. Visits and demonstrations can be arranged as part of on-the-job training.

Some types of on-the-job training can be very structured — for example, it is common for large organisations to have well-organised and lengthy training courses for the apprentices. Similarly, graduate engineers usually go through a very structured training programme to achieve professional status. For the individual concerned the quality of provision of such training can be a decisive factor when accepting or rejecting a job offer.

Coaching

This is usually used to instil particular skills, often of a physical nature; for example, coaching is often used for sporting or operating skills. It involves a coach who will take the students through the learning process and on to independent reproduction of the desired skill. For example, coaching would be used to teach an aircraft technician to perform an emergency shut-down of a jet engine at full power whilst beside the engine.

Coaching often involves teaching individual's to go against behaviours they currently hold. In general, the more deeply the skill or behaviour being coached conflicts with normal behaviour, the more coaching is required and the closer the relationship between coach and student needs to be. For example, a trainer will have to command a great deal of respect if required to persuade trainees to jump out of an aeroplane and be saved by a parachute they have packed!

Coaching is very effective at instilling skills and is of particular relevance when personal safety or credibility are threatened. It can seem expensive but the cost of the coaching should be compared with the cost of errors made by the untrained. When this is done the suitability of expensive coaching is often revealed. It is one of the few ways that behaviour in unusual or undesirable situations may be made automatic. In all cases of coaching the importance of excellence on the part of the coach cannot be overstated. Coaching is a skill in its own right and there is no point in using inappropriate individuals. Two special cases of coaching are worth noting: mentors and shadowing.

Mentor training is simply one-to-one coaching. It requires an experienced coach who follows the trainee almost everywhere during the period of training. The method is particularly appropriate where the material to be taught is hard for the trainee to learn. The technique is very dependent on the mentor. It is clearly, intrinsically expensive. For example, a trainee marketing executive might be mentored by a qualified training consultant who takes notes as the day progresses and discusses these with the trainee at the end of each day. The consultant will also deliver short bursts of training at times when poignant examples have just occurred. The mentor system is able to deliver one of the most specific and appropriate training systems of all. When given by a competent mentor it is a very effective training method but has a price tag to match.

Shadowing still involves one person learning from another but does not use the close teaching relationship of the mentor system. Shadowing means learning the job content of another by following them through a typical period of work, perhaps a day or two, or even a month. The shadow does not interfere with the person being shadowed, but observes all that goes on and asks questions when required to clarify confusion. It is appropriate for gaining understanding of job content but not particularly good for acquisition of skills — on-the-job training is used for this. It is especially appropriate for training individuals where there are many other departments about which understanding is required.

Roleplay

In roleplay the trainee uses their ability to see things from another's point of view and learns from the experience of playing another person. The trainee takes on a role which is provided and a scenario is played out. Since this is done away from real life, mistakes do not matter and learning can be made less risky. For example, a graduate engineer in a company might be sent on a business skills training course and roleplays a telephone conversation with a difficult customer. The roleplay is videoed and the small group watch afterwards and debrief the process. Another group member then tackles a slightly different situation using the skills learnt. After the experience the individuals may modify their behaviour and improve their effectiveness.

The process usually involves two or three people roleplaying in accordance with instructions. One or more members of the group are appointed to record what happens and try to see the situation from all sides. A video camera may also be used. The technique is particularly appropriate to interpersonal skills training since it revolves around feelings and how to deal with them — it is not appropriate to try and use the technique for factual tuition. One of the limitations of the technique is that not everyone is good at seeing things from another person's point of view and so may find the process difficult. It is also worth noting that when difficult scenarios are presented they may unknowingly upset trainees by dwelling on sensitive areas. An example might be someone recently widowed being sent on an appraisal training course and having to roleplay someone whose work has suddently deteriorated because of a bereavement.

Roleplay offers the only real way to study and develop personal skills in a safe environment. The trainees themselves, however, can limit the effectiveness of the technique. Any individual who cannot act at all, or who finds it difficult to assume the role of another person in front of others, is not likely to learn much because they simply cannot make the technique work. Interestingly, it is often exactly this sort of person who has the most to gain. The lack of this skill can itself be the source of the poor interpersonal skills the training is seeking to rectify. Consequently, those who have the most to benefit from the method can be the least able to do so.

The low cost, wide availability and instant play-back attributes of video cameras and recorders mean that they are extensively used for roleplay purposes. Seeing yourself on video is usually both surprising and informative, although for some people it can be embarrassing. The trainees first perform the scenario and are filmed by the trainer; they then view themselves. A trainer goes through the video making observations on specific relevant points and so debriefs the replay. Seeing yourself on video allows an individual access to a record of performance that cannot be questioned. Other means of providing feedback depend on the observations of others and therefore lack the objectivity of the camera. People are often amazed by the sound of their own voice which is very different when heard as others hear it. People become aware of many annoying idiosyncrasies or habits and some people can be demoralised by the experience. For this reason care should be taken with debriefing and it needs to be handled sensitively if it is to facilitate learning. Feedback should never be used in such a way that individuals are embarrassed by their performance. Videos can be particularly effectively used in 'before' and 'after' scenarios to make clear the improvements that the training has brought.

Study

Study is especially appropriate to the personal acquisition of factual knowledge; it is not, however, efficient for the acquisition of practical skills. When you need to learn about the history of France you reach for a textbook; if you want to juggle, you ask a juggler to coach you.

During study the individual often acquires the material alone, through books or correspondence courses, and is in complete control of the learning activity. If the study is in the interests of the organisation the commitment may be supported by study leave or use of company resources. The study might take the form of formal higher education courses or less structured study courses at other institutions. Occasionally the trainee will be given sabbatical leave to complete the course.

In some sorts of study, as in the formal higher education course, it should be remembered that an individual not only gains the factual body of knowledge from the study but also the methodology of thought associated with it; this indirect asset is hard to value. For example, many computer companies recruit extensively amongst classics graduates. This is clearly not because of their computing skills, but because their systems of thought and ability to analyse options are useful to computing organisations.

Formal study within a structured syllabus gives high-quality knowledge. Formal qualifications are portable and an organisation may find a good portion of its trainees marketing their skills elsewhere. However, low cost training outside of working hours, and wide applicability combine to make the option attractive. Private study might be used as a route for a marketing manager to gain a Master of Business Administration or an organistaion might offer a distance-learning course in German to an export clerk.

Studying for a formal qualification whilst working demands a high level of motivation on the part of the student. The pressures caused by formal study on top of a full-time job are not to be taken lightly by either side. Such a life-style leaves little recreation time for the student and the long study hours can reduce the student's on-site performance, especially at exam time.

Games, simulations and case studies

These are usually used within other training material such as books or courses to illustrate particular points. Training games usually take the form of a competitive game between teams who have some aim and must employ the tactics being taught in order to win. Often the games are based around imaginary companies engaged in competition and are similar to popular board games.

In simulations, the trainees are presented with situations in which they must devise strategies to solve the problem or achieve the aim. The simulation allows the trainees to practise in a safe environment away from the dangers of real life until they are confident of success. The simulators can vary greatly. Flight simulators are examples of how practical skills can be acquired in a situation so lifelike that the trainees are coached as if they really are in an aeroplane.

Case studies provide examples or illustrations of the issue being taught. They are sometimes used in a purely illustrative way. Often they may achieve their instruction through involved and detailed analysis of a lengthy case history. Case studies provide the trainees with access to material just as it was faced by those involved at the time and so brings them much nearer to the real world.

Internal training courses

It is quite common for organisations to run their own training courses. The larger the organisation the more likely it is to make economic sense. Internal training courses may be tailored to the individual organisation and can provide a good vehicle for learning about the company's culture. On the other hand they take a considerable amount of organising and there is the danger that they will become outdated and stale compared with external consultants whose constant exposure to the market-place keeps them up to date.

Organisation-specific training usually takes the employee off-site for a period of time, usually between one and ten days. Often the courses are residential and they are typified by long hours of intense study and learning. A great team spirit can develop on such courses and the absence of external factors leads to a sustained and intense atmosphere. Sometimes television, newspapers and incoming telephone calls are banned. The learning process is very rapid in such situations. Provided the courses are well constructed and relevant, they are very effective. Often such courses, especially factual ones, are supported by good documentation which can become an invaluable reference.

The trainee quickly gains highly relevant experience. The opportunity to develop a good working relationship is provided. It brings the disadvantage that the trainee is not available in his own department for normal duties and that colleagues may be resentful of the training and consequent career advantage gained. However, such courses are often expensive, especially when high-quality hotels or conference centres are used, and fatigue can interfere with the learning process.

We have now looked at six methods of training and development. The education of a professional person is something that continues throughout life — a qualified engineer can expect to experience most, if not all, of the above techniques during an engineering career. By understanding the different techniques and having a knowledge of them it is possible not only to select the best method, but also to get the most from it. Such improvements in training efficiency are clearly of interest to successful engineers.

4.6 Job design and payment systems

Employment is a relationship between an organisation and an individual — the individual provides services and in return receives two things. Firstly, payment of some form and, secondly, job satisfaction. Money is clearly an important part of the relationship but it is not all of it. Nobody is truly happy whilst engaged in work they find uninteresting. The simple fact that not all vocations receive the same remuneration, yet people still compete for jobs even in the poorly paid ones, shows that people attach great importance to the nature of the job as well as to the remuneration associated with it.

As engineers we may well be called upon to assist with, or even take responsibility for, the creation of new jobs. This might come about through a need to expand a department in which we work or because we have too much work ourselves and need assistance in the form of a junior. If we are to do this important job well and attract committed applicants for selection we will have to understand something of how a job may be designed to be motivating and how to arrive at a figure for the remuneration that the job should carry. In some organisations, particularly larger ones, there may well be a department with just this responsibility; alternatively it may be the responsibility of the individuals concerned. Either way, it is important that as professional engineers we have an understanding of what the process involves. We will now look at two subjects which address these issues: 'job design' deals with engendering job satisfaction and 'payment systems' deals with the provision of fair pay or remuneration.

4.6.1 Job design

Job design aims to make a given job required by the organisation motivating to the employee whilst retaining the benefits it brings to the organisation. In the first half of this century attitudes towards employees were very authoritarian. Employees were seen as obstinate and lazy, people who had to be forced to work. As time went on, it was realised that this attitude was quite wrong and that people are highly motivated if given responsibility for their work.

Modern job design takes note of these factors and the needs of the organisation are seen in conjunction with the needs of the individual. Designing a job well means finding a happy marriage between the needs of the organisation and the motivation of the individual so that both benefit from the committed attitude of the other. This task may involve the inclusion of obvious motivational factors such as developing responsibility. It may also include the avoidance of de-motivating factors. No one likes to be given inappropriate tools or insufficient facilities for a job and job design is a means of ensuring that this does not occur.

There are many aspects of job design. Below we will consider four particularly important ones: (i) ergonomics; (ii) work study; (iii) consistency; and (iv) job specification. The first two, ergonomics and work study, are large subjects in their own right. A trip to the engineering section of any reputable library will soon uncover many texts on these subjects. In this section we are considering job design as a part of personnel management, and we have therefore referenced ergonomics and work study since these subjects impinge greatly on the design of jobs; however, the sections should be considered only as a brief introduction to large subjects that might need further investigation if much job design is in hand. The second two aspects, consistency and job specification, also have a great impact upon job design. These subjects, however, are very dependent upon the nature of the job in question and the circumstances of the organisation. For these reasons the important issues are raised but these may only be addressed by considering them in the context of the organisation.

Ergonomics

Ergonomics is the study of the interface between man and machine. The aim is to make this interface an efficient one and allow the machine to be operated at speeds limited by the operator, rather than the interface. If we are designing jobs indirectly, that is by designing equipment that will be used whilst doing a job or designing an environment in which a job will be carried out, a knowledge of how best to design the interface between man and machine is desirable.

The design of control consoles or instrument panels for aircraft may draw heavily on the concepts of ergonomics. For any given situation the ergonomic approach is, firstly, to understand the way in which the operator acquires information and gives instructions or moves controls; secondly, to characterise this exchange; and thirdly, to design an interface that permits optimal exchange.

Different situations have different needs. A workstation for an assembly worker may require speed to be maximised; alternatively, accuracy may be paramount as with the control room of a power station, or a balance of both may be important as in the case of a fast jet fighter control panel.

To be successful in such a task the ergonomist must have a thorough understanding of the human body's capabilities. Data relating to humans is called anthropometric data. Books of it can readily be bought and it is usual to display the various data in percentiles — for example, a diagram showing the typical reach of the human arm will have the dimensions for the mean shown and will also have a table to show how the dimensions change for each percentile of the population.

The environment also affects the performance of the interface and so ergonomics is also concerned with optimal levels of lighting, shift lengths, working temperatures and so on. Again, many standard reference tables have been prepared of such data.

Work study

Work study means exactly what it says, studying work. The aim is to produce the optimal methods of performing a given job. This involves consideration of the work and the machines or other resources involved. There are clear benefits for the organisation in doing this well but it is also motivational for the worker. It is certainly satisfying to know that one has the best method and tools for the job in hand.

The study can take place on any scale — for example, in the assembly of a watch, during which the work study engineer will consider things such as the distance of hand travel in picking up the components and the optimal balance of the tools used. The engineer might weigh up whether two tools should be used for a particular task or if efficiency would be improved by the use of one slightly heavier tool which could do both tasks. Alternatively in a shipyard the engineer might consider the distance hull plates have to be carried, how many hot rivet stations are required, or how many times a particular point has to be visited and so on. These allow the preparation of optimal work schedules.

Work study has many areas which are often considered separately. Method study, motion study and time measurement are three important components. Frank B. Gilbreth did much work in these areas and is often attributed with being the founder of work study as a discipline in its own right. His wife was also a considerable force in work study and they jointly ran Gilbreth Inc. between 1910 and 1924. The company was hired to apply their scientific examination of the best way to complete specified jobs by many prestigious industrial organisations. Whilst conducting research into work study Gilbreth even gave his name, in reverse that is, to a symbol he invented. The 'Therblig' is a small symbol used to represent a specific body movement when writing down a particular task as a series of component motions. Once proficient in the 'language' of Therbligs one can write down complex operations with a view to manipulating them and finding improved ways of accomplishing the task.

The unique thing Gilbreth brought to his subject was his unwavering dedication to identifying the absolutely optimal method of performing even the simplest task. He was very clear and objective about his work and his attitudes extended into his private life. A novel, *Cheaper By the Dozen*, written by one of his twelve children, gives a humorous insight to the family life of the Gilbreths and does much to illustrate Gilbreth's own obsession for his subject.

Consistency

We have already considered various observations on motivation in Section 4.3.1. These need to be considered during the process of job design and in this section we shall not repeat the ideas, but rather make some observations on their application to job design.

Although jobs need to be individually designed, a job designer must also consider consistency across the organisation. One should not, for example, combat low pay in a company by introducing a few jobs with appropriate remuneration to start with. Attempting to rectify the situation in such a way will, in reallity, have a de-motivating effect on those whose pay is left behind. Individuals will naturally look at the fortunes of colleagues as well as themselves when judging issues like fair salaries or recognition. It is therefore vital to ensure that the design of all the jobs is consistent.

Consistency must also be maintained over a period of time. As skills develop, the individual must be able to move on to new tasks that assist in the quest for self-fulfilment. During this process the job designer must avoid the pitfall of failing to acknowledge advancing abilities and responsibilities. For instance, there is no point in increasing the motivational aspects of responsibility and authority without increasing remuneration to a level consistent with the new job.

Job specification

To specify a job completely one needs to specify all the things that impinge upon the job. Such a specification is of use both to the designers of jobs and also to applicants who are interested to know, as exhaustively as possible, the nature of the job. Section 4.2 contains much detail about the process of employing people. Times change and as circumstances move on the job description and job scenario that used to apply may also have to change. Jobs do change gently and it is important to monitor the changes to be certain that the job actually being conducted, as opposed to the original description which may be out of date, is still of optimum design from the employment relationship point of view. The following four dimensions may be used to define any given job.

Job descriptions and contract of employment: These two documents have previously been discussed in Section 4.2 and are referenced in this list since they are clearly documents that specify what a job entails. Of all the items included within a complete job specification, the job description is often the only one a job applicant is clear about before accepting a job.

Standard codes of practice for the organisation: Many organisations have their own guide-lines and codes of practice that affect jobs within them. Some organisations will have particular safety codes that have to be adhered to at all times. Others may have company meetings or regular departmental away-days. Such activities can shape the way communications occur and can affect jobs in all sorts of ways. For instance, in some companies the normal practice is to answer a ringing telephone within three rings regardless of what you are doing. Such a practice is clearly harder for some people to endure than others and therefore has a bearing on the nature of the job.

Law: Organisations that provide more than the legal minimum on issues such as redundancy benefit or pension arrangements will attract loyalty in a way that those who begrudge their employees even the legal minimum, will not. The law also makes a requirement to employ people in non-discriminatory ways and it is illegal (except for certain special cases) to stipulate, for example, the sex of an employee in an advertisement. Section 4.2.4 explains such issues. The whole issue of employment law is complex and beyond the scope of this engineering management textbook. A search of any reputable library will soon produce a list of titles on this one topic alone. However, it is clear from this paragraph that the law has the potential to affect the specification of a given job; it is therefore the responsibility of a job designer to be familiar with the law as it applies to the particular industry and to accommodate it when designing jobs.

Organisational culture: The culture of an organisation can have a profound effect on a job. The culture tends not to operate in writing and is much more a socially controlled issue. For example, in the contract of employment for many professional jobs, one will usually find a normal working week of less than forty hours specified. In such organisations, however, it is common to find employees working far longer than this stipulated minimum. This extended working week is soon taken as normal and those who work less are seen as uncommitted and are less likely to receive promotion. Unless such factors are accounted for and included in a job specification the resulting picture of the job is inaccurate.

4.6.2 Payment systems

Remuneration is the name given to the package of benefits that an employee gains from their work. It does not normally consist only of pay and the amount of other benefits included depends on the nature of the job in question. Such additional benefits may add up to as much as 20 or 30 per cent of the basic wage and form an important contribution to the employee's standard of living. For example, the employee might receive some or all of the following:

productivity bonuses
personal insurance
healthcare cover
share issues
pensions
Christmas hampers

- unpaid holiday entitlement
- subsidised meals
- subsidised social club
- company car
- company petrol
- crèche facilities
- book allowances
- reduced prices on company goods
- free sponsorship tickets
- expense account
- sports facilities

This list is by no means exhaustive but clearly shows that remuneration means more than a basic wage. The organisation will usually have its own policy concerning the allocation of such benefits. For instance, it is common for all employees to have healthcare cover but only a few to have a company car. In most cases the organisation assigns a monetary value to each element of the remuneration package and the wage is calculated by subtracting the value of the other benefits from the monetary value assigned to the job. Of course, different individuals may rate the value of the benefits differently. For example, a car will be valued much more highly by someone who has a long way to drive to work than by a local employee who rarely leaves town.

The central aim of a payment system is to define the way in which the employees of an organisation will be remunerated. There are two extreme views that may be taken with regard to such a system. In the first, everyone receives the same wage. This rather ideological system has been tried but never seems to survive. Such systems are soon undermined by the inevitable movement of personnel. The senior people leave the organisation to earn more elsewhere and the resultant skill imbalance ruins the organisation.

The other approach is to say that everyone should receive remuneration in proportion to their contribution to the organisation's success. This proposal is normally accepted as the most equitable method and many organisations freely state it as their aim to remunerate employees in this way. The central question is how actually to rate the jobs within an organisation. For example, how does one decide the relative importance of a draughtsperson and a managing director? Below are various important issues that must be considered when preparing an assessment of the contribution of each job to the organisation.

External comparison

A comparison is made with other jobs available in other organisations. This is often more difficult to do than first it seems, since companies are in general reluctant to disclose details of their own payment systems. Salary surveys are regularly performed by independent journals and can often help in providing raw data. However, these rarely consider the value of benefits other than salary.

Job ranking

For a payment system to be fair the relative value of each job must be known. A list of the jobs within the organisation is compiled which reflects the importance that the organsiation places upon each one. Since the jobs are in order it follows that the remunerations should also be in order and a glance down the column of salaries will easily reveal any anomalies.

Job assessment

A comparison is made of the content of all the jobs and their specifications within the company against the people who fill the posts. This is done to ensure that there are no inconsistencies within the organisation. People of a certain standard should have similar levels of complexity and remuneration across the organisation. Nothing is more likely to cause trouble than having two people doing the same work and receiving different levels of remuneration. To ensure that employees are fairly rewarded personnel assessments may be conducted. In this process personnel are assessed by their qualifications and skills and given a ranking relative to others. The more highly qualified and skilled the employee, the higher the remuneration should be.

Once these investigations are complete, the organisation has a list of the jobs in order of importance and various bench-marks of external comparison against which to judge the level of pay. There is both relative and absolute information and so a remuneration value for each employee can be defined. Two commonly used systems for choosing the final numerical value of a salary are described below.

Graded system

The jobs are divided into 'grades' and within each grade there are various points. Each grade corresponds to a given level of seniority and there is a considerable difference in the job description of people between grades. To each point a salary is assigned. Everyone in the organisation has a salary somewhere on the chart. An example is shown in Figure 4.5.

Changes in salary are then brought about by three separate means. Firstly, through an annual increase of the number associated with every point. In this way the salaries of the company may be moved as a whole to allow for general economic effects such as inflation or recession. Secondly, an individual may be moved up the points ladder within the grade in response to good work. This may be done following an appraisal or simply after a given length of service, often each year. Thirdly, the salary may be increased through promotion, which involves moving on to the next grade up and into a more senior job altogether.

TIC-TOC COMPANY LTD PAYMENT SYSTEM					
POINT	GRADE 1	GRADE 2	GRADE 3	GRADE 4	GRADE 5
1	10500	14500	17900	21780	25780
2	9300	13100	16100	19964	24124
3	8700	12400	15200	19056	23296
4	8100	11700	14300	18148	22468
5	7500	11000	13400	17240	21640
6	6900	10300	12500	16332	20812
7	6300	9600	11600	15424	19984
8	5700	8900	10700	14516	19156
9	5100	8200	9800	13608	18328
10	4500	7500	8900	12700	17500
INCREMENT	600	700	900	908	828

Figure 4.5 Salary structure for Tic-Toc Co. Ltd

Free position systems

In companies using free position systems individuals are permitted to take on any salary level the organisation sees fit. The advantages are that it permits people with quite similar jobs to be paid very different salaries if their performance merits it, since there is no scale against which everyone compares themselves. The disadvantages are that employees can feel that the organisation is paying the least it can get away with in each case and the more secret approach leads to increased salary speculation. Payment systems must be fair and be seen to be fair. The system does allow considerable flexibility and provided the employees feel that it is indeed the greater contributors that receive the greater rewards the system will endure. The system makes extra demands on the appraisal system if this is used to assess performance and therefore salary increases.

4.7 Summary of Chapter 4

We started this chapter by looking at the case of Hilary Briggs, a Logistics Director in the Rover Group. We saw that in a few years since graduating Hilary now has a very considerable personnel management task. She has five employees reporting directly to her but manages a department of 365, and she has to motivate her staff, appraise their performance, and manage development and training for them. In addition, as with all companies, the staff have to be recruited and paid.

In this chapter we have looked at each of these aspects of personnel management. We saw that in order to employ someone there are a number of legislative and economic factors that have to be taken into account, and we saw that there was a process that had to be followed of recruitment, selection and induction.

We then considered how you might encourage your staff to work to their best and considered some of the ideas put forward by people such as Maslow and Hunt and how you can use these to determine motivation factors for individuals.

Appraisal systems were reviewed in Section 4.4 and we saw that the process of appraisal, whilst costly, can have great benefit for both employer and employee.

One of the main reasons for appraisal is to identify potential and determine training and development needs of employees. In Section 4.5 we looked at the training methods that could be used to meet these needs.

Finally, in Section 4.6 we looked again at some factors that affect motivation, but from the viewpoint of job design and methods of remunerating people for their work.

Personnel management is obviously an enormous topic and many people specialise in particular aspects of it, such as motivation or recruitment. As engineers you are unlikely to have to follow a specialised route in this area, but you will undoubtedly have a need for a general understanding of the topics. This text has provided an overview of those topics and in Sections 4.9 and 4.10 we have directed you to some more detailed texts.

4.8 Revision questions

1. How can companies using a free position system for determining the level of pay of its managers ensure that it still complies with the Equal Pay Act (1970)?

2. (i) List the main features that you would expect to see in a contract of employment.
(ii) What is the legal position with regard to the preparation and issue of job descriptions?
(iii) What role do job descriptions have in terms of personnel management?

3. A friend at work, also a graduate engineer, has asked for your advice on how she might resolve what she feels is a clear case of sex discrimination. You both work in an office with fifteen other graduate engineers — she is the only woman.

The company that you work for publishes no information on pay scales, all pay negotiations being done on an individual basis. Generally no one discusses their pay; however, two weeks ago your friend overheard a colleague talking to someone about pay in the pub and she got the impression that this person was being paid more than she was. She feels this is particularly unfair becasue: (i) she has been in the company longer than the other person; (ii) she feels that she has a much more responsible position than he has, as she works on a much more expensive project.

During the last two weeks she has thought of nothing except this and is now also convinced that she has been discriminated against in other ways. For example, she was not given the supervisory job that she had expected. She is now determined to walk into the manager's office and hand in her resignation, having made her views on the situation clear.

You had noticed that she had seemed particularly upset today and when you asked what was happening she told you the story and finally asked, 'What would you do?'

What advice would you give, taking into account both legal and personnel implications?

4. As a personnel manager consider the legal and personnel aspects of the following two cases. How would you deal with them?

(i) An engineering apprentice has a history of using bad language at supervisors. He was verbally reprimanded for this three months ago and until this week there had been no recurrence. Yesterday he was asked to do a particular job by his supervisor and responded with a stream of abuse which culminated in him walking out of the factory; he has not returned.

(ii) Due to the introduction of new machining centres your company wants to make a fifth of the people in the machine shop redundant — there are currently 50 people working there.

5. What is the role of a job description for (i) employer, (ii) employee?

6. Why should personnel specifications be prepared when planning recruitment?

7. Outline the process of finding and taking on a new employee.

8. What costs are involved in recruitment and selection?

9. What options are open to a company that fails to recruit a suitable candidate to a key managerial post?

10. What is the purpose of the contract of employment?

11. For what reasons can an employee be legally dismissed?

12. Under what circumstances can a company lawfully discriminate when recruiting?

13. What do you understand by the term 'motivation'?

14. What motivates you now, and how do you think this will change over the next five years?

15. How can a manager find out what motivates his/her subordinates?

16. How does leadership affect motivation?

17. How can a leadership style be developed?

18. What benefit does the manager gain by having a knowledge of the 'motivation theories'?

19. What are the benefits of an appraisal scheme for an employee?

20. What costs are involved in operating an appraisal scheme and how can these be justified?

21. What is the purpose of an appraisal form?

22. What are the requirements for carrying out an effective appraisal interview?

23. Why is brainwashing not appropriate to training in the workplace?

24. Difficulties can arise when trying to monitor the effectiveness of training — what are these, and how can they be overcome?

25. Management training makes much use of roleplay. What benefits can be achieved from this training technique rather than study?

26. On-the-job training is widely used in industry — what are the requirements for this to be successful?

27. What advantages can a company gain from using on-the-job training rather than any other method?

28. What are the aims of job design?

29. How can a knowledge of job design benefit a design engineer?

30. If a company has an unfair allocation of responsibilities between its managers how could it determine a fairer apportionment?

31. What are the advantages for (i) the company, (ii) the employee, of using a graded payment scheme rather than a free position one?

32. What are the effects of using a single graded payment scheme in a company that operates on a number of sites nationally?

4.9 References

Motivation and leadership

Herzberg, F., Mausner, B. and Snyderman, B.B. (1959), *The Motivation to Work*, New York: Wiley.

Hunt, J.W. (1986), *Managing People at Work — a manager's guide to behaviour in organizations*, London: McGraw-Hill. (See Section 4.10 also.)

Hutchins, D. (1985), *Quality Circles Handbook*, London: Pitman.

McGregor, D. (1960), *The Human Side of Enterprise*, London: McGraw-Hill.

Maslow, A.H. (1987), *Motivation and Personality*, 3rd ed., New York: Harper & Row.

Tannenbaum, R. and Schmidt, W. (Mar/Apr 1958), *How to Choose a Leadership Pattern*, Harvard Business Review.

Training and development

Landy, F.J. and Trumbo, D.A. (1976), *Psychology of Work Behaviour*, Homewood, IL: Dorsey Press, p. 222.

Job design and payment systems

Gilbreth, F.B. and Gilbreth Carey, E. (1949), *Cheaper By the Dozen*, London: Pan Books Ltd.

4.10 Further reading

Bentley, T.J. (1991), *The Business of Training*, London: McGraw-Hill (training series).

This book discusses the role of training and its impact on business. It looks at the need for training and the way in which training strategy can be developed, and training designed, to meet those needs.

Hunt, J.W. (1986), *Managing People at Work — a manager's guide to behaviour in organizations*, London: McGraw-Hill.

This is an excellent book which is easy to read. It fully explains Hunt's theories on the use of goals as motivators in such a way that their practical application can be understood even by people without industrial experience.

Sidney, E. (ed.) (1988), *Managing Recruitment*, Aldershot: Gower.

This book is a compilation of articles from different contributors covering all aspects of recruitment and selection. In particular it discusses techniques for recruitment of special categories of people such as graduates.

Torrington, D. and Chapman, J. (1963), *Personnel Management*, Hemel Hempstead: Prentice Hall International.

This is a comprehensive text covering all aspects of personnel management. It is recommended by the Institute of Personnel Management and is an invaluable reference.

Chapter 5
Teamworking and creativity

Overview

In the first chapter of this book we introduced engineering management. We explained the sort of skills that the engineering profession demands of its members. One of these skills is the ability to work in teams. Successful teams greatly extend the abilities of the team members and allow the team to achieve far more together than they can singly. The ability to work successfully together in teams is a prerequisite of creative engineering. We do not work in isolation — we are all in teams of some sort. In this chapter we shall take a detailed look at the important science of teamworking.

5.1 Introduction

The history of science and engineering is filled with great advances made by brilliant individuals. We all know of Newton's laws of motion, Stephenson's Rocket, or Brunel's civil engineering achievements. These pioneers seem always to be from the past — why is this? It clearly is not because progress has slowed down. In fact quite the reverse is true, progress has never been faster than it is today. The reason for an absence of such individual pioneers is simple. It is no longer possible for individuals to accomplish the vast intellectual challenge that is associated with even a small advance today. People cannot do it alone, but they can do it in teams. Teams of scientists unravel the structure of matter. Teams of engineers design and manufacture airliners and teams of professionals bring together the many skills that are necessary to introduce new products like compact disks to our markets. As engineers, we are clearly interested in studying this productive co-operation and in this chapter we will examine its operation.

The chapter is divided into three main sections. In Section 5.2 we shall examine how and why teams work. We still start by looking at the problems and advantages that teamworking brings. We shall then investigate teamworking from a theoretical point of view and introduce a particular theory together with the experimental evidence that supports it. Section 5.2.3 will examine teamworking from a practical point of view and make clear some of the limitations that apply to team theory in practice.

In Section 5.3 we will introduce the subject of group dynamics which describe the actions and results of the interpersonal forces that exist within any collaborating group of people. We will then consider the needs of a group. In Section 5.3.1, we will see how various human behaviours meet these needs and so make humans natural teamworkers. Such behaviours are the dimensions of group dynamics.

In Section 5.4 we will introduce the most important techniques for managing a process that is often thought of as being unmanageable. Teams are often formed to solve problems and so need to be creative. Some teams are much more successful at this than others and in this section we examine what actions may be taken to improve the effectiveness of the creative process. The section starts with planning innovation, which introduces some techniques for planning tasks whose progress depends on the unknown outcomes of future work. The next area considered is problem-solving. This section includes material on brainstorming and lateral thinking. Finally, in Section 5.4.3, we investigate and explain how good decisions are made, and finish with a look at how the team issues of Sections 5.2 and 5.3 apply to decision-making.

None of the issues above can operate effectively in isolation. Many of the management and psychological principles in other sections of this book are therefore relevant to this chapter. Section 6.3 is of particular importance in ensuring the team is directing its efforts to the actual task and not wandering off the point. All aspects of Chapter 4 apply to the management of, and participation in, teams as well as to individuals interacting with others to solve problems. Chapter 7, and in particular Section 7.5, is relevant to teamwork where financial autonomy has been delegated to the team along with task-oriented objectives. Chapter 8 is again important for teams which are autonomous and are therefore responsible for achieving their objectives without assistance. Finally, Chapter 9 applies to teamworking just as it applies to all management issues.

It might seem from the above that practically everything that applies to the management of an organisation applies to a team, and in general this is true. It is not surprising, however, when one considers that a team is an organisation within an organisation.

The learning objectives for this chapter are as follows:

1. To examine the benefits that teamworking brings.
2. To review some theories of team composition and to see how these theories may apply in practice.
3. To understand group dynamics.
4. To present some contemporary problem-solving ideas.
5. To understand how the creative process may be planned.
6. To understand the various aspects of decision-making.
7. To explain several decision-making techniques.

5.2 Teamworking

For mankind, teamworking is a lifelong habit. We enter this world in the team of our family. We are schooled in the team of our class-mates. We choose our own team of friends and we join teams for sport and recreation. Some teams last a long time, others do not. Team composition is often beyond our control. In employment, we choose a team with a common interest but have little choice over our team-mates. Most adults have been members of many teams and know well the fortunes and dangers of life in teams. In becoming parents we rejoin our first team in a new role and pave the way for another generation to live their lives in a world filled with teams.

5.2.1 Holistic teams

The adjective 'holistic' describes an entity whose whole is greater than the sum of its parts. We are all familiar with the concept applied, for instance, to colonies of termite ants. These colonies are amongst the oldest of all life forms. By good management, the colonies are capable of far exceeding the puny abilities of the individuals. This example from nature illustrates vividly the benefits that holistic teams can bring.

In life, we are used to seeing teams of people whose group abilities come to more than the sum of their individual ones. When we observe this, however, we must be careful to distinguish between the simple numerical advantage that teamworking brings and true holism. The ability to lift heavier objects or perform more operations simply by having more people is an example of the former. The group's acquisition of new abilities, that the individuals could not have provided alone, is the latter. A simple holistic example might be a song-writing duo in which one member writes the lyrics and the other the words. Separately they are nothing, but together they are everything. Another example comes from a case study of engineering students:

Some final-year undergraduates had to present a design to their colleagues after a term's work of preparation. One team had to design a rubber-powered buggy to negotiate an obstacle course defined at the start of the term. One of the team members describes the work:

> Our team worked really well together. In the end we came second and it wasn't really very difficult. It wasn't even that we were particularly good at design or anything, we just seemed to find it easy. Joe was better at ideas than the rest so he did most of the actual design, I suppose. But then Angela did loads of reading and chasing things up; I hated that bit of the project so I was really pleased when she volunteered to do it. Dave was good too, he isn't really good at anything in particular but he did make sure we all had our say and knew what we were all going to do. He gets on with everyone. In a way he was probably the boss but it didn't really feel like that and he never got bad-tempered or anything like that. Then there was me, I like pulling things together, I made up the report from the information the others gave me. I had to chase them up over a few loose ends and get them to explain what some of it meant but that served to tie up all the loose ends. We really enjoyed it, our marks were high and the buggy worked. It was great!

Of course not all teams are so productive. Most of us can think of examples when we have been in teams that just didn't work together. Perhaps a group of friends who went off on holiday and things didn't go too smoothly; perhaps a group at work charged with some special task which never quite got done; or perhaps a meeting or committee in which many people gathered, much was said, little came of it and nothing was done. Most people can think of such cases. There are many reasons why a meeting can be ineffective. Section 9.6 explains the mechanics of meetings but even when everything has been properly prepared group meetings can still go wrong if the team does not have the capacity for holistic behaviour. It is easy to look at the above case study and see how it could have been different, how the members might have fittled less well together and how if any one of the tasks performed had been done badly, or there had not been someone to do it well, the whole thing would have gone wrong.

The potential benefits of being able to put good teams together are enormous. Managers need to know how to bring about the benefits of holistic teams and how to avoid forming ineffective teams. We will now examine team composition in theory and then make some observations on it in practice.

5.2.2 Optimising team composition — theory

The ultimate question for a theorist in teamwork is: 'How can one form the best possible team?'

Since good teams must have good people in them, one might try to make a good team by choosing someone good at each of the roles the team needs. For instance, to form a winning football team, simply select a star player for each position. However, no one with any experience of team games would actually try such an approach since it takes no account of how the team members would operate together. Potentially great football teams composed of star players at each position occasionally do perform well but they are by no means a guarantee of success.

A good team theory must therefore describe not only how to select individuals, but how to select groups that will work effectively together — this process is called 'team balancing'.

Simple theories

In the next section on advanced theories, we will explain the work of R. Meredith Belbin who developed a very good theory which is of enormous practical value. Since this theory is so useful there seems little point introducing other, less capable theories except for the fact that the reader may come across them in professional life. For this reason the way these simpler theories work is described briefly in the next paragraph.

In most models, separate team roles are identified: these may be functions that the team require for success, such as 'leading', or they may be attributes of people that are helpful in teamworking, such as 'respect for colleagues'. According to the theories, for a team to work effectively together, each of the roles of attributes must be present. This does not imply an absolute minimum number of people in a team, however, since one person may be strong in more than one role. Such models have many interesting features. The idea that the team needs balancing is catered for, and the fact that people are different is recognised. Usually, each theory will describe a way of identifying the attributes a given individual can bring to the team. Often this is done through a questionnaire. The ultimate teamworking question we are trying to answer is: 'How can one form the best possible team?' If we are to use a theory to answer this question we will have to make advances on these simple theories. Such theories generally do not provide any experimental evidence to support their predictions, nor do they provide a way of measuring the role each individual takes in a team with accuracy; we are not even sure that the roles described are 'pure', meaning that they are indeed the elemental factors of teamwork that must be present. Whilst such theories are useful in as much as they point us towards an understanding of teamwork they cannot generally be regarded as theories in the scientific sense of the word. To really prove a theory we require unassailable experimental evidence of its predictions. The next section describes just such experimental vindication of a particular theory.

Advanced team theory

There is one team theory which has the characteristics we are seeking. It has been experimentally proven and it does provide predictive results.

This celebrated theory is the work of R. Meredith Belbin MA, PhD and is described in his writings. What follows below is a brief summary of Belbin's work, together with some direct quotations from it. His description of the eight team-roles he identified is reproduced, together with the questionnaire he produced to identify an individual's team-roles. Also reproduced is his description of certain disastrous combinations of roles within a team. All the material described and quoted comes from R. Meredith Belbin MA, PhD (1981), *Management Teams: Why they succeed or fail*, London: William Heinemann Ltd. (ISBN 434 90127 X). (Reproduced by kind permission of the publishers.)

The theory was developed over a period of about nine years and was predominantly researched at the Administrative Staff College, Henley, England, by the Industrial Training Research Unit from Cambridge. Additional experiments were conducted in Australia.

Belbin's research was centred around the performance of teams at a particular management game, 'Teamopoly'. The teams played the game competitively against each other. The game, loosely based on 'Monopoly', produced winners and losers after many rounds of play. The game was developed to eliminate the usual weaknesses of luck and chance from having a major effect. The game not only placed the teams under stress but afforded them the opportunity to use imagination and skill in winning. Above all the game was played by teams whose collective decision selected their strategy for the next round.

Belbin and his research colleagues used proven psychological classifications to generate their team-roles. They started with the dimensions of introvert vs. extrovert and anxiety vs. stability. This immediately produced four team-roles although by the end of the work the researchers had identified eight 'pure' team-roles.

Belbin's team model features a questionnaire-based analysis of team-roles. These psychometric tests are used to produce a numerical rating of each individual. The experimental verification of the theory is the most impressive aspect of the work. After many years of developing the theory, Belbin and his researchers were able to predict, with indisputable statistical correlation, the likely outcomes of given team compositions. They were able to do this from the results of the psychometric tests alone and did not even have to meet the team members.

Interestingly the researchers found it easier to predict 'losing', rather than 'winning' teams. This led to the preparation of a list of disastrous combinations which is quoted below. Often these problems stem from a missing role, or from two members that clash.

An important empirical fact supporting the value of the theory is that it has already found its way into the practice of many organisations who use it for all sorts of tasks from making sensible choices about the composition of the Board of Directors through to general recruitment. The fact that such organisations can justify using the theory in these ways is an important indicator of the theory's value. The eight team-roles developed by Belbin and his colleagues are described below.

Company Worker (CW): As a team-role, specifies turning concepts and plans into practical working procedures; and carrying out agreed plans systematically and efficiently.

Chairman (CH): As a team-role, specifies controlling the way in which a team moves towards the group objectives by making the best use of team resources; recognising where the team's strengths and weaknesses lie; and ensuring that the best use is made of each team member's potential.

Shaper (SH): As a team-role, specifies shaping the way team effort is applied; directing attention generally to the setting of objectives and priorities; and seeking to impose some shape or pattern on group discussion and on the outcome of group activities.

Plant (PL): As a team-role, specifies advancing new ideas and strategies with special attention to major issues; and looking for possible breaks in approach to the problem with which the group is confronted.

Resources Investigator (RI): As a team-role specifies exploring and reporting back on ideas, developments and resources outside the group; creating external contacts that may be useful to the team and conducting any subsequent negotiations.

Monitor Evaluator (ME): As a team-role, specifies analysing problems; and evaluating ideas and suggestions so that the team is better placed to take decisions.

Team Worker (TW): As a team-role, specifies supporting members in their strengths (e.g. building on their suggestions); underpinning members in their shortcomings; improving communications between members and fostering team spirit generally.

Completer Finisher (CF): As a team-role, specifies ensuring that the team is protected as far as possible from mistakes of both commission and omission; actively searching for aspects of work which need a more than usual degree of attention; and maintaining a sense of urgency within the team.

Belbin's Self-Perception Inventory

Directions: For each section distribute a total of ten points among the sentences which you thing best describe your behaviour. These points may be distributed among several sentences: in extreme cases they might be spread among all the sentences or ten points may be given to a single sentence. Write the point distributions down beside the questions.

SECTION I. What I believe I can contribute to a team:

(a) I think I can quickly see and take advantage of new opportunities.
(b) I can work well with a wide range of people.
(c) Producing ideas is one of my natural assets.
(d) My ability rests in being able to draw out people whenever I detect they have something of value to contribute to group objectives.
(e) My capacity to follow through has much to do with my personal effectiveness.
(f) I am ready to face temporary unpopularity if it leads to worthwhile results in the end.
(g) I am quick to sense what is likely to work in situations with which I am familiar.
(h) I can offer a reasoned case for an alternative course of action without introducing bias or prejudice.

SECTION II. If I have a possible shortcoming in teamwork, it could be that:

(a) I am not at ease unless meetings are well structured and controlled and generally well conducted.
(b) I am inclined to be too generous towards others who have a valid viewpoint that has not been given a proper airing.
(c) I have a tendency to talk a lot once the group gets on to new ideas.
(d) My objective outlook makes it difficult for me to join in readily and enthusiastically with colleagues.
(e) I am sometimes seen as forceful and authoritarian if there is a need to get something done.
(f) I find it difficult to lead from the front, perhaps because I am over-responsive to group atmosphere.

(g) I am apt to get caught up in ideas that occur to me and so lose track of what is happening.
(h) My colleagues tend to see me as worrying unnecessarily over detail and the possibility that things may go wrong.

SECTION III. When involved with other people:

(a) I have an aptitude for influencing people without pressurising them.
(b) My general vigilance prevents careless mistakes and omissions being made.
(c) I am ready to press for action to make sure that the meeting does not waste time or lose sight of the main objective.
(d) I can be counted on to contribute something original.
(e) I am always ready to back a good suggestion in the common interest.
(f) I am keen to look for the latest in new ideas and developments.
(g) I believe my capacity for cool judgement is appreciated by others.
(h) I can be relied upon to see that all essential work is organised.

SECTION IV. My characteristic approach to group work is that:

(a) I have a quiet interest in getting to know colleagues better.
(b) I am not reluctant to challenge the views of others or to hold a minority view myself.
(c) I can usually find a line of argument to refute unsound propositions.
(d) I think I have a talent for making things work once a plan has to be put into operation.
(e) I have a tendency to avoid the obvious and to come out with the unexpected.
(f) I bring a touch of perfectionism to any team job I undertake.
(g) I am ready to make use of contacts outside the group itself.
(h) While I am interested in all views I have no hesitation in making up my mind once a decision has to be made.

SECTION V. I gain satisfaction in a job because:

(a) I enjoy analysing situations and weighing up all the possible choices.
(b) I am interested in finding practical solutions to problems.
(c) I like to feel I am fostering good working relationships.
(d) I can have a strong influence on decisions.
(e) I can meet people who may have something new to offer.
(f) I feel in my element where I can give a task my full attention.
(h) I like to find a field that stretches my imagination.

SECTION VI. If I am suddenly given a difficult task with limited time and unfamiliar people:

(a) I would feel like retiring to a corner to devise a way out of the impasse before developing a line.

(b) I would be ready to work with the person who showed the most positive approach, however difficult s/he might be.
(c) I would find some way of reducing the size of the task by establishing what different individuals might best contribute.
(d) My natural sense of urgency would help to ensure that we did not fall behind schedule.
(e) I believe I would keep cool and maintain my capacity to think straight.
(f) I would retain steadiness of purpose in spite of the pressures.
(g) I would be prepared to take a positive lead if I felt the group was making no progress.
(h) I would open up discussions with a view to stimulating new thoughts and getting something moving.

SECTION VII. With reference to the problems to which I am subject in working in groups:

(a) I am apt to show my impatience with those who are obstructing progress.
(b) Others may criticise me for being too analytical and insufficiently intuitive.
(c) My desire to ensure that work is properly done can hold up proceedings.
(d) I tend to get bored rather easily and rely on one or two stimulating members to spark me off.
(e) I find it difficult to get started unless the goals are clear.
(f) I am sometimes poor at explaining and clarifying complex points that occur to me.
(g) I am conscious of demanding from others the things I cannot do myself.
(h) I hesitate to get my points across when I run up against real opposition.

Enter the points assigned to each question into the table below and then add up the vertical columns to give a score for each team-role.

SECTION	CW	CH	SH	PL	RI	ME	TW	CF
I	g__	d__	f__	c__	a__	h__	b__	e__
II	a__	b__	e__	g__	c__	d__	f__	h__
III	h__	a__	c__	d__	f__	g__	e__	b__
IV	d__	h__	b__	e__	g__	c__	a__	f__
V	b__	f__	d__	h__	e__	a__	c__	g__
VI	f__	c__	g__	a__	h__	e__	b__	d__
VII	e__	g__	a__	f__	d__	b__	h__	c__
	__	__	__	__	__	__	__	__
TOTAL								
	__	__	__	__	__	__	__	__

Interpretation of total scores: The highest score on the team-role will indicate how best the respondent can make his or her mark in a management or project team. The next highest scores can denote back-up team-roles towards which the individual should shift, if for some reason there is less group need for a primary role.

The two lowest scores in a team-role imply possible areas of weakness. But rather than attempting to reform in this area the manager may be better advised to seek a colleague with complementary strengths.

A successful team is a balanced team, one in which all roles are present, and an unbalanced team will be a losing team. The process of using the analysis of the individuals' performance to choose a team in which all the required roles are present is called 'team balancing'. This result is contrary to many first opinions of how to produce a good team. One might suppose that the best way to form a team is to pick an expert in each field required and even double-up on experts in the most important fields. Belbin and his researchers did construct such teams, much to the annoyance of the other competing teams who thought these teams certain to win, but to the researchers' surprise they nearly always fared badly. So spectacular were their failures that they earned the description 'Apollo Syndrome'. These highly intelligent teams were noted for their in-fighting, each member seeming to gain more satisfaction from using their intelligence to defeat the proposals of other team members in a sort of 'fight to the death' approach; before long intellectual supremacy became their unacknowledged aim and the objectives of the group fell by the wayside. Belbin and his colleagues produced a list of the dangerous group combinations and this is quoted below.

> A Chairman along with two dominant Shapers both above average in mental ability. (The CH will almost certainly fail to get the job of chairman.)
>
> A Plant together with another PL, more dominant but less creative, and no good candidate to take the chair. (The Plant will be inhibited and will probably make no creative contribution at all.)
>
> A Monitor-Evalulator with no PL and surrounded by Team Workers and Company Workers of highly stable disposition and good mental ability. (The team is likely to generate a climate of solid orderly working and not foresee any need to evaluate alternative strategies or ideas.)
>
> A Company Worker in a team of CWs with no PL and no Resource Investigator. (The company will lack direction and the organisers will not have much to organise.)

A Team Worker working with TWs, CWs and Completers but no Resource Investigator, Plant, Shaper or Chairman. (A happy conscientious group will be over-anxious to reach agreement so that the presence of another TW merely adds to the euphoria.)

A Shaper working with another Shaper, highly dominant and of low mental ability, a 'superPlant', anxious and recessive, plus two or more Company Workers. (The SH will find that any display of drive and energy is likely to increase provocation and aggravation and disturb further an already unbalanced team.)

A Resource Investigator with other RIs and PLs but no TWs, CFs, MEs or CH. (A formula for a talking shop in which no one listens, follows up any of the points, or makes any decisions about what to do.)

A Completer-Finisher with ME and CWs but no RIs, PL or SH. (The CF, if he intervenes at all, will probably only help an already slow-moving group to get bogged down in detail.)

5.2.3 Optimising team composition — practice

So far we have met and can now apply an experimentally substantiated theory of team balancing. The question we must now ask is why is this theory not used all the time and teams always balanced in this way. There are two basic reasons. Firstly, it is not always appropriate, and, secondly, a number of practical reasons often make it difficult. We will now look at each of these points.

It is not always appropriate to use the team-balancing approach. The well-balanced team is a formidable weapon with an enormous capacity for creative work. It is only appropriate to create such a team if there is a task available to challenge their abilities. A balanced team will not be fulfilled with tasks beneath their abilities. It may well be an expensive group, providing another incentive to give it complex problems with potentially valuable solutions. In general the technique will work well for selecting teams to face rapid change, competition, a need for innovation, and for action. The converse is also true. There is little point in using the model for line-management applications or where relatively repetitive work is in hand. On the one hand, team balancing is extremely appropriate when establishing a new group or project team to undertake ambitious and challenging work. On the other hand, it is not appropriate for repetitive work or work without a need for innovation performed by groups with loose working affiliation and who do not face rapid change.

It is often difficult to apply team balancing for logistical reasons. The opportunity to recruit the majority of a team at once does not often arise. It is common for existing employees to take precedence and so limit flexibility in team balancing. Even when recruitment is taking place there are the two problems of not necessarily having applicants of an appropriate profile and of having those whose team profile is acceptable but who cannot be selected for other reasons, perhaps being too weak technically. However, even when these problems apply, team balancing still has benefits to offer. It is certainly better to make some effort at balancing a team than making none. Perhaps the disastrous team combinations can be avoided even if the ideal ones cannot be achieved. Also, a team that applies the model to itself can use the analysis to assist performance without necessarily changing its composition. If a weakness is identified it can be compensated for. Either the person who is best at the missing role can take it on, or the group can have a constant eye for the problems brought by the missing role and act before the weakness lets them down. If a team cannot be balanced by changing its composition at once, it may be possible to change it slowly over a period of time, perhaps after several rounds of recruitment. Having a policy on the sort of person to be recruited next time the opportunity comes up is clearly better than taking each case on an *ad hoc* basis.

In these ways, many of the benefits of team balancing can be realised in situations even when it is impossible to form a new team.

5.3 Group dynamics

The word 'dynamics' is familiar to engineers — it is used in mechanics to describe the effects of forces in motion. The term 'group dynamics' is well chosen, though it is often misunderstood. It means the forces in motion within a group. Of course here the dynamics refers to personal forces, instead of mechanical ones. Group dynamics is therefore the study of personal forces within groups. Much work on group dynamics has been conducted on groups in the widest sense of the word, such as families or gangs. In this book we are studying management and our interest in the subject comes from the fact that engineers normally work in groups to achieve their aims.

The most basic question that one can start with is 'Why do groups ever form in the first place?' The answer to this question is very straightforward. A group together can achieve very much more than the individuals can alone. In Section 5.2.1 we saw how this can be true for teams of people in their work. In group dynamics, we take the examination much deeper and realise that even from a zoological point of view, this benefit of group work still applies. Wolves hunt in packs, whales live in schools, most animals exhibit group behaviour of some sort — the difference with humans is that there is a lot more of it.

This programmed teamworking instinct is there to make each species more successful. Inherent in the genetic coding is much more than a desire to be in a group. Many special behaviours, such as the acceptance of authority or the need to conform, are aimed especially at life in groups. We are not just conditioned to operate in groups, we are actually designed to.

Group dynamics is the study of personal forces in motion within groups of people. It deals with the behaviour of individuals and their interactions with others. These are very complicated things and making accurate models of them is not easy. Understanding always precedes rational control. Therefore, as engineers, if we wish to benefit from rational suggestions to improve team weaknesses, or to engender those factors that assist our teams, we shall have to start by understanding group dynamics.

In the next two sections we shall introduce group dynamics firstly by looking at the needs of the group. We have seen that the basic motivation for group working comes from the much increased capacity for activity that it brings. This benefit does not just happen; many problems must be solved in order to reap the benefits. We will then see how various group behaviours exist to meet these needs and overcome the problems they bring.

5.3.1 The needs of the group

In this section we shall consider the needs of the group. These are things that the group requires in order to provide the benefits of group membership to the society of individuals that constitute it. We are studying group dynamics in this section and so our examination of group needs does not include material things such as shelter or finance. We are, however, interested in things that the group must have before the interplay of personal forces that is group dynamics can start to operate. There is no definitive list of group needs but the following six are quoted in many studies in one form or another.

The need for a task

The whole purpose of forming a group is to achieve some task. The nature of the task can have a great impact upon the life and times of the group. If the group task is continued prosperity, as might be the case with a team of company directors, the perpetual nature of the task shapes the group. They take a long-term view, are a strongly bound group and there is much scope for powerful political in-fighting. By contrast, if the task is short, such as a ten-session, one-hour-per-week training course, the opposite applies.

The need for the group to have a focus is very powerful. It is responsible for creating substitute tasks where no real ones exist, as is the case with the imagined enemy of the 'boy's gang' or the instinctive recourse to competitive behaviour between groups engaged in similar tasks. The task is also responsible for influencing the type of person attracted to the group. Technologically advanced companies engaged in rapidly changing markets attract one sort of person whilst traditional craft industries, practising skills noted for their constancy rather than their transience, attract another.

More than anything else the task colours the group. It pre-selects group members and favours those with certain values or aims. Often even a particular background and intelligence are favoured. When you ask people about themselves they say what they do, not what groups they belong to — such is the importance of the task.

The need for creativity

We now have a team that has something to do. Depending on the nature of the task there will also be a need for some level of problem-solving skills. The team has been formed to provide greater benefits to the members than they could provide for themselves, therefore a mechanism is required for ensuring that good ideas are plentiful. This in turn requires that each individual is able to put forward their ideas and that the best amongst them are selected. The concepts explained in Section 5.4 will be of use to the team here. The important thing is that the group are able to generate a solid bank of good ideas when necessary.

There are two ways in which the group can improve the likelihood of producing the ideas it needs. The first way is to ensure that the group has members with the skills to produce the ideas it needs. These might be intellectual skills and may be provided by the team-role of 'Plant' that we met in Section 5.2.2, which allows the group to produce the ideas for themselves. Alternatively the ideas may be found outside the group and a member to perform the 'Resource Investigator' team-role is required.

The second way that the group can improve their chances of creative output is to have an environment that is conducive to good work from the 'Plant' or 'Resource Investigator'. Some humans exhibit a process called facilitation. This describes the improvement in work quality and creative thought that can occur when an individual is surrounded by other productive minds. Some individuals almost have to be surrounded by others before they can produce much in the way of creative ideas. The question of what consistutes a conducive environment is a matter for the group to decide for themselves. Some things that are frequently associated with the creative process are enough time to think, freedom from petty pressures, a lack of deadlines or strict plans, and the opportunity to work on one thing at a time. These needs are almost exactly opposite to the overall management needs of the group where a constant eye for progress against plans, pressure to achieve and constantly working on more than one thing at once are good ways to pursue the group's task.

The group therefore has a need to be creative, either by having creative people or by using the creativity of others; in either case an appropriate climate is required before group members can perform these tasks to the best of their abilities.

The need to resolve disputes

A group engaged in using new ideas to solve a task will very often have to choose between alternatives. One member of the group may favour one course, another will favour the opposite. If the issue is a very important one, such as how much profit to keep back in order to fund next year's work, or which way to design the main components of the next new product, one can expect the debate to be intense. However, the group can only follow one path and a decision has to be made.

It is clearly the aim of the group to ensure that the most rational decision is made. There are many ways in which the rationality may be assisted and Section 5.4.3 describes good decision-making practice. From the group dynamics point of view, however, rather than from the purely logical one, the important issue is that personal motives and interpersonal contests do not allow the group decision to drift into the irrational. Plenty of decisions taken by groups are actually only the product of one member. Whilst that member's judgement is good the situation may be tenable, but it clearly exposes the group to an undesirable risk.

Where the decision is dependent upon the special skills of one team member it makes sense to question the decision and, if found acceptable, under examination by the rest, to adopt it. This might mean delegating the design of various components to certain experienced engineers. Often delegation is achieved through the group leader who, although ultimately responsible for the decision, delegates the authority to choose to one of the group.

The need to survive

For a group to be successful, it must survive for a period of time that is consistent with the task in hand. The structure of the group must be dynamic — if a member leaves, a replacement must be produced and the completion of the task interrupted as little as possible.

In primitive societies the contest for group positions is driven by the ambitions of individuals and this ensures that there are a steady stream of contenders pushing upwards. Usually they are selected against some ritual or custom that aims to assess their suitability in some way. In modern organisations the process is formalised into promotion and recruitment. Without the improved status and achievement that such a hierarchy brings, there would be no fierce contest for the top roles and the task would not have champions to ensure it is completed.

The need for optimal resource deployment

To be at its most effective, a group must have a means of ensuring that its resources are deployed in an optimal way. In the case of special skills or equipment, optimal deployment is usually obvious. In other situations it is less so. For instance a junior engineer might be given a task which is at the upper limit of that individual's abilities, and such a task would almost certainly be more competently accomplished by a more senior engineer. However, in doing the task the junior engineer has the opportunity to develop and so increase the resources of the group. In such a situation, comparison between the advantages of deploying a junior engineer and improving the skill base of the group against the advantages of the senior engineer's speed and quality must be made. Such comparisons are constant in the optimal deployment of group resources.

The need for unity

After a course of action has been chosen for the group there is no benefit in individual members directing efforts in other directions. This reduces the resources that can be employed to achieve the task. The group needs unity if it is to survive and it comes from either rejecting those who do not support it or convincing members of the suitability of the chosen course.

If previously sceptical individuals are to be converted to the chosen route, respect for the group decision-making process and a commitment to the team are vital. A united stand maximises the group's chances of survival and achievement. It makes the group less prone to defeat from rival groups and provides assurance to the group members of their security. However, it should be remembered that making a good decision is the most important thing. Even a strongly united team heading in the wrong direction will eventually fail.

The need for unity applies to the long term and the short term. In the short term unity prevents the group from becoming ineffective under pressure and being prone to exploitation from other groups. In the long term unity is needed simply to ensure that the group continues to be committed to paths of action that are consistent with the group's objectives.

5.3.2 Meeting these needs — group dynamics

So far in this chapter we have seen that a group can greatly extend the abilities of its members and put them at a considerable advantage.

We have also seen that in order to provide these benefits there are various needs of the group that must be met.

Below, we will introduce some of the basic dimensions of group dynamics and explain a little of how these provide for the needs of the group. Each of the dimensions is an attribute that may be observed in a group. Group dynamics is an all-pervading subject and applies to ancient tribe and modern office alike. After reading this section you should be able to identify the group behaviours in your own social groups and observe them within your family, sports teams, friends, societies or associates.

Norms

Norms are standards of behaviour. They describe the 'normal' way in which the group behaves. It is possible to imagine all sorts of different ways in which these norms might come about.

Some norms stem from sheer practicalities. We all drive on the left in the United Kingdom, not because the left is inherently better than the right, but because you have to choose one side to avoid a lot of crashes and we've chosen the left. Other norms may be quite covert and unspoken. Examples of this might include a meeting that routinely considers topics in a certain order, or a group of employees who use their own made-up terms or expressions. The reluctance to explain such a system to newcomers enforces their oddity and in this way norms can be used to draw lines between those who are 'insiders' and those who are 'outsiders'.

These various types of norm support the likelihood of group success in many ways. Norms bring familiarity which protect individuals from feelings or fear or threat that might inhibit their contributions. Our natural desire to avoid insecurity therefore makes us take quickly to group norms and accept them in accordance with our level of personal security.

Some norms extend to preserving the social hierarchy and so illustrate to all the orgnasation of the group. It is common, for instance, for the order in which a formal dinner is served to reflect status, though in fact any order would be just as practical. Where norms engender sound practice they may serve to make the group more long-lived. Certain cultural food restrictions are good examples of this. They may well have formed sensible hygiene practices when introduced, but such is the power of norms that they have survived long after the invention of the refrigerator and are now unnecessary.

Within a group of engineers it may be an office norm that the engineer who performed the majority of the initial project work gives a presentation to the Board and takes on the project. Such a norm might support optimal resource deployment since the jobs are probably best done by the same person. However, if an engineer was not asked to follow the route this would almost certainly lead to anxiety and the thought that the initial work performed was not acceptable. In this way norms may be used as vehicles of judgement.

Norms simplify things, they serve the belonging need we all have and they can be used as a ready reckoner against which to judge behaviour.

Group culture

All groups have their own culture. In group dynamics the term is used to describe the atmosphere or ambience of the group. Group culture has a major role to play in producing a climate conducive to the production of ideas. In Section 5.3.1 we saw that there was a need for creativity and that this in turn required a climate conducive to idea generation. Different groups might find different climates successful but all groups will want an appropriate one. Group culture provides this creative climate.

It is easy to see that at least in some way the culture of the group must be the product of the individuals that compose it. If the group in question is a design team, it may be that one senior member of the organisation has a potent character trait which before long permeates the group. For example, if one person in a team is really enthusiastic about the work in hand, it can raise the spirits of the whole group. The whole team becomes bright and optimism prevails. Such outlooks can become entrenched as part of the culture. If we compared this group with a rival group in a different organisation we may well find the cultures to be completely different. The culture of a group is a dynamic thing and will change with time. One cannot, of course, decide to engender a particular type of culture and simply create it. The culture of a group is a result of the group, not the other way round. Group culture can be thought of as an extension of the factual group norms into the more abstract. Issues such as beliefs, values and opinions shape group culture and they are not so easily defined.

Leadership

Leadership is a vast issue. Its many facets assist the group in every one of its needs. It is the most potent role of all within a group and it is the kingpin upon which the rest of the group depend. If you ask someone what their job is like, they will usually include a description of their boss. Difficult questions get referred to the boss. The science-fiction cliché, 'Take me to your leader', alludes to the overwhelming power of the leader's position.

Leadership is often misunderstood and seen as anything from the romantic swashbuckler leading a merry crew to fortune and adventure through to the lonely isolation of a compassionless military hero. Such fictional distortions have no place in engineering management. Leadership is an essential ingredient of group success provided by ordinary people.

When did you last hear of a successful human endeavour that had no leader?

An important distinction must be made between leadership and a leader. Leadership is defined as those tasks which need to be done in order for the group to be led. A leader, on the other hand, is someone who is actually doing the leading at that time. Leadership tasks can be performed by anyone in the group, even by more than one person at once. Usually, one person is the leader for the majority of the time and is the person named when people are asked who the boss is. Much of the time, other group members take the lead with the full approval of the 'main' leader.

The style of leadership has a great influence on the rest of the group. In the long term the leader is judged by the other members against the group success in achieving the task. Unsatisfactory leaders are then impeached or abandoned. Many people regard leadership as the only group position worth holding because they have been conditioned by their culture to aim for the top. This does nothing for the individual's personal happiness and everyone has their own ideal group position with very few actually preferring leadership. One cannot change one's basic preferences much and it is important to remember that whilst skills can be improved through good training one can never acquire abilities that were previously completely lacking. To some extent, the saying that good leaders are born and not made is true.

The leader of a group can affect the task. The leader is the ultimate authority and it is only the leader that can ultimately take the group away from the task. The leader's good judgement on group issues is vital for the continued success of the group. Group dynamics provides a safety valve on the authority of the leader and it is only while good performance endures that the group permit the leader to stay. It is the group's respect that regulates the leader's authority. The respect of the group gives the leader a mandate to lead and it can be withdrawn at any time. There may be no ballot boxes but a leader is democratically empowered. It is this need for a leader to be in position at the command of the group that undermines the authority of a leader empowered only by organisational decree or appointment.

The leader is the ultimate judicial authority in a group. In this role the leader settles disagreements and resolves disputes. A competent leader must already have shown fitness to lead and therefore commands the respect of the team. When disputes occur the team members have to be prepared to respect and trust the leader's decision. So powerful is the group need for a decision-making authority that when a new group is formed, choosing a leader for it is almost the first thing it does. Often this is done subconsciously. A group of three or four people asked to move a heavy object forms a good example. They normally sidle up to the object and start a reasonable but unco-ordinated attack, then one of them says something organising and constructive and is instantly made the leader. Usually it is this person who counts the group into the lift.

The leader is often the one who guides important discussions and makes the final decisions following group debates. In this way the leader has more influence on the fertile generation of ideas than any other role. A group may have a very good ideas base but if the leader fails to seek the opinions of other group members, or is not convinced by a correct argument put forward by another, the ideas might as well not be there. Again there is a safety valve on the authority of the leader. If a leader subsequently finds a correct option put forward by a group member was discarded, the respect of the leader may suffer greatly. To command respect, the leader must constantly be making excellent choices. Because of this, it is harder to hold leadership than to challenge it and so potentially superior leaders are always eager to take over. Of course, when the authority of the leader comes by appointment, as is the case with modern employment, social norms prevent a rival leader from carrying out a challenge, but this does not mean that a loss of respect has not occurred. It is possible for a leader to lose command in all but name.

Finally the leader is the task master and figurehead of a group. The leader is ultimately responsible for who does what and is the person put forward to represent the group.

Conformance

People who have joined a group have already demonstrated some form of common aim by their very presence. Conformance in group dynamics refers to the group need for unity. The need is met directly by the individual's need to belong or rather their need to conform. Of course, different individuals have different ideas about conformance and one group may be ideal for one person but appear hopelessly identity-crushing to another. From the group's point of view conformance is a force that must be appropriate to the needs of the particular group — if it is too weak in its binding effect the group runs the risk of falling apart. If it is too strong it runs the risk of inhibiting the creative abilities of the group members and so frustrating the very reason for the group's existence.

All individuals have conformance needs and all groups demand them in some form or other. Even the socially rebellious groups of teenagers, found in one form or another in all societies, cannot escape. Such groups claim to reject conformance and yet wear a uniform as recognisable as that of any army.

In general the greater the commitment to the common aim the more personal differences can be suppressed and the more powerful the conforming force. In a group at work, the conformance of individuals to the group's task can become a contentious issue. The members of the group are bound together by the cohesive forces of the group and these must be maintained in balance in order for the group to stand the best chance of success in the task. Interpersonal compatibility, however, can interfere with this cohesive balance either by becoming too strongly attractive between two individuals or by becoming negative and repulsive. In both cases the mutual feelings may distort rational judgement and lead to a de-optimised decision-making process. This effect will be greatly exaggerated if the leader is affected. To prevent these problems from reducing group effectiveness, people tend to limit their social interactions within a group at work. Business is not usually mixed with pleasure and in-fighting is not normally tolerated.

Conformance has beneficial effects on the generation of ideas, and two compatible minds can work together to form a formidable intellectual entity. Having like-minded members brings solutions to many group needs, the acceptance of norms, similar judgement of leadership, common purpose, all make the group more likely to succeed in its task. Conformance should not be confused with similarity or lack of independence. A group of very similar and yet conforming people is very strong. Such a group will enjoy a good match of differences and is well equipped for all circumstances.

Conformance is the group behaviour that directly answers the need for unity. Conformance also provides benefits to the group by making the individuals relaxed within it and confirms that they are accepted. In this way, conformance can answer the deepest of all psychological needs, the need for identity. By wearing the uniform of your group, you state your identity as one of them. By using their expressions you pass the test of conformance. By resisting those who are different, you separate your identity from theirs.

5.4 Managing the creative process

Contemporary creativity comes from group work. The days of individuals providing the majority of breakthroughs in science and engineering have past. Modern problem-solving is an organised, rational approach aimed at producing the required breakthroughs at the times they are needed. The techniques described in this section are all aimed at managing the creative process of engineering by assisting groups to plan their efforts.

In Section 5.4.1, we will meet some methods that can be used in an overall way to plan large blocks of creative work. In the next Section, we follow with the techniques of lateral thinking and brainstorming. Both are techniques for assisting the production of new ideas. Finally, in Section 5.4.3, we investigate a particular skill in the creative process, that of good decision-making.

5.4.1 Planning innovation

There seems to be an inherent paradox in planning innovation. Innovation is associated with inspiration, imagination and originality. Yet none of these things seems amenable to planning.

We have a very stereotyped view of a creative person at work. We imagine a great engineer thinking things over until inspiration strikes and a brilliant invention is born. In reality, even when individuals do make breakthroughs, they usually come after much hard work. People create the environment for innovation and breakthrough by constantly working at the problem. Of course, no one would suggest that people can plan which ideas will be masterpieces and which will not but this does not preclude the use of planning to assist creative results. By planning their work, so that they are constantly in fertile areas, people affect their 'luck'. There is, in fact, no paradox involved in planning innovation.

Chapter 8 describes planning techniques for determinate problems. The 'special' techniques in this chapter are used when coping with indeterminate issues and should be used in conjunction with the methods described in Chapter 8.

In general, the techniques that we have at our disposal to assist in planning innovation may be divided into two categories. In the first category are the techniques of planning itself. These are ways in which good plans may be prepared. In the second category are the different approaches to the work that may be taken. These are the underlying principles upon which the plan depends. We will start by looking at the different approaches and follow with the good planning techniques.

Approaches to planning innovation

The following techniques may be applied when planning the innovation process. They may be used singly, or in conjunction with each other. Once a principle upon which the work is to be based has been selected, the planning techniques of the next sub-section, together with the planning techniques of Chapter 8 may be used to provide a complete plan.

Multi-directional attacks: In certain circumstances a solution to the critical problem might come from more than one direction. In such cases it may be sensible to embark on developing parallel solutions based on different principles even though only one will ultimately be required.

This 'insurance policy' approach can greatly increase the confidence of a successful outcome as well as doing much to ensure the solution chosen is indeed the most appropriate. The amount by which these two benefits off-set the cost of the additional work must be estimated and a decision made on the best information. This approach is particularly prevalent in the placement of military contracts where companies may compete for a tender long before the state finally chooses the winner of the contract. In the meantime each of the companies will have invested a great deal of time and money in demonstrating their ability to meet the needs of the contract in their chosen way.

Fixed resource research: In this approach a fixed amount of resource is directed at a problem with the intention of identifying the best solution available within a budget. This is particularly appropriate when the definition of performance is as yet unknown, perhaps when research is undertaken as part of a large programme or if the project is relatively small and is of an investigative nature, rather than a manufacturing one. A negative result is also useful in that the organisation may then be sure that no competitor can solve the problem without spending more than the organisation's own research investment. Some organisations attempt to pre-empt the creative work of others employing levels of expenditure that are known to exceed those of their rivals. In this way, such organisations hope to be first to achieve the next hurdle, whatever it may be.

Rolling plan: This technique can be applied where success is more important than timing. As the project progresses and the results of key decisions are known new objectives for the next phase of the work are defined. The managers employing such a principle must exercise control in order for it not to degenerate into an unguided block of work in which the management has become mere reporting. The end goal of the whole programme must not be forgotten and time-scales must always be imposed for each section. The technique is particularly appropriate for projects in which the future path is completely dependent on results that are as yet unknown, or to situations where a relatively large number of alternatives must be tried in turn until an acceptable result is found.

Undirected research: In this approach the researchers are given a problem but left to see what ideas and results they can generate. The free environment fosters their creative output and their natural enthusiasm leads them towards useful work. The technique can be very effective in providing those with creative skills an opportunity to make the most of their skills in a way that benefits the organisation. However, the technique can degenerate into an unproductive ramble through the interests of the researchers and it should be managed in terms of desired goals and resources to ensure direction but otherwise left alone. Such work should be scheduled into existing commitments.

Key result research: Clearly there is no point conducting research into all areas of a particular problem if there is still a chance that the project will fail on some fundamental principle that has not yet been proven. For this reason it makes sense to start with 'proof of principle' research. Care must be taken with such research projects, as the temptation here is to perform very cursory investigations whose results look positive. The research is then taken further and it is discovered that the principle does not really work in the situation in which it is required because of other fundamental design needs. This problem may be overcome by having further grades of research programme; after the initial surface investigation the work is taken a little further and may go through several stages before becoming a product. One should avoid the trap of thinking that simply because the work has gone a certain distance it must continue — many successful companies claim a project drop-out rate of nine out of ten. Such organisations are still comparing the objectives of the projects with their strategic aims even in the last stages of a project's life, and if a mismatch is found there is no point continuing. Individuals don't pour good money after bad, nor should organisations.

Planning techniques for the innovative process

The following techniques may be used in conjunction with the planning techniques of Chapter 8 to plan the innovation process.

Upper and lower estimates: No matter how complex, or dependent upon future outcomes, it is always possible to put estimates on the upper and lower limits of cost and time. Such estimates can initially be used in a crude way to identify whether further planning is sensible. The first phase of examination can be done almost on the back of an envelope, giving order-of-magnitude estimates of the resources and principles involved. This will suggest whether it is sensible to consider the option further and will also make plain the critical milestones in the project. Once the first estimates have been prepared, further refinements may be made to increase the accuracy of the estimates.

Milestone reviews: Even though a new idea or device may depend on certain critical problems being solved, they rarely depend on a great number of critical issues. It is therefore possible to pick a few milestones in the life of the project. These are stages at which, if the project has not achieved certain targets, completion within the original budgets is no longer possible. Milestones break unmanageably large blocks of work into manageable packages. Progress through the milestones is a good indicator of overall success.

Vested disinterest: Individuals who are champions of their own projects or ideas are not necessarily good planners of the work. Their input may be essential to the process since they are the ones with the detailed understanding, but they are also ambitious for the work and want it to succeed. Such people tend to minimise the difficulties and overstate the benefits. When planning is undertaken people with opposite points of view should be invited to examine the proposals.

Strategic comparison: Great benefit can be derived from comparing how the organisation performed in similar previous ventures. Clearly if a task is to be completed in an uncharacteristically short period of time an explanation of how this can be achieved is required. There is, however, a danger that people will use this 'experienced viewpoint' to decry perfectly viable options. We are all familiar with the phrase 'we've tried that before, it's no good'. When memories of previous tasks are recounted, questions need to be asked to establish rationally whether the new idea should be tried: 'What is different *now*?', 'Why did the previous project fail?' — such questions deserve answers. This makes abundantly clear the complete waste of time and money that can result from an undocumented research project. If no write-up of a previous project exists no benefit from the experience can be fed into future plans. Regardless of success or failure, valuable planning experience can be gained from every single project an organisation undertakes if it is documented.

In Section 5.4.1 we initially saw that there seemed to be a paradox in planning innovation — how can one plan to have an inspirational thought or choose which piece of work will be ordinary and which brilliant? However, we soon realised that there is no paradox in planning innovation and the most important thing is to keep attacking a problem in order that eventually we achieve the break we are looking for. People make their own breaks by working steadily at problems. The difference between planning innovative processes and planning other processes is that innovative processes involve indeterminate outcomes, in some way the path that the plan must take cannot be known until the result of some future activity is known. This indeterminacy makes planning the innovative process difficult — but not impossible. In this section we have also seen various underlying planning approaches that may be used when designing a plan to deal with an innovative process and some good planning techniques that may be used to ensure that such plans are effective.

5.4.2 Problem-solving

Engineers solve problems. The organised production of food was an early achievement of the first agricultural engineers. As societies advanced new problems were encountered. Engineers solved them by providing, amongst other things, buildings, sanitation, energy, lighting, armaments, transport, flight. The list grew and grew as engineers strove to meet mankind's indefatigable appetite.

Today engineers continue to propose solutions to solve these problems as well as many new ones. Cars straight out of yesterday's science-fiction are in every garage showroom. For less than the average weekly wage machines can now be bought that effortlessly wash the clothes of a large family; such a wash used to take two whole days of hard work — today people complain about just having to hang it out!

In all areas of human endeavour problem-solving is the gateway to progress. It is not surprising then that the very process of solving problems has been studied in a rigorous manner. Researchers attempt to reveal dependable strategies for success.

There have been many attempts to produce reliable problem-solving techniques. Often these take the form of a short-list of tasks that must be performed before the problem can be solved. Any good technique will include the following steps in some form or other.

Examination: One cannot start to solve a problem without being absolutely clear what the problem actually is. For instance, an engineer working on the design of a component in an engine may be trying to improve the life of the component. At first the problem seems well defined, but in reality the problem might actually be that the oil pressure needs increasing or the material needs altering, or the speed of the component should be reduced. It is possible that what is required is another method to be used for the same function. It may be that all these things are important but which must be tackled, what actually is the problem? In some cases the problem may not be correctly defined at all. A company wishing to have more graduate employees may implement a recruitment policy lasting several years to correct the perceived imbalance. However, the shortage may be caused equally by high levels of resignation as low levels of recruitment. The real problem may actually be how to stop the graduates leaving rather than how to attract them more effectively. Clarity of thought and the application of rationality are the keys to being accurate in identifying the real problem.

Proposal: Good problem-solving depends on being able to identify the best solution to the real problem. This might be achieved by having the ability always to think straight away of the best solution.

Unfortunately this approach is practically impossible for all but the simplest of problems. The ideal approach is to produce lots of different ideas and then weigh one against another to find the best one. Often, thinking of new ideas makes one rethink and improve old ideas and so the process leads to an improvement in quality. In any case one cannot know that the best idea has been selected unless all other possibilities have been presented and evaluated.

Implications: A proposal designed to solve a problem of any complexity will have implications. The implications may even be significant enough to render a proposal unworkable. For example, if an organisation finds that a valuable employee is resigning there is a temptation to offer more money as an incentive to stay. However, the implication of doing so is that the organisation was not paying the individual as much as it felt the employee was worth in the first place. Clearly this means the organisation was guilty of underpayment. In addition, how will all the other employees react? They cannot be blamed for thinking that the organisation is underpaying them too. In such a case, the implications of the proposed course of action mean that it should not be taken up.

Implementation: No solution to a problem is complete without due consideration being given to the implementation of the solution. A workable plan is required to demonstrate that the solution is indeed feasible. Once a problem has been defined, the best solution identified and its implications thought through, it is always possible to produce a plan for implementing the solution. Such a plan can then be used to manage the work. Implementing a solution to a problem is, by definition, a management issue and depending on the solution in hand any of the topics covered in this book may be appropriate.

Below we consider two methods for improving our abilities to propose solutions to problems. The techniques are excellent at unjamming thought processes that have become hindered by deductive reasoning and have reached a situation from which there seems to be no way forward. The techniques are used in conjunction with deductive and rational reasoning to ensure that the best solution is found. These processes work in all problem-solving situations but are especially appropriate to a process as creative as engineering. They are called brainstorming and lateral thinking and are familiar, in name at least, to many people.

Brainstorming

Brainstorming is a method by which a flood of new approaches to a problem can be generated. In most cases the majority of ideas are discarded but from the generous flow only one idea may be needed to solve the problem and so the wastage rate does not really matter.

The technique of brainstorming relies on the assembled group seeing the problem with fresh eyes and using their imagination to produce new or novel approaches to the problem. A purely rational approach is used after the brainstorming session to assess the ideas but during the brainstorming phase the most productive individuals are those who let go of rational constraints and use their imagination to bring entirely new views to bear on the problem.

Successfully managing the brainstorming session means creating the environment in which this torrent of ideas flourishes. Various things can be done to foster good brainstorming conditions:

- Have a leader experienced in brainstorming.
- Have a group of less than about ten and more than three.
- Include some people who have little or no technical knowledge of the problem.
- Don't permit negative comments.
- Don't allow suggestions of solutions or evaluation of ideas whilst the ideas are being generated.
- Produce a group atmosphere that puts everyone on the same level; people tend to be inhibited by having more senior people present.
- Encourage positive statements. A negative statement like 'the problem is that we can't . . .', is a lot less stimulating for participants than saying 'what if . . .', or 'how may we . . .', or 'hey, how about . . .'.
- Keep the session short, about half an hour.
- Record the ideas in sufficient detail to allow examination later.

After the ideas-generation phase is over the more technically qualified may consider the practical aspects of each suggestion and the implications of them. The reviewers should be aware that it is in the nature of brainstorming to produce some quite large and general ideas that normally require a considerable amount of work before they can be implemented. There may even be new ideas in quite different directions emerging.

The brainstorming process is particularly useful when an impasse has been reached or when the physical principle used in an attempt to solve a problem has been found to be unsatisfactory. The process tends to throw up rather wild ideas, many of which are interesting or even amusing but of little practical value, so consequently the use of the process is sometimes limited. In general the technique is only applicable on an infrequent basis but when it is used the results can be spectacular.

This technique is really an extension of a thought process we all go through when we are problem-solving. Whilst the problem is comprehensible and understood we apply rational thought to produce the optimal next step. However, we can often reach an impasse or get so close to the problem that we cannot take an overview and jump to fruitful alternatives. At this time we may have the sudden, inspirational thoughts that unblock the log-jam. These thoughts often come at unexpected times and are often jogged along by association with some unrelated event or coincidental encounter. Brainstorming tries to manage and optimise this process.

Lateral thinking

We are all educated in classical problem-solving. Our schooling makes deductive, logical thought habitual. It is certainly true to say that its application has brought impressive advances in our knowledge. In 'vertical' thinking one starts with some known condition and then applies steps of reasoning aimed at reaching a goal. The only rule is that the steps must each be true and self-consistent with the system of thought being employed. After many steps have been employed one is in a new place that might not have been foreseen without the logical process. As long as each step is valid we may go as far as we like. Engineers make great use of this type of thinking and think nothing of starting with two or three physical laws, applying them to a physical system and producing, through vertical reasoning alone, an experimentally verifiable model of the system. It is little wonder that the process has earned respect.

Lateral thinking is the opposite to vertical thinking. In lateral thinking, one is not constrained to follow the formal deduction process — one has an inspiration or flash and sees a new angle on the problem which, if locally self-consistent, is a new model in its own right. The human mind is very good at seeing patterns, maintaining conformity of ideas and reinforcing concepts. Lateral thinking is used to free us from the limitations that these cause: it is not a replacement for our existing system of thought, it is a powerful tool for use at certain times. In many ways, lateral thinking is brainstorming on your own. A great deal has been written about lateral thinking, much of it by one of its most famous exponents, Edward de Bono. The reader will find his book *Lateral Thinking for Management* listed in Section 5.7. Lateral thinkers make jumps to new ideas and views, they are creative and imaginative, often supplying ideas that vertical thinking alone could not provide.

5.4.3 Decision-making

Being consistently better at making decisions will place anyone at an advantage over their colleagues. Good decision-making is common to all professions and engineering is no exception. Good engineering managers are good decision-makers.

In this section we will look at the problem of decision-taking with the aim of understanding the process and how to optimise it. Decisions vary greatly in their magnitude — some involve trivial issues such as choosing an item from a restaurant menu, others are vast. When President Kennedy announced that America would undertake the Apollo moon-shots he is thought to have undertaken the most expensive commitment ever made in a simple statement. At first sight these two decision have little in common. However, they both involve choice and they both have consequences. In some ways at least, all decision-making is the same.

We shall now investigate how to improve our decision-making by dividing the process up into independent, elemental sections and optimising each one.

Start with objectives

A prerequisite of making a choice is that one must have some objective against which the choice is to be judged.

Objectives facilitate clear and corporately beneficial decision-taking at all levels. If the objectives are not clear, the decisions that result cannot be guaranteed to support the original intention of the objective. Worse still, if an individual is supplied with conflicting objectives rational decision-taking becomes impossible when a decision that cannot serve both objectives is faced. Section 6.3 explains how good objective-setting always avoids such conflict and produces clear, unambiguous objectives which assist the decision-making process.

Collecting the right data

Clearly a rational decision made in the knowledge of all the relevant facts will be a good decision. Sadly, we are rarely in possession of all the facts we need. In life, all but the most ordered and the smallest of decisions are made in a state of partial ignorance. The aim of the data-collection phase is to reduce the level of ignorance.

Whenever a decision is to be made there are always options. The two important goals of this data-collection phase are therefore, firstly, the collection of a sufficient amount of facts about each option and, secondly, to ensure that all options are known. During the first phase the data is assessed to estimate the error in decision that may result from the error in the data. If this error is unacceptable, more data is collected until a more reliable decision can be made. There is no point gathering unnecessarily accurate information to permit a decision; the important issue is that the accuracy of the information is consistent with the measurable effects of the decision.

Brainstorming sessions or more formal strategy meetings are often held to be certain of covering all options.

Decision-making techniques

When the decision-maker is in possession of an acceptable quantity and quality of information, and has an objective to achieve, the process of selecting an option can begin. We will now look at some of the evaluative techniques that may be used to assist the process.

It should be remembered that making occasional wrong decisions is an inevitable consequence of making decisions at all — no one can expect to get it right all the time. The well-adjusted person will put the mistake down to experience and learn from it. No one can do no more than make their best effort, one must accept one's own fallibility and not bear grudges.

Search for extrema: In some cases the objective may require the decision-taker to head for an extremum (maximum or minimum) of some variable. If a project is required to be completed in the minimum possible time, for instance, the decision-maker need only consider critical-path plans which allow this, rather than fixed-capacity or fixed-resource plans which may be more appropriate but take longer. Alternatively a camera designer may have the objective of minimum possible weight and so will have to select plastic for the lenses rather than the optically superior glass available.

In general, extrema are very poor ways to specify objectives and almost always lead to unclear choices. The questions of 'how light must the camera be?' and 'what project duration is acceptable?' are unanswered in the above examples. The use of extrema has been included here for two reasons: firstly, because very occasionally it is the appropriate way to specify an objective; and secondly, because whilst it is bad practice, it is also common practice.

Penalty costs: All decisions come not only with potential benefits but also with costs associated with the decision being wrong. These are called *penalty costs.* Occasionally, even though there is almost no chance of the decision going wrong the consequences are so dire that an almost certain gain is left untouched. No one in their right mind would gamble with their lives for modest financial gain no matter how favourable the odds.

Matrix assessment: This is a very simple and effective way of choosing from amongst alternatives when many different attributes have to be weighed against each other. The attributes are first listed and assigned a maximum score to reflect their relative importance. The items that are most important are assigned the largest maximum scores. Each alternative is then marked out of the maximum score for each attribute,

Attribute	Max. score	Glass	Plastic-N3	Silica	Plastic-N2
Clarity	20	19	17	15	17
Low cost	20	5	15	10	18
Weight	20	2	19	13	16
Refractive index	15	14	8	9	12
Uniformity	10	10	6	8	6
Stability	15	15	11	13	11
Total		65	76	68	80

Figure 5.1 Matrix assessment for selection of lens material

a high score indicating that the option can successfully provide the desired attribute. When this is complete the columns for each option are added up to determine the best choice.

In the following example, Figure 5.1, a designer is selecting a material to use as a camera lens. The different options open are listed across the top. In the column down the side are the various attributes considered important by the designer. The next column shows the maximum score available for each attribute. The designer has allocated these scores and has, for example, decided that clarity is twice as important as uniformity and so given it twice the maximum score. The body of the table shows the score for each option. The bottom line clearly shows that plastic-N2 is the best choice.

In the example above a decision has been made on technical grounds but the method is equally appropriate to any type of decision in which options have to be selected against several criteria.

Overriding constraints: In some cases, overriding constraints significantly reduce the number of options available. This serves to make the decision-making process easier and so it is helpful to look for such constraints at the start. Sometimes the constraints may be the quest for extrema as described previously but there are other constraints too. For instance, a manufacturer may decide to upgrade an existing camera model and include various new features without changing the original body shell. The design engineers will therefore have overriding space constraints imposed by this decision. In reviewing the design options, the engineers can reduce the number of alternatives significantly by applying this overall constraint.

Use of maths: Nearly all decisions involve an evaluation of probabilities. Wherever it is possible mathematically to select an option it is best to do so. Maths is an unbiased and dependable assistant; even when the uncertainties are significant error bars may be allotted and maths still applied. No issues are able to defy the laws of probability, and understanding them gives one a significant advantage in decision-taking. For example, when asked how many people must be gathered in a room before two of them will have the same birthday, most people will give an answer somewhere in the hundreds. Only statistics can easily predict the answer — that you stand a fifty-fifty chance with only about nineteen people.

Consequence analysis: In some cases the consequence of a decision interacts with the very issue upon which the decision was made. For example, a company which reduces its prices in an attempt to undersell its competitors cannot be surprised if they react by doing the same and thus restoring the status quo. The decision the company faces in the first place is not whether to reduce prices, but rather who can ultimately reduce their prices the most and for how long these reductions can be sustained.

Anticipating all the consequences of a given action is not usually possible and so efforts must be directed in the most important directions. One of the best-known and most comprehensive uses of this technique is the war game, in which certain scenarios are fought out by staff taking the roles of all sides and the consequences of certain actions on the whole course of a war may be examined. Conducting such analyses becomes more difficult and less accurate the further into the future the scenario is taken.

In Section 5.4.3 we initially looked at what a decision is and saw that whilst the substance of decisions may be very different, the processes involved are common. All decisions involve choice and if implemented they all have consequences. There are clear advantages to being a consistently good decision-taker and the section progressed with a look at the three key areas of decision-taking. In the last section we introduced several very useful techniques for making evaluations that can be applied to all manner of decisions.

5.5 Summary of Chapter 5

In Chapter 5 we have studied three major areas. These were team working, group dynamics and managing the creative process.

In Section 5.2 we examined the potential benefits of teamworking and met holistic teams, teams whose whole is greater than the sum of their parts. This interesting and useful phenomenon was then studied and an experimentally verified model developed by Belbin and his colleagues was presented. The application of the theory in practice was then considered and it was clear that even when the opportunity for team balancing is limited the model has useful management contributions to make.

In Section 5.3 we discussed various aspects of group dynamics. The section started with an introduction and we then considered group dynamics by firstly examining the needs of a group and then seeing how the behaviours studied in group dynamics are able to meet these needs. We discovered that a great many behaviours are designed to answer the needs of a group. We do not just prefer to operate in groups, to some extent we are actually designed to. Man is a social animal and this result is in good agreement with experimental observations.

In Section 5.4 we examined the apparent paradox of planning a creative process and found that in reality there is no paradox. Planning creativity is possible. We described the problem-solving process and found that creative output can be increased by using the techniques of brainstorming and lateral thinking in conjunction with rational thought and deductive reasoning. After this we investigated the decision-making process and explained various techniques for ensuring that good decisions are made.

Taken as a whole, the material in this chapter has the potential to make a massive impact on our effectiveness as engineers. By using the team theories we have introduced, holistic teams can be composed whose productive output will far exceed that of unbalanced teams. We have introduced group dynamics and as engineers can improve our teamworking success through understanding more of our team-roles, why we have them and how they work. Finally, an ability to manage the creative process is clearly a benefit to the individual and so Section 5.4 is of use to the individual. However, when these skills are used in conjunction with good teamworking skills a staggering level of effectiveness can be achieved. A team of engineers who competently apply the concepts of this chapter will never be limited by their own abilities.

5.6 Revision questions

1. Describe in your own words the following terms from this chapter:

   Apollo Syndrome
   Brainstorming
   Group dynamics
   Holistic
   Norms
   Team balancing
   Matrix assessment

2. How did the team roles developed by Belbin come about?

3. What governs whether a team is more successful together than the members are singly?

4. Under what circumstances is team balancing particularly appropriate?

5. List five differnt 'teams' to which you belong.

6. Why might team balancing be of interest to engineers?

7. What features of Belbin's team model distinguish it from other such theories?

8. For one of the 'teams' to which you belong listed in question five, get each member to answer Belbin's self-perception inventory.

   How well does your team fit together?
   Do you each agree that the team roles the questionnaire has assigned to you fit yourself?
   Do the others agree?
   What strengths does the questionnaire have?
   What weaknesses does it have?

9. What factors make improving the balance of your team difficult?

10. How does a group's task affect the group?

11. Why must successful teams have creative skills and the ability to resolve disputes?

12. How are these needs met?

13. For one of your teams listed in question five, describe behaviour or incidents that are examples of the following:

   Norms
   Group culture
   Leadership
   Conformance

14. Why is there no paradox in planning to be innovative?

15. Think of an example of a creative process you have been involved in and describe either a process that was used to plan it, or a process that would have helped it to go better.

16. Get together in a group of five or so and 'brainstorm' the following problems. Use the techniques described in this chapter and ensure that a timekeeper is appointed and used.

(i) What sort of businesses could your team start?
(ii) Choose a local traffic problem — how may it best be solved?
(iii) How many different purposes can your team think of for an electric kettle?

In your team think of another similar problem to brainstorm that is of benefit to the team or a team member.

17. What three steps are essential in decision-taking?

18. Think of a decision you have recently made. What techniques did you use to arrive at your decision? Would using any of the techniques described in the chapter have helped or changed the decision you made?

19. In a group of three or four, choose a technical engineering problem with which you are familiar (perhaps the bending of beams or the design of a circuit). Choose a goal to be achieved and then brainstorm four or five different approaches to the problem. Use the 'matrix assessment' technique to determine the best solution.

5.7 Further reading

Meredith Belbin, R., MA, PhD (1981), *Management Teams: Why they succeed or fail*, London: William Heinemann Ltd (ISBN 434 90127 X).

A complete look at the work of Belbin that led to the famous theory of team balancing. A must for anyone serious about team balancing and a good read.

Stewart, D.M. (1987), *Handbook of Management Skills*, Aldershot: Gower (ISBN 0566 02635X).

A book in three parts: Part I, managing yourself; Part II, managing other people; and Part III, managing the business. Chapter 17 is a good general review of team-building.

Torrington, D. (1985), *The Business of Management*, Hemel Hempstead: Prentice Hall International (ISBN 0-13-1049283).

Chapter 14 deals with team-building. The rest of the book forms a comprehensive investigation of the essential management skills required for work in business.

de Bono, E. (1971), *Lateral Thinking for Management*, Maidenhead: McGraw-Hill (ISBN 0-07-094233-1).

Chapter 6
Personal management

Overview

Charity begins at home, so does good management. One cannot expect to be good at managing other people whilst unable to manage oneself. Personal management is more than simply knowing where one's things are, or having a tidy desk. It extends to planning one's career, setting objectives and being able to assess oneself objectively. In Chapter 1 we saw some examples of advertisements for engineering jobs. Personal management is particularly easy to scrutinise at interview time. Individuals who are weak at it are characterised by being poor timekeepers. They do not appear clear in their goals and are uneasy with questions that probe their reasons for wanting the job on offer because they do not truly know for themselves. The opposite characteristics apply to those with good personal management skills. Which do you want to be?

6.1 Introduction

Some people always seem to be organised and have their arrangements under control whilst others are muddled, often break engagements and are late for things. However, our organisational skills control more than just our arrangements– we use them to achieve our goals in life. If our personal organisation is good we are more likely to be successful in achieving our life goals. The poorly organised drift, hoping success will come to them, the well organised name their destination and head straight for it.

This chapter is about improving our personal management and ensuring we achieve the great things we are capable of. It is of relevance to everyone since, whatever your level of personal organisation, an improvement in it will always bring results.

The chapter is divided into three sections.

In Section 6.2 we will introduce time management, good desk-keeping and boss management.

In Section 6.3 the need for objectives and the process of writing objectives are considered. Objectives are the most basic building brick of management and their use from a personal point of view is covered in this section.

In Section 6.4 we shall look at appraising oneself. This section will look at career planning, the preparation and use of curriculum vitae, and finally will look at one's general well-being.

The learning objectives of the chapter are as follows:

1. To understand how to be personally organised.
2. To understand time management and good desk-keeping principles.
3. To realise the importance of managing one's boss.
4. To develop the ability to set objectives.
5. To understand the process of effective self-appraisals.

6.2 Personal organisation

6.2.1 Time management

Time is a physical dimension that has puzzled scientists since humans first asked the question. 'What is time?' Even today the question still cannot easily be answered by physicists. The average person is left with a concept of time as some medium through which things pass, changing as they go. One cannot affect time's passage, you cannot know the future and the only certainty is that eventually one's allotted span runs out.

People are very different. Some people achieve a great deal whilst others appear to achieve almost nothing. Some manage to find success and fulfilment, others do just the opposite. Making oneself happy and fulfilled in life is surely everyone's aim and yet some achieve it easily whilst others find it elusive. Although there are many differences between these types of people, one difference stands out above the rest. The way they use their time is different. Time is unlike any other resource. It cannot be stopped, it cannot be saved up in periods of plenty for use in shortage and absolutely everybody gets the same amount, just twenty-four hours a day. The difference is that the achievers use all their time effectively. People who do this are characterised by knowing what they are trying to achieve and managing their time to ensure that they get it done. They never seem to be involved in things that do not build towards their goals, and they frequently review their situations, making new plans and using opportunities to achieve their aims. People who are unable to achieve happiness have the opposite characteristics.

Time management is the rational way to ensure that our limited time is always used effectively. We shall now look at the mechanics of it.

Time management principles

As a first step to effective time management we have to identify those things that make us happy, the life goals we are currently seeking. These goals should include all aspects of life — private, social, sporting, work — and finances should certainly be included. It is important that the statement of these goals is very clear and Section 6.3 will help greatly with the preparation of these. For example, a person moving to a new town to start a new job may want to do well in the new post but also meet new people and make friends, perhaps buy a house and run a car. These things potentially conflict: money must be saved for a house but spent on the car. Time must be spent on the job yet it is also needed for socialising. Such a person will have their own ideal balance between these needs and it will not just happen along — they must be managed, time included.

The following paragraphs describe the essential steps required to manage your time. These are having clear objectives, prioritising tasks, sticking to plans, making time to manage time, dealing with the unexpected, and controlling time-wasters.

Clear objectives: Time management starts with these overall goals which are then broken down into more manageable tasks. There may be smaller, individual objectives formed to deal with specific issues or ongoing tasks that require regular blocks of time. Tasks that are sequential are divided into manageable chunks, planned and deadlines set.

These are needed before time management can begin. One cannot organise oneself to achieve something without actually knowing what one is trying to achieve. To use a navigating metaphor, you cannot plot a course until you know your destination. If you do not have your goals clear, sort them out first.

Prioritise tasks: First, break down the overall objectives into manageable chunks of work and collect up all the tasks that are in progress. Second, assign priority to each task. Everybody has some tasks that simply have to be done and others that are less vital. The tasks on a 'to do list' should be prioritised and generous allocations of one's most productive times made to the most important tasks. At the other end of the scale, desirable tasks that are not vital can be scheduled into the background. Then, if other jobs permit, they will be tackled; otherwise they are left. It is inevitable that some low-priority jobs will not be tackled — this does not mean that time management is not working, rather that it *is* and you are achieving important tasks at the expense of unimportant ones. Good time management means that only jobs of a low priority do not get done. People who are bad at managing their time often complain that they do not have enough time to complete the important tasks. This cannot be — after all, they have just as much time as everybody else; what is usually happening is that some low-priority tasks are completed instead of the important ones.

Stick to scheduled tasks: No system of organising your time will actually do things for you! You have to follow your own plans, stick to them and get things done within your own deadlines if your system is to work.

Allow time to manage your time: Setting out your time in an ordered way takes time. One must allocate a slot of time for time management. Even for very busy managers five or ten minutes a day is enough for the task. You spend money managing your money; you should spend time managing your time.

The unexpected: Just because you cannot know what will happen does not mean you cannot plan for it. Individual jobs vary in size and complexity, of course, but nothing we do is so unlike anything else that we cannot estimate time by some form of comparison. An organised manager will allocate periods of the day to background tasks which, though beneficial, can be delayed if the need arises to make way for today's crisis, and the undone work made up when the unexpected interferes less. If no such periods occur then the day is filled with too many tasks and it is time for a serious look at whether all the tasks should be under way — perhaps too much is being attempted or tasks that do not support the objectives have crept in.

Managing time-wasters: All sorts of normal daily activities are time-wasters. It would be madness to think that they can be eliminated; indeed, it would not even be good time management for two reasons. Firstly, because everyone needs to vary their mental activity rate and some periods of apparently wasted time, for example a conversation in the corridor or the time spent photocopying, can be beneficial due to the very diversion that they bring. The second reason is that sometimes these apparent time-wasters turn out to be goldmines of information. How many times do you pick up something new in passing or as idle comment thrown in whilst talking about something else? For these reasons it is not appropriate to remove all the traditional time-wasters. What is sensible, however, is to try and manage them so that the beneficial aspects are retained but the negative ones lost. We will now look at five of the worst time-wasters: procrastination, the telephone, meetings, the drop-in visitor and poor delegation.

Procrastination is, by definition, a waste of time, yet who is responsible for it occurring? One of the hardest lessons of time management is that it is not everybody else who wastes most of our time, it is ourselves. We choose what we do and so the responsibility is ours; other people do impinge on our time but only as much as we let them. Procrastination is easy to do and hard to combat. Often people procrastinate because they do not relish the task in hand, because they find it difficult, or because it will bring them into contact with people they don't like. The best way to combat procrastination is firstly to recognise that you are doing it and secondly to be strict with yourself about the time allowed and the results required for the awkward task. Procrastination can extend to long time-scales as well as short — often people have 'pet tasks' that have been hanging aroung for months or even years and never get done. Some people procrastinate over major issues and the indecision can have a large effect on their lives.

The telephone is often cited as a time-waster, which is surprising really, since when used well it is a very great time-saver. The important issue is to control the use of the telephone so that the communications performed with it are all contributing to the achievement of one's goals. A telephone call that brings understanding of how a goal may be achieved is clearly a useful call and may very well have brought the information more efficiently than by another method of communication. However, a half-hour call to explain something that takes seconds to understand in diagrammatic form or in which you achieve someone else's goals is clearly a waste of your time. The use of the telephone is a skill in its own right and is often taught on skills training courses to new recruits. The use of the telephone is discussed further in Section 9.5.

Meetings are another perennial offender for wasting time. As with the telephone, an effective meeting can be an excellent time-management aid or a formidable waste of time. Meetings in which decisions are reached rapidly and actions are agreed by attendants who have the authority to do so, can be as effective as they are brief. Conversely, meetings where the required authority for a decision is not present, where the purpose of the meeting is not clear, or where the attendees are unprepared can be as ineffective as they are long. The individual can ensure time is not wasted in meetings by insisting on the good meeting practice explained in Section 9.6.

The drop-in visitor can be controlled at the times when they are unwanted by simply and courteously turning them away. Drop-in visitors generally waste time either by consuming time or by the very disturbance that turning them away causes. Strangers should have to make an appointment and a reason for the call asked for. There is no need to be curt in such situations: the ubiquitous 'Hello, What can I do for you?' is both polite and to the point.

Poor delegation sounds more like an issue for personnel management than time management but in fact poor delegation is one of the greatest enemies of good time management. Poor delegation means not being clear what outcome is required from a particular task. It brings three ways in which time is wasted. Firstly, if a task has been poorly delegated its definition may not be clear, resulting in effort wasted in directions that do not relate to the task at all. Secondly, it may be that the task was not well chosen and does not contribute to the overall aims at all. In this case, even if the task is achieved, the time has been wasted since the task itself is of no use. Thirdly, if a person does not delegate when really they should, they waste their own time since they end up doing the task themselves. An organisation can lose substantially here since if a long chain of delegation is affected every person involved is engaged in a task that truly belongs with their subordinate.

Time management systems

Many commercial systems are available for time management. It is important to realise, however, that you do not actually need specialised stationery, when a loose-leaf diary is enough. However, the commercial systems do offer very practical solutions and many people who manage their time use them. The different systems all have their own particular format but typically they contain the things listed below. The price varies considerably — they start at very reasonable prices but specialised time-management companies tend to offer rather more expensive solutions. Choosing a system is a matter of taste.

Binder: It is essential that the system has a mechanism to allow the removal and insertion of pages. This is because as time progresses past pages of the diary need removing and new ones inserting, and also since all the other items within the system are constantly being updated, the old versions must eventually be discarded.

Diary: The largest section in any time-management system is usually the diary. Diaries can be bought in almost any configuration from showing a year to a page through to showing one day to an A4-sized page. Most people prefer to be able to see a week at once but the important thing is that the one you have suits you by being large enough to contain the typical number of entries you wish to record per day. For people to whom this means a day to a page it is advisable to have another, synoptic diary which shows how things will go over the next few weeks, to keep an overview.

Address section: A section for keeping the names and addresses of contacts is essential.

Blank paper: Blank paper is often included for use in making up lists, notes, sketches, 'to do lists', writing notes for others, and any other activity best served by blank, rather than pre-printed, paper.

Report sheets: Many manufacturers have available all sorts of report formats including, for instance, monthly budgets or expense account forms. Some people find these useful, especially since the binder keeps them all together; often the use of the forms is restricted by the need to use the pre-printed stationery of your own organisation.

General: Many commercial systems include all sorts of other useful information such as maps, telephone area codes, common words in foreign languages, even puzzles and diversions for idle moments.

Objective sheets: Inserts that are used for recording the objectives and overall aims. Separate sheets should be used for 'to do lists' or 'this week's tasks' since these will change whilst the objectives are relatively constant. These pages should be referred to from time to time to be sure of overall aims, especially every time a new task has to be done, when the objectives should be used to decide whether to accept the task or not. Saying no to new tasks that do not assist the overall aim is the golden rule of time management.

Case study: Time management

A service manager for the subsidiary of a large manufacturing company selling washing machines has four service engineers working for him. The normal daily pattern is for the service engineers to perform the day's work and at the end of each day the team gather and arrange tomorrow's work. The meeting takes over an hour.

After his annual review the manager becomes aware that similar service departments in other companies service more machines per head. He is also alerted to the fact that his engineers feel over-stressed; certainly their long hours of work must be tiring. Some of them complain that they are untrained for the jobs they meet. The manager's policy is to make the customers happy but recently this seems to involve more and more areas of work. He discusses the situation with the service department personnel and the daily meeting comes under fire. After a long day on the road no one wants to end the day thinking about tomorrow's headaches. The boss has a very long day because of taking his children to school first thing and is always run down by the day's end. The service engineers also feel that the meeting is in some senses a waste of time since it is impossible to plan a visit: 'Service isn't like that, you just have to get there and see what's involved', said one of them.

After talking it through with the group and his own boss the manager introduces a new daily order. The department have the objective of solving all problems within two days. Overtime is to be used to provide more effort in time of need. Now each engineer phones into the manager first thing in the morning to collect the first job of the day. The manager gets in early, which suits him, and collects the messages on the answering machine and plans as much of the morning as he can. The engineers phone in between jobs and by lunch-time he usually has the remainder of the day planned for each of them. He lets the answering machine take over after about three o'clock with a message that the call has been logged and a return call to arrange a visit can be expected the next morning. The service engineers continue until the job is done and then get the customer to sign a form which includes the time at which work stopped. The boss uses the end of the day to do his office administration, catch up on the reading that goes with his job and plan possible future developments of the department. These less stressful tasks are done at the end of the day when he is least effective.

The manager now actually has time to keep up with the organisation's developments and has time to install measures of effectiveness. Amongst other things he realises that the department is solving many more problems than rivals normally do and charging the same. He discusses these and other strategic issues with his boss and they agree a strategic business plan for the department to make service support become a marketing tool for the company. At the next annual review, the quantified success of the department is clear, the service engineers are happy and promotion looks imminent. ■

Case study observations

In this case study various important management issues can be examined. We shall only consider the time management issues.

The case study illustrates a common thread in management problems. Initially there seems to be no way forward. The manager knows that other departments in other organisations are more efficient yet everyone in the department seems to be over-worked, including the manager. As a group they correctly percieve the daily meeting to be ineffective. However, the fact that they have no overall objectives, no measuring systems, they seem unable to control the flow of work to their own satisfaction, and do not seem to know what is required in order to become efficient are much deeper criticisms. However, even if these big issues could not be solved — in fact they didn't even discuss them at the meeting — at least they can now manage their time better. The time-wasting meeting is replaced with a more appropriate communication system: each engineer hears only what they personally need, not what everybody else needs to know too. The objectives for each service engineer are improved since they receive them during the day, when they already know how the work is going, rather than the day before, based on pure guesswork. The change in communication method here is crucial and most service departments of this nature issue their staff with portable telephones. Time has also been set aside for different tasks by the manager. Early in the day the engineers' workload is being managed; later the answer-phone takes over and the manager is free to take on other tasks. It is not long before the manager is able to address the underlying issues of overall objective and measurement systems and is soon implementing them successfully. The new regime of 'management from above' has replaced the old regime of 'management from below'. The manager now has departmental objectives with which to control the direction of the department, while the needs that the organisation has of the service department can be converted into objectives for it to achieve. Whereas the manager used to spend all his time serving the needs of his subordinates — *they* used to manage *his* time — he now does so for himself. A clear picture of what must be achieved and how it may be achieved has replaced the old order in which the group only responded to problems and were constantly battling against a tide of work without ever getting control of it.

The difference in management between the 'before' and 'after' case is very great and yet the way in which it was brought about was not radical or heavy-handed. Good management practice was introduced and it made way for more improvements. It is certainly true that scrapping the meeting saved time but, as these observations have shown, that was a small change that eventually led to a completely new departmental time-management system coming into operation.

6.2.2 Good desk-keeping

We all know messy colleagues. Some people run a desk so untidily that the desk surface cannot be seen. Such people would be outraged if they visited a supplier to find an order of theirs lying in a muddle on a messy desk, yet this is exactly how they deal with other people's jobs on their own desk.

A desk is the workshop of a craftsperson. Like all good craftspeople, the good desk user should only get out the tools necessary for the job in hand. The job will be completed and then tidied away before the next is begun. Tools and equipment are not left lying around to become damaged or muddled and at the end of every day the workshop is cleared away completely. Most people judge a workshop from the way it is kept and you can do the same with a desk and its owner. After all, if the owner finds working with so many things muddled up easy perhaps they have a muddled mind, perhaps they don't think the things are important, or maybe the organisation overloads the person to the point of them not being able to cope. In any case the muddled desk indicates muddled work going across it.

Fortunately there is a simple system to follow which will clear the mess. Once these rules become habit the desk is always used effectively and the craftsperson has more time to spend on creative work. We shall now look at the clean desk rule.

The 'clean desk rule' states that the desk is to be clear of all things at the end of the working period. The only permissible exceptions are the telephone and the empty 'in-tray'. One must be ruthless in despatching the clutter of a messy desk and even hardened managers find it nearly impossible to accept that there are only four options for every item.

1. Bin: Just because you have received something it does not mean it has value. The bin is the most underestimated aid to good desk-keeping we have. If an item does not fit one of the categories below then the bin is the only place for it. Periodically, scour the filing cabinet for bin-worthy material. Judge everything against your objectives and anything that does not fit in must go in the bin. Look at material not from the point of view of how useful it is but how it may be justifiably binned. Try to imagine the worst thing that would happen if you binned the item. Only reluctantly refrain from using the bin — try to wear it out with over-use!

2. Delegate: If you cannot put something in the bin, maybe it belongs to someone else. Ask yourself whether you are the correct person for the job. Delegation can be upwards, downwards or sideways. Just because someone else gave you the task it does not mean that you should do it. Only start to do something if your are sure that it is within your objectives and you are the best one for the job.

3. Act immediately: Some items can be acted upon instantly and should be — there is no point putting off a small action that might as well be done now or filing a trivial task, noting it on the 'to do list' and doing it later. A surprising number of things can be dispatched in almost no time at all, for example a single phone call, one letter, a payment, or an instruction.

4. File it: Many things will need filing. Some will be for information only, to be used later when action is required, others will require action too complex to do immediately. These tasks are defined on the 'to do list' and then filed ready to be worked on at the allotted time. Some items may be of interest only, perhaps reading about your subject, which should be done but will probably not produce particular actions. Whatever the situation, the first question is where to put it. People who cannot file things away to be completed later usually run an ineffective 'to do list'. They cannot trust the list and so dare not put something away for fear of forgetting it. Such people have the number of tasks they can simultaneously administer limited by the size of their desk. A succesful manager will need to avoid such a limitation and must employ good filing.

There are many ways of filing — alphabetic, chronological, by size for example — but for your own desk you need a retrieval system appropriate to your personal needs. The most effective way to accomplish this is to have an objective-based filing system. Start with the list of tasks that you are working on and have a file, or group of files, for each. Various benefits result from this system. Firstly, all the information for a particular task is in a particular file and so the desk is not covered with many files for just one job. Secondly, when an objective is completed it is easy to archive, keeping the number of active files to a minimum. Thirdly, since the files are laid out in accord with the objectives it is easy to find the things you are looking for.

Some items for filing may belong in a library of data relevant to the job, some may be reading material. To cater for these items two additional files, one for each should be kept.

Finally, having set up a system do not be tempted to let it lapse. Set aside a regular slot once a day to maintain the workshop and do the time management. In life there are few things more satisfying than a tidy desk after a productive day.

6.2.3 The boss–subordinate relationship

No one shapes our view of work more than our boss. Bosses allocate work, they make decisions about us and our work, they have authority. All the things that motivate us in our work come via the boss. Is it any wonder that for most people, the image they have of their employer is really the image they have of their boss?

Bosses comes in many different shapes and sizes. Some people are natural managers with much skill whilst others may have accepted managerial responsibilities reluctantly in order to further their career. Some bosses need support from above, others are fiercely independent. Normally people do not choose their boss and bosses rarely choose all their staff. Consequently, neither side is usually able to choose someone best for them.

A boss–subordinate relationship is a two-way thing. Bosses clearly want the relationship to be a productive one and so reflect favourably upon their own management. Similarly, the subordinate wants the relationship to be successful and to be fulfilled in the work done for the boss. Both parties are responsible since both contribute to its productivity and both can wreck it.

The following three areas of the boss–subordinate relationship are, sometimes wrongly, thought of as the responsibility of the boss. As can be seen they are influenced by both parties.

Control of objective-setting: It is important the objectives are clearly understood and Section 6.3 describes the mechanics of producing good objectives. Such techniques help enormously in ensuring clarity and are even more effective when done as a collaboration between boss and subordinate. The first step in managing your boss is to ensure that only tasks which have a minimum acceptable standard of objective clarity are tackled. After all, 'What, exactly, do you want me to do?' must always be a perfectly reasonable question to ask your boss.

If objectives are poor, the reasons should be examined. If they are simply poorly expressed, then the situation can easily be corrected. If, however, the weakness comes from a lack of clarity in the overall aims, or fundamentally weak premises, then there will be rather more work in correcting them. It is better to face the problems at an early stage and decide exactly what must be done rather than leave it till later and realise that unusable work has been in progress and so much valuable time has been wasted.

Manage expectations: Much job dissatisfaction comes from a mismatch of expectations. The mismatch may occur if the boss has an expectation from the subordinate who does not meet it or vice versa. Such mismatches are often associated with future career prospects or job content.

An employee who is ambitious and who performs well may feel ready to move upwards. It is vital that the boss in such a situation does not create false expectations of promotion. If a person is given to understand that success today will bring promotion tomorrow, the organisation must honour its offer. You cannot blame someone for being disappointed if they feel they have earned a promotion but it does not come. There are two issues here. Firstly the expectation, not just the promise, should only be given where the organisation has the ability to meet it. Secondly a clear definition of success must be understood. If objectives are used in the career management of employees there will be no disagreements over whether a given standard has been achieved.

Bosses and subordinates alike have ambitions and hopes, but only communication between them and with the rest of the organisation will ensure that a realistic view endures. Neither side is telepathic and it is madness to assume that the other party can know something if you don't tell them.

If a mismatch occurs it can be resolved when it is confronted. The longer it takes to face an issue the harder it is to resolve. For this reason the management of expectations must be tackled whenever a divergence of opinion is suspected either by the boss or the subordinate.

Maintain the relationship: There is no evidence to prove that getting on well with your boss as a person is central to a good working relationship. The converse is, however, not true — it definitely has a de-motivating effect if you get along badly. From this we can conclude that it is necessary to avoid quarrelling, but not necessary to go further. In general, attempts to generate artificial friendships with bosses go down very badly. Of course, there are many friends who work for each other but there are clearly potential dangers and such people must constantly guard against the obvious conflicts of interest that might result.

It is not being likeable that makes a good boss, it is the quality of the boss's personnel management that is important. You can be happy working for a boss who makes absolutely clear what is required but whom you do not particularly like; whereas it is not easy working for someone likeable, but who never actually tells you what is required and cannot guide you.

6.3 Objective-setting

Objectives feature in all branches of management, they are one of the most basic and fundamentally important tools of management and they are constantly in use within organisations. Since objectives are so central, a good question to ask first is why do we actually need objectives at all?

6.3.1 The need for objectives

There are two powerful influences that bring the need for objectives in engineering management.

The first reason for needing objectives is because of the practical limits of communication. When a task is delegated one individual hands a task to another. The two must communicate the task and this allows room for misunderstanding. Objectives exist to eliminate misunderstanding from the delegation process. During the management of organisations complex pieces of delegation become necessary and if during the process of delegation we are to avoid explaining every job within the task it must be explained unambiguously and succinctly. There must be no hidden assumptions. For example, if a manager says to a subordinate 'Run the overseas section of the sales office for a while please', there is no way for the subordinate to know what is actually required. They may have a good idea from experience but that is not enough. If the person delegating had said, 'Over the next three months log ten million yen's worth of Japanese orders using existing resources', the picture would have been a little clearer. Objectives are also a desirable way to be managed from a personal point of view since they allow room to do things the way you want to. You are not forced into methods that you do not necessarily approve of and you have the opportunity to bring your own ideas to bear on the problem.

In summary, the first need for an objective is to be an unambiguous answer to the question 'What, exactly, is it you want me to do?' This is a perfectly reasonable question to ask and a boss who cannot answer it should not anticipate admiration.

The second reason for objectives is to fulfil a deeply rooted part of our motivational psychology. Humans are 'goal-seeking mechanisms'. In Section 4.3.1 we reviewed some theories of motivation, and found that having something to aim towards is a very deeply rooted need. It seems we cannot be satisfied without pursuing a goal. This is often illustrated by the experience of people who strive for some particular goal for many years. The objective becomes all-consuming and when in the end such people either achieve the goal or have to give up they often experience depression or even emotional breakdowns. Some may wonder how such people, who often have impressive abilities, can possibly fall apart so dramatically. The answer is, of course, that their objective system has been destroyed. For years every minute has been filled with work aimed at one goal, then there is nothing. Although such extremes do not probably apply to you, most people are familiar with the lack of direction and anti-climax that follows a great achievement — this is the same thing but on a smaller scale.

People who are good at personal management will have many balanced goals. Some will be short term and others long term, they will have plans for today, next month and often for years ahead. In general, individuals who have such objectives and actively pursue them control their destiny, steer their lives, achieve much and are happy; those who don't, aren't.

We have now seen that there are two reasons for objectives. Firstly, to provide unambiguous delegation and secondly to serve a fundamental drive of humans, the need to achieve. By working towards an objective, these two things are brought together in one statement.

6.3.2 Writing objectives

An objective is an absolutely clear and unambiguous statement of a desired outcome. Objectives are used by people who delegate or by individuals who are managing their time. They may be corporate objectives applying to many people or they may be departmental objectives, applying to only a few. In all cases, there are various important characteristics of a well-specified objective that prevent ambiguity and foster accurate interpretation. There are many popular lists of attributes that well-written objectives should exhibit; in general they include being quantified, achievable, compatible, time-based and measurable. We will now look at these properties in detail.

Quantified: Objectives should be quantified, numerically if possible. If a given objective cannot be quantified then you can never know whether it has been achieved. Who would want to head for a goal such as 'reduce the failure rate' or 'make the price acceptable'. These goals do not explain what is required and consequently are de-motivating for the individual and useless for the organisation. It is important that the qualification is appropriate: words like 'all' or 'never' are special words, and should only ever be used to quantify objectives to which they genuinely apply. For example, the objective 'reduce the waste heat generated to zero' may be desirable and is certainly quantified but the use of zero is unlikely to be appropriate.

Achievable: This component of an objective is aimed at two things. Firstly, the motivation of the individual. It is clearly motivating to be given an objective that is exciting, and although tough, still achievable. The important factor is the individual's belief that the task is achievable. Motivation is a personal thing and it is therefore personal opinions that count. If an employee does not believe an objective is achievable, their boss will not sway them by simply stating that it is. The reasons for disbelief must be made clear and dealt with. The boss may be overlooking something that worries the individual, self-confidence or lack of experience perhaps.

Secondly, achievable objectives make planning possible. Where complex, interrelated tasks are in progress planning is vital to optimise the use of resources and predict overall lead-times. If the individual tasks that have been delegated prove unachievable, the overall plans will collapse as each failed task affects the progress of the next. If any of the benefits of planning are to be realised, there must be a known likelihood that each of the tasks will be achieved.

Compatible: This attribute is related to feasibility but is distinct from it. Compatibility relates to a collection of objectives and describes the need for them to fit together as a whole. For instance, a design department manager might have the objective of 'answering ninety-five per cent of telephone calls from customers with design queries within half an hour'. The same engineers in the department may also wish to improve their understanding of their customers' needs and have the objective of 'visiting fifty per cent of home land customers over the next year'. Both of these objectives may individually be feasible, but no one person can do both of them. One cannot always be in the office and always be out of it. Such objectives are not compatible. A set of objectives must be compatible no matter whether they are for an individual or a whole department. For this reason, objective-setting always starts with one, central goal that is supported by others that follow from it. Progressive subdivision ensures conflicting objectives do not occur.

Time-based: An objective which is not limited in time has no use whatsoever. Every result that is desired has some appropriate deadline. Open-ended objectives waste resources in rambling activities unrelated to central goals. If there is no particular time by which you need something, you don't actually need it at all.

Measurable: This is related to being quantified. However, for an objective to be measureable you have to have access to an appropriate measuring system. For instance, it is quite possible for a computer software manufacturer to quantify the objective of halving the number of operating problems experienced by customers during the first year of operation of a particular piece of software. It may even be a very sensible objective to have chosen. But if the company does not acquire information from the customer about the operating problems the objective cannot be measured. Waiting for complaints or estimating the errors in tests are clearly not useful substitutes for the real data. Consequently, the engineers will not know if their efforts have been successful. If you cannot measure the changes you are making you cannot know if they are having the desired effect.

6.3.3 Maintaining progress

When pursuing an important objective it is not surprising to find considerable interest in the progress of the work. The athlete aiming to win a race monitors progress in detail. Daily changes are used as a short-term pointer to the long-term success. This interest is much more than simple curiosity, it is used as a control mechanism to make success even more likely. If the athlete's progress is not fast enough, for instance, the training is stepped up, different training is used or advice sought from specialists.

Maintaining progress is a closed-loop control activity which starts with measuring progress, proceeds to comparing it to the desired amount and ends with action to correct any deviation. The loop is constantly in operation and has the effect of perpetually steering the work towards its objectives. The process has the same principle as control theory.

Measurement: Various techniques are used to measure progress. Reviews are often held where the achievements of the last period are discussed and data from the measurement systems are presented. Progress meetings may be organised for suppliers at which demonstrations of progress are called for. Whenever the individual is acquiring data that indicates in some quantified way how the task is proceeding, progress is being measured. It is important to bring the data together from time to time and ensure that the whole picture is satisfactory.

The measuring of progress is not over until it has provided unbiased observations upon actual progress whose accuracy is quantified. Only when you are in possession of sound data can you make a reasoned analysis.

Comparison: This stage of ensuring progress involves the comparison of the data against the plans made to achieve the objective. In Chapter 8, and Section 5.4.1, we discuss different planning techniques. The comparison stage of maintaining progress is when the data of the measurement stage is compared against the plan to quantify both the progress made and the work remaining.

Corrective action: Using the information gained so far it is possible to formulate rational paths of actions that offer the best route towards the objective from the current position. Once a certain position has come to pass the management question is not 'How did the situation come about?' but 'How best to proceed from here?'

These three steps of the closed-loop control process should be operated at appropriate intervals for the goal in question. It is through operating control systems that the decision-takers can be alerted to danger or opportunity in advance of its arrival. The benefits of getting such information at the earliest possible opportunity are self-evident.

6.4 Self-appraisal

People vary greatly in their ability to manage themselves. Those that are good at it and are clear about their objectives manage their time to ensure they achieve their objectives and they monitor their progress towards them. One of the key processes of good personal management is the ability to review yourself and see impartially how you are progressing. Such a process is called *self-appraisal.*.

The main difference between appraising yourself and appraising someone else is that it is even more difficult to be objective about your own achievements and failures than about those of another person.

In Section 4.4 we looked at the appraisal process as it is used between a boss and subordinate. Self-appraisal is very similar and still involves the review of the previous period, comparison against objectives and new targets for the next period. It is very easy to fall into the trap of selective memory and post-rationalisation where our own performance is at stake. We do not find our own failure palatable and so we tend to remember only the good things and to think up rational reasons for our behaviour after the event to cover up any irrationality at the time. Consequently, serious self-appraisers will record it for future reference. Particularly where objectives for the next period are concerned it is important to be clear and have a written record.

Of course, a self-appraisal may consider any item the individual feels appropriate but there are three topics that are of perennial interest. These are career planning, the curriculum vitae and general well-being.

6.4.1 Career planning

Good career plans have two aspects to them, the long term and the short term.

The long-term planning should start with what sort of person you are and what you want to achieve. Section 4.3.1 examines the things that motivate people and they should be considered from your own point of view. Your long-term plans should provide you with a life that is fulfilling and that involves you in work you find deeply motivating. As with many aspects of personal management, doing this successfully depends greatly on your ability to be objective in assessing yourself and being able to separate out into different areas the reasons for having certain views. Those who can be objective about their feelings are much more likely to be able to make clear and rational decisions about themselves and their future than those who cannot.

The short-term planning process involves tactical moves towards the long-term goals and selecting the best next step towards them. A short-term plan answers the question 'What am I going to do when the present career phase is over?'

Between these two plans we can be sure of constantly steering ourselves towards a fulfilling life. Clearly taking the view that we can bring such achievements about and actually doing so is more likely to lead to fulfilment than simply hoping that it will happen. One must recognise that as life progresses views may change and so the plan should too. If your outlook has changed, you must look for the basic change in motivation that caused it. Perhaps new long-term aims are called for. Many people do have changes of career, not because they choose wrongly in the first place but because they themselves have changed. It would certainly be a shame to miss out on such opportunities by simply not bothering to think about it. The important thing is that we always know where we are heading. For instance, an engineer may well place importance on becoming a member of a professional body and perhaps becoming a chartered engineer. If so, then planning will be required to ensure that the necessary postgraduate training is available from prospective employers, and choices of employer may be affected by this ambition.

6.4.2 Curriculum vitae

The curriculum vitae (CV) is a description of yourself, prepared for the benefit of another who wishes to view a synoptic account of you and your skills. Your CV is your envoy, it goes ahead of you and it represents you. It is particularly useful and should find frequent use when applying for jobs. Whenever a self-appraisal is conducted, it should include a look at the CV. Ask the question, 'Will my CV impress the next person who uses it?', if not, then plans can be made to gain the appropriate skills and experience to improve it. One's CV should always be balanced, meaning that it has a good cross-section of the skills needed in your field of work. Breadth of knowledge and experience are just as important as depth. From these points we can draw several conclusions about the way in which a CV should be prepared.

1. It should be appropriate for the organisation to which it is sent. CVs should be tailored for each job application.
2. It should be synoptic, there should be no unneccessary verbosity, but it must provide the reader with the information required. People receiving CVs have many to read. The long and boring ones get skipped over, just as you would if you received them. If your CV is

not a clear and complete summary of yourself the reader will infer that you are incapable of preparing such a piece of writing or will assume that there is something to hide.

3. It should be comprehensive and up to date. If you received a CV from an applicant that was ill-prepared or out of date you would be disposed to reject it. If the applicant takes little care over something so important, how much less care will be taken with tasks the organisation might give out?
4. The CV is your envoy. Until you meet, the organisation knows nothing of you except your CV, it goes ahead of you and represents you. It should therefore be neat, tidy, short enough to read fairly quickly, well enough written to make the reader want to know more. The spelling, grammar and punctuation should be beyond reproach.

Although a CV should be prepared for each application, all CVs have points in common and so it is possible to use a format that will be generally applicable and then update the content as required.

All people who receive a CV will want to know some of your personal details to form a picture of you. It is therefore normal to include a section on your personal details and explain at least something of your education. In the case of a CV prepared for employment the recipient will want to know whether you have the ability to do the job and whether you are likely to be a personable employee who will fit well into the organisation. To provide for these needs your CV should include a description of what you are doing at the moment and what you have done in the past. Early in your life you will have far fewer things to talk about than later on and so, in order that the CV stays at a manageable size, it is normal to say less about your early achievements as time goes by. For instance, in the CV shown in Figure 6.1, there is very little said about the time immediately after graduating since much has been achieved since. In earlier days this section would have been much larger and might have included things that the person learnt then or more details of the work undertaken. Personal skills and experience are of interest to someone receiving a CV since they say a lot about how someone views themselves. After all, why should you be considered for a post if you do not even have something to say about why you might be particularly suited to the post? Finally, to complete the picture of what you are like the reader will want to know something of your interests or pastimes. If you read some CVs it is more likely that you would remember those people that had something interesting and human about them. Not everybody has dangerous or exceptionally exciting hobbies to put down, but everybody does have something that interests them, or that they enjoy doing. Put it on the CV and the reader will have a better chance of knowing what you are like.

The example CV in Figure 6.1 is of an engineer with several years'

CURRICULUM VITAE

Name: Anne Example BSc (Hons)
Date of Birth: DD.MM.YY
Address: XX Your Road, Your Address, Your Post Code.
Status: Married with one child aged eight.

EDUCATION

Date	Place	Qualifications
XX–XX	Crompton Park School	Eight O-levels, three A-levels (Maths, Physics, Chemistry)
XX–XX	University of Lorrowbridge	BSc (Hons) in Electronic Engineering Class II(ii)

CURRENT POSITION
YY–Present day
Senior Designer in High Current Division at Mainwaring Peylon plc.

I joined Mainwaring Peylon as the project engineer for the high current thermo-intelligent switch programme. I am responsible for all aspects of the group's management. I have seven people reporting directly to me including three graduates.

The remit of the group is the complete design and manufacture of a new product range. This includes the design of the switch amplification cavity and the three-dimensional multi-parameter finite element analysis. I personally designed the resonator control system. I have negotiated all aspects of the contract with the sales and marketing division which involved several trips abroad to provide technical support for the division and to visit customers. I am also responsible for all quality assurance aspects of the programme. The programme is nearing completion and I am engaged in identifying new work for the group at present.

PREVIOUS WORK EXPERIENCE
Last job YY–YY: Project Engineer at Halton Freionics Ltd: Employed as a Project Engineer reporting to the group leader. I was responsible for all phases of product development from initial customer visits to commissioning. Projects undertaken included a Muon-catalysised fusion power cell for specialist Hi-Fi systems, and a portable 27 TeV, 10A electron accelerator for pest control.

I also handled the commercial account with the company's second-largest German customer. This included price setting, hosting visits, product support and the handling of all communications.

Job before that YY–YY: Dagaba systems Inc. (UK): Employed initially as a graduate trainee and ultimately as a salesperson. In my training period I was introduced to various functions of the business and learnt how the organisation functioned as a whole. I liked travelling and was successful in my application for promotion to sales. In my four years in sales I travelled extensively in the UK and made three trips to the United States and the Far East. I became interested in the process of product development and spent more and more time on new product generation. Eventually I left to expand my opportunities and joined Halton Freionics Ltd.

Job before that YY–YY: Immediately after graduating I spent a year abroad travelling. I worked in Australia for nine months whilst travelling. The time included two months as a hospital assistant.

PROFESSIONAL TRAINING
Since graduating I have attended various courses in professional skills:
Time management (5 days)
Communication skills (7 days)
Management organisation and teams (10 days)
Appraising employees (1 day)
Leadership skills (5 days)

SKILLS AND EXPERIENCE
I have a broad range of personnel skills including dealing with people at all levels within organisations. I have briefed a board of directors on many occasions but have also trained apprentices. I have interviewed candidates for selection and have conducted performance-based appraisals.

Upon delivering equipment I have presented training in its use and written instruction and service manuals for most of the equipment I have designed.

I have experience of various computer packages and have supervised Computer-Aided Design draughtspeople on both electrical and mechanical systems.

INTERESTS
I am interested in the philosophy of design and have given various guest lectures at my old university college. I enjoy sailing and in 19YY I was entered in a national ocean-going yacht race. I was second rigger on an all-women's crew. I attend night school in art at a local college where I study photography. I enjoy both listening to and performing music, I play the piano and also sing in a choir. I enjoy cooking and entertaining for friends.

PUBLICATIONS
In 19YY I co-authored a paper on the magnetic confinement probems of muon-catalysised fuel cells and presented it at a conference in New Jersey.

REFERENCES
The following people may be contacted in connection with my application and will be happy to answer any questions.

Name : Address : Relationship

Name : Address : Relationship

Figure 6.1 Example of a curriculum vitae

experience, applying for a job as a chief design engineer in the electronics industry.

6.4.3 General well-being

Every self-appraisal should include a review of general well-being: health, enjoyment of life, whether you are getting enough exercise, sleeping well and all the other things that compose a wholesome life. A healthy body and healthy mind will always go together.

People whose general well-being is good are characterised by having a happy home life, a group of friends they regard as permanent, good health, enough sleep, leisure pastimes and a satisfying vocation to fill the week days. All these things are equally important and they should all be reviewed, not just the last one. Any development that deprives the individual of their required balance causes stress. The more the imbalance, the greater the stress. Good personal managers are clinically accurate in identifying the source of their stresses. Their good self-appraisal skills give them the earliest warnings possible and they rapidly execute strategies either to defeat or to avoid the stress. They undertake actions that build up their general well-being.

Interestingly, people's ability to maintain their general well-being is often affected by stress and the situation is unstable. Those who are doing well find it easier than those who are not. Stress causes a lack of well-being and the most common causes of stress are not having enough time, not having clear objectives, and not being able to take a clear, objective, look at your own progress. In this chapter we have discussed ways of managing these three things and therefore have provided material of use to people trying to reduce their stress levels.

6.5 Summary of Chapter 6

In this chapter we have introduced personal management. You can use personal management to help achieve life goals and be fulfilled.

In Section 6.2 we introduced time management, effective desk-keeping and boss management. Through these activities you can be sure that everything you are doing is assisting you to achieve your life goals; that you use your desk to its best advantage; and that you have the most productive relationship possible with your boss.

In Section 6.3 we introduced objectives. We examined the need for them and found that there are five attributes of well-written objectives. The section ended with a look at how progress may be assured and the overall operation of objectives was seen to be similar to the operation of a closed-loop control system.

In Section 6.4 we discussed three important aspects of self-appraisal. These were career planning, curriculum vitae and general well-being. In the section we say how these may be reviewed. It is clearly a matter for you to decide whether you will use these tools to ensure that you achieve your goals.

To find deep satisfaction in life, you have a choice. Either you can just hope that someday you chance to come across it; or you can use self-management to head straight for it. The choice is yours.

6.6 Revision questions

1. Why is personal management important? What advantages does it bring? What characterises people who are good at it? How may it be achieved?

2. What is 'time management'? How does it apply to the following groups of people?

Students
Qualified Engineers
Salespeople
Shop-floor Apprentices
Executive Managers

3. Keep a log of your activities over the next week accounting for periods of time no smaller than one hour. At the end of the week total the amount of time you have spent on each category of activity. Does the way in which you allocate your time look sensible when compared against your objectives? How much time has been wasted? Could you manage your time better? How?

4. Could your filing system be improved? Can you always find the material you are looking for? Explain to someone else what problems you have with your filing system and how you will change it to overcome them.

5. What can you do to control the effectiveness of the working relationship with your boss?

6. What attributes do well-written objectives have? Find an objective of your organisation and compare it against these attributes. Does it have them? Re-write the objective and improve the quality of writing, taking the required attributes into account.

7. What are your objectives for the next year and for the next five years? Write down five things you want to achieve over the long and the short term. Work on the statements you have prepared until they have all the important attributes of a well-written objective. Show them to a friend and ask for comments.

8. How is progress towards objectives maintained? How might this occur in the workplace of engineers?

9. Prepare a CV for yourself. Ask a friend to go through it with you and compare it with the example in the chapter. Will it impress someone to whom you might send it? What things do you think you should do in order to improve the balance of it?

10 Explain the following terms:

Clean desk rule
Compatibility
Time-wasters
Objective

6.7 Further reading

Adair, J. (1988) *Effective Time Management: How to save time and spend it wisely*, London: Pan Books Ltd (ISBN 0 330 30229 9).
A well-written and informative account of time-management principles. The book is easy to read and includes sensible guidance to improve your time-management abilities.

Chapter 7
Finance

Overview

It doesn't matter what you think about money, you need to be able to manage it. Corporate business and church fêtes both need to balance income and expenditure. No part of management goes without financial implications and professional engineers have needs for financial understanding just as other professionals do. We have seen how an engineering organisation has need for various functions and so far we have seen how the needs for personnel, teamworking, and personal management may be met. In this chapter it is time to examine how the financial needs of an engineering organisation may be met.

7.1 Introduction

To introduce this chapter we will start with a question: 'Would you like to make some money?' Not just earn a wage, everybody can earn some sort of wage, but actually turn over money and keep some of it for yourself.

Those who apply themselves to the task of making money have one thing in common. They all realise that, amongst their many other skills, they will have to understand money. Most people know that to make money goods have to be sold at a price that will cover materials, labour and overheads, and still leave some profit. However, very few people can look at a business and say whether a sensible asset distribution prevails or say whether the accounting system is facilitating optimal decisions. Few understand that a healthy business can be rapidly bankrupt by simply not distributing its money correctly. Once we get into the realms of shares, debentures, mergers, acquisitions, taxation, investment policy or corporate financial planning all but a few are left behind.

Finance pervades nearly all activities in some way. A musician arranging a concert has to pay attention to its financial viability. A retail shop has more complex financial problems and needs to know, for example, how to price larger goods which take up more space, and therefore consume a greater proportion of the cost of this space, at the exclusion of other products. For a trading organisation that takes in goods, processes them and then sells them, the financial complexities are immense and cannot be tackled by the novice. The most commonly quoted reason for the failure of new business is a lack of financial acumen.

This chapter will give you an introduction to finance. Engineers do not need to become skilled financiers in their own right but they do need to understand basic financial axioms in order to make good decisions. In the same way that a good farmer is knowledgeable about soil science and meteorology, a good engineer is knowledgeable about finance.

We will start by examining the need for monetary control. Before we embark upon an investigation of finance in detail we must examine what it is we expect to achieve through having a system at all. Section 7.2 examines this need.

In Section 7.3 we introduce the subject of financial accounting and explain how this branch of finance serves one of the needs identified in the previous section, to provide financial information about the organisation to those in the outside world.

In Section 7.4 we examine the need for budgeting and explain how budgeting is performed and consider how financial management may be achieved through the use of budgeting.

In Section 7.5 we introduce the subject of management accounting and explain how this exists to provide financial information to those inside in the organisation who have responsibility for its management.

The learning objectives of this chapter are as follows:

1. To explain why financial control is important.
2. To explain financial accounting.
3. To be able to understand the major financial documents.
4. To explain management accounting.
5. To develop an understanding of the budgeting process.
6. To appreciate the overall structure of organisational budgeting.

7.2 The need for monetary control

Money is required in the administration of all organisations. Some organisations seem to have making money as their sole aim whilst others, such as charities, seek to encourage plentiful contributions and then distribute them in accordance with a perceived need. Even though charities are not trading, they still have to be skilful handlers of money. It doesn't matter what an organisation is trying to do it must be able to control its finances. The more complex the situation the greater the need for good financial control.

We shall now consider a simple business to see how the need for a clear financial management system comes about and requires more than a simple personal finance system to be effective.

7.2.1 Inadequate Financial Systems — A Case Study

Jake has set up his own business operating out of his back yard. He is renting his house and has finished studying for a higher degree in music and so has some spare time. He is convinced that with his hobby-based knowledge of digital electronics he can develop a home hi-fi business based on combining proprietary supplies of hi-fi equipment with his own creations to produce individual systems for each customer. He calls the enterprise 'Jake Sounds'.

He soon has an order from a friend. He discusses it, and produces a design. Jake buys the components from various suppliers and assembles the equpment. Some of the components have to be sent off for machining, painting, silk-screen printing and other processes with local companies, others arrive ready for use.

To keep track of finances Jake has a filing system consisting of three files. One file contains a record of all purchases, another contains a record of all sales. The third contains a summary of the other two; this one is the most important one to Jake, since it contains the total amount received and the total amount paid out. He is looking forward to seeing the first number minus the second get very big! The first job is finished in only six weeks and the total amount spent was £1,200. Jake notices that he has a few components left over but they cannot be sold back to the supplier and so he decides not to give the customer any discount. He reads a textbook on pricing policy and discovers that it is normal to charge 2.4 times the cost of materials and labour to the customer in his industry. He costs his own time at £8 per hour to give £960, adding this to £1,200 and multiplying by 2.5 gives him £5,400 which terrifies the customer so he charges only £4,500. Filling in his books he calculates in the third file that he has made £4,500 minus £1,200 minus £960 = £2,340 clear profit. As this is after paying his own wages of £960 Jake is delighted. The rent and bills for the premises can easily be covered for the whole year with just a few such jobs, and after that it's all profit in the bank for the business.

After some time operating on a similar basis Jake makes the following summary of his business. The order book is growing and he can no longer do all the work himself. Nearly all customers want a mixture of standard equipment and his own custom designs but these designs are practically never the same as something built previously. The accounting system is adequate.

Because of these factors he hires three people and divides the company into two sections, though they are all physically still in one room. The first section handles all the custom equipment and the second purchases all the standard equipment and combines it with the custom equipment to form the complete system. This is then tested, and corrected if necessary, prior to shipment. Each department orders the parts it needs which, when delivered, are placed on shelves covering one wall of the room. All purchases are recorded in the file, as are all the sales, just as before.

Two orders are processed sequentially in this way to allow everyone the opportunity to follow the system. There is some confusion sometimes over what belongs to whom on the shelves and it has become necessary for the three of them to look up the cost of goods used on each system in order to set the selling price accurately. These points are not considered to be serious and they continue operations, having had a profitable three months.

They now have many orders and some have to be worked on simultaneously. Jake hires a buyer to take the load off the engineers, leaving them free to concentrate on the equipment. This works very well and whenever the engineers need something they ask and it soon appears on the shelves. The buyer is always helpful and always seems to know what's going on.

Before long the next quarter is over and it is time for Jake to assess how things are going. The company have completed seven systems this quarter, which makes the productivity less than the 'one system per six man-weeks' that he used to average on his own. None the less, he calculates that at about £2,300 profit per system, that should make £16,100 clear profit to invest back in the business after wages have been paid. By the time he starts making up the books, Jake is already planning how to make his business grow faster with a new marketing initiative. He plans a profit-related bonus scheme to reward the industry of his employees, and a bottle of Chateau De La Tour 1961 for himself. Imagine then his disappointment when he finds that the company has only made £1,225 — he cannot believe his eyes and promptly calls a meeting to discuss the matter.

Several things come up in the discussion, starting with the pile of things on the shelves which seems to be bigger than before. Two particular jobs were especially difficult to get right, and while the components used were not extraordinarily expensive several sets were blown during manufacture. No one can put their hand on their heart and say accurately how much each system cost. Jake can certainly find out how much all the systems cost together by totalling the 'purchases ledger' but he cannot now find out how much was spent on materials for each one. He looks up total expenditure and finds that much more than he expected has been spent.

One of the employees comments that it is no longer possible to know exactly how much time is spent on each job. Before, when there were only one or two systems it was easy to keep track, but now he freely admits that he divides his time between all the jobs in progress and hopes it all balances out. The others comment that even though the two difficult jobs consumed the majority of their time, they still allocated their time evenly amongst the five jobs in progress at the time. Jake is very concerned at this and the unknown material costs since it means that the selling prices were fixed using the wrong information, and so the profit margin for each job will be quite different from his original estimates.

The team turn to the stores and purchasing system. The buyer explains that where possible he has been buying in bulk for economy. This is obviously good but has caused some difficulty. Kate mentions that she was about to ask for some components to be ordered when she saw some in stock, and so used those. The components were much the most expensive in the system and because of the discount the system must have been much cheaper to make but the selling price was based on the old price. On hearing this another engineer is enlightened: he ordered a matched set of output Stegatrons for one difficult system and later found some of them missing and had to order even more, and they were really expensive. It transpires that any profit that might have been made on Kate's system will have been lost by filling it with expensive and unnecessarily matched Stegatrons.

The engineers also noticed that there is room for improvement in the accounting and stock control system. They even found an example of the same delivery of ten components (worth a considerable fraction of the material cost) being included on the costing of two different systems even though each system used only one of the components. Another engineer had looked at this information and thought that all the components must have been used up, and so ordered some more. There were then eighteen units in the store. This was a really poignant example since the company had only orders needing six more of the particular component and the remainder would probably become scrap since the manufacturer would very soon be replacing the unit with a new version that the customers would want used in their systems.

Some chartered accountants who have been hired to audit the company's books send a letter to Jake asking the following questions:

- What is the value of your stock?
- How much profit did you make on each of the jobs?
- What is the component wastage rate?
- How much money is outstanding in unpaid bills?
- How much money is outstanding in uncollected debts?
- How much of the purchases were for components and how much for tax-deductible assets or equipment?
- One particular payment is questioned being much bigger than any other to a supplier.
- How much profit have you made overall?

Jake realises that he cannot answer any of the questions, not even the last. Jake lowers his face into his hands and makes manifest his heightened state of aggravation and woe. 'What are we going to do?' he asks. He calls a consultant to help him understand what has happened. The consultant confirms that the business was suffering from an unacceptable accounting system. 'If there had been a half-decent one in place the problems you now face would have been obvious to you as they occurred rather than all this time later. Think about this idea,' the consultant explains, 'the buyer orders the items and pays for them out of the company bank account. When the goods are delivered they are "booked" into stores meaning that a record is made of the items in a file. When the engineers need a component they go to the store, take the part required and then record in the file what has been taken, its value and the job on which it is to be used. They also decrease the stores count so that the next person knows accurately what there is and what it cost. The stores becomes a control system, not just some alphabetical shelf space. If an engineer knows of a future requirement, existing stock may be ear-marked in the book for use and so not used by someone else. If the engineers all use a time sheet to record how much time is spent on each individual job you will then be able to easily work out how much each job is costing.'

The consultant says that he found £6,000 of material in the store. Jake is delighted until he hears that £6,000 is what was paid for it and not what it is worth now. 'The value of the stock in terms of resale value is only about £1,000,' comments the consultant. 'Great savings could also be made by organising the timing of your purchases as you have already noted. You need to schedule your production to ensure that you take advantage of the discounts available. Make a list of all the components in each of the systems and get the buyer to order them sufficiently early to be sure they arrive in time without paying any premium for rapid delivery if the order is placed a little late. You could even note down the typical delivery times of your suppliers and order parts in a scheduled way so that they arrive only shortly before they are needed. In this way you could minimise your stock. The buyer would then know the best time to order a given component and would only buy early if any discounts or special terms that the suppliers may be offering make it worthwhile.'

The consultant's face darkens as he turns to Jake and says, 'Worse than this, however, and an unforgivable error in a manufacturing company, is the way the selling price has been based on the wrong data.' The consultant estimates that the two really difficult jobs were actually sold at a loss of about £500 each. He guesses that the profit was nearly all made on three other jobs whose selling price was set too high. This was because the selling price was based on estimates which turned out to be wrong. Jake remembers that one of these customers was rather concerned at the price and will probably not now do repeat business.

Jake is worried about how he can answer the questions of the auditors but is also resolved to sort this out so that his company may become highly profitable once again. He asks one last question of the consultant: 'But how come I was all right before, the accounting system was fine and I always knew where I was?' The consultant replies 'Yes you did, but that was because you had all the information you needed. The company was simple so the accounting system could be too. Your company has changed and just because something served you well under one situation it doesn't mean that it will continue to serve you well under a new one. You used to know everything about each job and that is still what you need to know. All you have to do is have a stores system, have a file for each job into which you book all stores requisitions for that job and into which you book all time spent on each project. Then you will have all the information your company needs to satisfy the tax man and make the best business decisions for the future. No one likes paperwork so have only enough to meet your needs but do make sure you have sufficient for the needs of the business.'

■

7.2.2 The ideal financial system

We know that money is the life-blood of all organisations: it does not matter what the organisation's purpose is, it will still have to manipulate money in order to meet it. There are good and bad ways to handle money and the ideal financial system will clearly be the one that enables the best manipulation of the organisation's finances to be achieved.

For instance, imagine a manufacturing company that purchases all its raw materials with sixty days' credit, the assembly and shipment takes twenty days and the product is sold within five days, on thirty days' credit. This leaves five days clear after receiving the customer's payment before the bill from the suppliers arrives. Such a business would not have to borrow money to start production, could cope with any number of orders and would have no financial limitations on its rate of expansion. If the same company had to pay the suppliers well in advance of receiving payment from its customers the situation would be very different. In order to produce each product the company would have to borrow the materials cost of the product for the period of manufacture, and the number of products it could make at once would be limited by how much it could borrow. The rate of growth would be limited by having to borrow against tomorrow's larger profits in order to purchase today's larger material bill. These differences do not necessary have anything to do with the design of the product, its success in meeting the customers needs or the production arrangements. They do, however, depend directly on the way in which the organisation negotiates and operate its finances. It is clear which sort of arrangement is preferable.

An ideal financial system is defined as 'the most cost-effective method of ensuring that the organisation can operate its financial affairs in such a way that the organisation's overall goals may be best served'. Such a system must consume an acceptably small portion of the organisation's resources whilst providing for its financial needs. These needs may be external, as in the requirement to produce annual audited accounts for the state, or they may be internal. A manufacturing business needs to know, with sufficient accuracy, how each area of the business is performing in order that good areas may be left alone and bad ones rectified. These two needs are often separated into two separate subjects.

The first, provision of information for people outside the organisation, is called *financial accounting*, the second, the provision of information for those within the organisation, is called *management accounting*. Both are a balance between providing too much information or too little. Too much will cause unnecessary expense, be confusing and probably be wasted. Too little information, or the wrong information, will compromise the quality of decision, lead to irrational actions and might even risk the organisation's survival.

7.3 Financial accounting

Financial accounting is the name given to the preparation of financial reports predominantly for external needs. There are two main bodies interested in this information.

The first are people who invest in the organisation in some way. Such people are clearly interested to know how their investments are performing. In the case of new investments, an assessment of the likely profitability will be prepared and the financial performance of the organisation will be of key interest.

The second body is the state, and its interest is two-fold. Firstly, because it requires tax to be paid on the profits of trading, and secondly, because the state regulates trading and seeks to prevent unscrupulous organisations exploiting investors and employees. It is for these reasons that there are legal requirements to produce audited accounts.

To meet these needs various financial documents have been developed and in this section we shall consider some of them. Although there are many financial documents we shall restrict ourselves to three of the most important: the balance sheet, the profit and loss account and the cash-flow projection.

7.3.1 The balance sheet

The balance sheet is a very important financial document. It not only lists all the different areas in which money belonging to the organisation currently resides but also specifies the amount in each. Since money is constantly flowing around the organisation the balance sheet can only apply to one particular instant. The best conceptual understanding of the balance sheet is gained by thinking of it as a financial snapshot of the organisation or, perhaps, as a single frame from a movie film of the organisation's finances. Because of this the balance sheet is prepared for a particular day, usually for the day that ends the organisation's financial year.

This snapshot nature sets limits on what may be inferred from a balance sheet. For instance, imagine walking into a business one day: you might see masses of money being made using very little stock, and this snapshot would make the business look profitable. Watching on another day, however, you may observe that things are quite different and that now that the company actually has a very poor rate of converting stock into profit. You would then have to revise your optimistic appraisal of the organisation. For this reason the balance sheet is not used on its own when assessing an organisation but is considered together with other documents such as the profit and loss account.

The balance sheet takes its name from the fact that it has to balance two amounts of money. It operates on the principle that if you have something it must have been paid for somehow. For instance, if you own a building you much have once had some money with which you bought it. To represent this on a balance sheet one side would show where the money came from in the first place, perhaps a loan, while the other side of the sheet shows what has been done with it, in this case converted into an asset. The basic axiom of the balance sheet is therefore:

Assets equal the sources of the funds

The balance sheet: preparation and understanding

To understand how a balance sheet is actually prepared in practice we shall now follow Jake Sounds, from Section 7.2.1, as Jake prepares a balance sheet to sort out the mess he got into. To do this he went back to the beginning when he started the company.

Jake invested £1,000 of his own money in the company. His company therefore had a source of funding and an asset. Jake makes up the balance sheet as follows:

Balance sheet for Jake Sounds on Day 1

Assets	£	Source of funds	£
Cash in account	1,000	Own capital	1,000
	1,000		1,000

At the time Jake felt that this was not enough and persuaded his bank to invest a further £800. This happened on Day 5 and the balance sheet was updated.

Balance sheet for Jake Sounds on Day 5

Assets	£	Source of funds	£
Cash in account	1,800	Own capital	1,000
		Loan from bank	800
	1,800		1,800

Although there is £1,800 in the account the balance sheet makes it clear that the asset is made up of two parts. His own investment of capital, which need not be repaid, and the bank loan, a liability, which will have to be repaid.

On Day 9, Jake bought some stock: £450 was spent with a local shop. The effect was to reduce the company's cash whilst increasing the value of the stock; there was no effect on the sources of finance.

Balance sheet for Jake Sounds on Day 9

Assets	£	Source of funds	£
Cash in account	1,350	Own capital	1,000
Stock	450	Loan from bank	800
	1,800		1,800

Jake then bought a further £850 worth of materials with his credit card. This meant receiving the stock, and paying the bill later. Of course the 'creditor' was owed money and the balance sheet must reflect this. Jake decided that the creditor must be a 'negative asset' because a consequence of acquiring a creditor is that money from the bank account will be required to pay off the bill and so it should be 'ear-marked' for this use immediately.

Jake then noticed something interesting: a balance sheet will still mathematically balance, regardless of whether an existing item is included as a positive value on one side or as a negative one on the other. One could, for instance, show a creditor as a positive source of finance and still have a balance. In fact this would be an equally fair way of looking at the creditor since, like the bank who gave the loan, the creditor supplied an asset on condition that its value would be repaid later.

Jake realised that once a convention is adopted one must stick to it throughout the balance sheet if it is to remain consistent and he decided to regard creditors as negative assets.

On Day 15, stock increased by £850 but this was off-set by £850 worth of creditors and the total assets of the business remained unchanged.

Balance sheet for Jake Sounds on Day 15

Assets	£	Source of funds	£
Cash in account	1,350	Own capital	1,000
Stock	1,300	Loan from bank	800
Trade creditors	(850)		
	1,800		1,800

The first customer took delivery on Day 31 and the business acquired its first trade debtor. The selling price was £4,500 and this amount must be entered on the left-hand side since a debtor, being someone who owes you money, is clearly an asset. Jake almost forgot that the transaction also resulted in a loss of stock equal to the value of the components in the system.

£1,200 worth of materials were used in the system. The balance sheet is now made up to take account of these transactions, assets have risen by £4,500 and stock has fallen by £1,200. However, both these items are on the left-hand side and since they are not equal the balance sheet will no longer balance. Jake cannot see what has gone wrong but is convinced that a source of finance must be found to balance the change in assets. Finally he realises that the product has been sold at a profit. Profit is a source of finance, the very source of finance he is trying to bring about. The amount of profit is easily calculated by making up the left-hand side and then entering the required amount to make the sheet balance. In this way Jake is sure that all the deductions that should be made have been included and the profit figure is accurate. The result of these actions is to increase assets by £4,500, reduce stock by £1,200 and to create a profit to balance the sheet. A new and important category must therefore be entered on the balance sheet to display this profit and is shown below as 'Profit and loss account'. Clearly it is possible for a loss to be made and in such a case the figure entered would be negative. This entry in the balance sheet should not be confused with the separate financial document of a similar name described in Section 7.3.2. Although the two things are related they are, as we will see, different.

Balance sheeet for Jake Sounds on Day 31

Assets	£	Source of funds	£
Cash in account	1,350	Own capital	1,000
Stock	100	Loan from bank	800
Debtors	4,500	Profit and loss account	3,300
Trade creditors	(850)		
	5,100		5,100

Jake's wages were £960 for the first six weeks. The above balance sheet was made up for Day 31 and the wages were not yet due and so the profit looks larger than it truly is, illustrating a management opportunity for 'dressing' the books. This can involve more than simply choosing a particular day on which to prepare the balance sheet. It can extend to running down stock near the year end or squeezing credit arrangements together with an aggressive round of debt collection in order to make the balance sheet unusually favourable on the day of preparation. Because of such possible ambiguity in meaning, anyone assessing a company never depends solely upon the balance sheet but uses other information to support the assessment.

One Day 40 the business receives payment from the customer (a debtor) after acceptance testing is complete. The supplier (a creditor) is also paid. At the same time, an appraisal is made of the remaining stock, which proves to be useless leftovers, changing the value of the company. An asset previously valued at £100 turned out to be worth nothing. If the value of stock has changed, something else must change too in order for the balance sheet to remain in balance. The simplest way to visualise this stock revaluation is to imagine that a product with a material cost of £100 has been sold for £0. The profit and loss account must therefore be reduced to pay for the loss.

The effect of all this is to increase the cash in hand by £4,500, reduce the debtors by an equal amount, reduce the creditors by £850 and reduce the cash by the same amount. Stock is reduced by £100 and so is the profit and loss account.

Balance sheet for Jake Sounds on Day 40

Assets	£	Source of funds	£
Cash in account	5,000	Own capital	1,000
		Profit and loss account	3,200
Stock	0	Loan from bank	800
Debtors	0		
Trade creditors	0		
	5,000		5,000

Finally on Day 43 the wages are paid. The effect of this is to reduce the amount in the bank by £960, which must clearly come from a source of finance, the profit and loss account.

Balance sheet for Jake Sounds on Day 43

Assets	£	Source of funds	£
Cash in account	4,040	Own capital	1,000
		Profit and loss account	2,240
Stock	0	Loan from bank	800
Debtors	0		
Trade creditors	0		
	4,040		4,040

We have now followed a balance sheet for a particular company through a process of money creation. Initially the company purchased raw materials, work was performed with the expectation of payment, a sale was made and, finally, cash received. The cash was used to pay the creditors off and the remainder was kept as profit. In reality organisations do not make up a new balance sheet every time a transaction is conducted and the balance sheet contains much more information than in the simple example above.

Five years later Jake Sounds is doing very well and Jake has prepared the balance sheet for the annual accounts. It is shown in Figure 7.1. Like all balance sheets, assets equal the sources of their finance. This balance sheet shows last year's figures too for an easy assessment of performance. There are many legal requirements imposed upon the preparation of an audited balance sheet and the books listed in the 'Further reading' section at the end of the chapter include an explanation of them.

Balance sheet: Jake Sounds

		YEAR 5 £k	YEAR 4 £k
	Fixed assets		
(a)	Tangible assets	35.6	28.6
(b)	Investments	4.1	2.6
(c)	Total fixed assets	39.7	31.2
	Current assets		
(d)	Stocks	45.6	41.3
(e)	Debtors	15.6	14.3
(f)	Cash at bank and in hand	8.9	6.9
		70.1	62.5
	Current liabilities: amounts due within 1 yr		
(g)	Debenture loans	0.5	1.2
(h)	Creditors	20.5	17.4
(i)	Net current assets	49.1	43.9
(j)	Total assets less current liabilities	88.8	75.1
	Liabilities: due after more than 1 yr		
(k)	Finance loans	1.6	1.8
(l)	Creditors	0.8	0.6
(m)	Provision for liabilities	7.5	6.8
(n)		78.9	65.9
	Capital and reserves		
(o)	Called-up share capital	17.3	16.1
(p)	Reserves	0.5	0.3
(q)	Share premium account	7.6	7.1
(r)	Profit and loss account	53.5	42.4
(s)		78.9	65.9

Figure 7.1 Balance sheet for Jake Sounds

Jake explains what each section means in the lettered paragraphs that follow the balance sheet.

(a) Tangible assets: These are under the general heading of 'fixed assets' and so list the physical assets such as plant, machinery or buildings. Depreciation is allowed for and the value of a given asset falls with time, the depreciation being paid for from the profit and loss account. Occasionally, particularly when a company is being sold, a financial term called 'Goodwill' is listed is a fixed asset. Goodwill reflects the increased worth of a business, over and above its asset value, by virtue of some particular benefit. This might take the form of an especially loyal customer base who are prepared to pay over the market rates for the organisation's products. The consequence of such an advantage is that the business is really worth more than the basic asset value since it has the capacity to generate more profit than would normally be expected for such a business. Goodwill is therefore defined as the difference between the sale price and net worth (net assets) of a company. In non-accounting terms this is the difference between what a buyer is prepared to pay and what the books indicate the company is worth. This difference is quantified at the time of sale and is recorded as 'Goodwill'.

(b) Investments: Listed under 'fixed assets' these relate to long-term investments, usually meaning longer than one financial period, such as loans made to other companies. Short-term investments, also known as 'money market investments', are shown as current assets.

(c) Total fixed assets: The sum of (a) and (b) above.

(d) Stocks: The stock of a company is all the useful components and raw materials used in manufacturing the product, and owned by the company. Some stock items devalue with time and, where applicable, correction must be made for this. Work in Progress (WIP) means materials and equipment that are not contained within the stores area because they are actually being worked on at that moment. Material in this situation must be accounted for and it is included in the value for stocks.

(e) Debtors: Someone who is shortly to pay you money is clearly a current asset. It is common to find a small percentage rate of non-payment, so-called 'bad debts' and a 'provision' for this may be made (see (m) below).

(f) Cash at bank and in hand: No possession can more clearly be a current asset than cash in hand or at the bank!

(g) Debenture loans: A debenture loan, like a share issue, is a way for the company to raise money. The company enters into an agreement with the lender to repay money at a fixed rate over a period of time in return for a lump sum. Where an asset of the company is offered as collateral to the lender, the debenture has become a mortgage; in this case the lender takes possession of the asset in the event of a default on the loan repayments. A debenture is issued for a specified period of time. The length of time will decide whether the debenture should be listed under current liabilities (as here) or below (see (l) and (m)). Although debenture loans are normally long term, they can also be bought and sold on financial markets, just as shares can.

(h) Creditors: Creditors are people to whom the organisation owes money. The normal period of credit is thirty days. Creditors, together with debenture loans, are normally regarded as 'negative assets' in the sense that they must soon be paid for and will consume cash. Such 'negative assets' are called *liabilities*.

(i) Net current assets: The net current assets are calculated by adding or subtracting from the current assets of the business the amount currently outstanding. This figure is of interest since it represents the pool of finance from which creditors can be paid. A company can have great wealth tied up in its long-term assets but still find difficulty meeting its operating expenses if its net current assets, sometimes called 'working capital', are too small.

(j) Total assets less current liabilities: This number is the total assets of the company less the liabilities falling due this financial period. It is the sum of (i) and (c).

(k) Finance loans: One common source of funds is a long-term finance loan. Such a loan is classed as a negative asset, rather than a source of finance, since it has associated repayment. Often such a loan will have scheduled repayments lasting many years.

(l) Creditors: It is quite possible to acquire long-term creditors in the same way as short-term ones. The purchase of major machinery, for example, frequently involves paying the supplier over some years.

(m) Provision for liabilities: Prudent management will make allowance this year for impending liabilities next year. This entry on the balance sheet is used to show money set aside for future liabilities that have arisen out of past transactions and are already known. Provision for liabilities is also called 'contingent liabilities'.

(n) Net assets: (sometimes called 'total assets') The title of this entry is self-evident given the nature of the balance sheet and is not usually given a heading.

(o) Called-up share capital: In this section of the balance sheet the source of the finance used to purchase the assets above is made clear. A common way for an organisation to raise money is to issue shares. The money so raised is shown as called-up share capital. This account contains capital raised from the sale of shares at their nominal value.

(p) Reserves: The total reserves of a company consist of the share premium account and the reserves transferred out of profit at the discretion of the directors. Together they comprise the shareholders' funds. This, along with the share capital, is the value of the shareholders' investment in the company. Reserves also include balancing amounts of money caused by, for example, revaluation of property.

(q) Share premium account: This shows the amount of money raised by issuing shares at a price above their nominal value. Such shares are 'issued at a premium', hence the name. Double-counting must be avoided and so the money raised by share sales at the nominal value is included elsewhere.

(r) Profit and loss account: Once the shareholders have ratified the decision of the Board of Directors concerning the amount of any share dividends which are paid for in cash, there will be an amount left to balance the balance sheet. This will be the profit and loss figure for that particular day. This is calculated as (n) minus the sum of (o), (p) and (q). This number is distinct from the separate financial document called 'profit and loss account', sometimes called the 'profit and loss, and appropriation account'. In the first financial year the value on the balance sheet will be the same as the retained profit figure from the profit and loss account. In subsequent years, the figure in the balance sheet will show the cumulative amount of money raised by the organisation over its trading life, whilst the profit and loss account will show the profit generated and retained during only the previous year. Just as the share entries show the total sources of finance raised by share issue, the profit and loss entry of the balance sheet shows the total sources of finance raised by profitable trading.

(s): The total sources of finance. Like the net assets the title of this entry is not usually printed out. It is, by definition, equal to (n).

PROFIT AND LOSS ACCOUNT	Y5	Y4
	£k	£k
Turnover	90.7	83.8
Net operating costs	71.5	66.6
Profit before taxation	19.2	17.2
Taxation	6.2	5.8
Profit for the financial period	13.0	11.4
Dividends	1.9	1.7
Profit retained	11.1	9.7

Figure 7.2 Profit and loss account for Jake Sounds

7.3.2 Profit and loss, and appropriation account

It is the aim of most traders to make a profit. They aim to do this by buying things, doing something to them to make them more valuable and then selling them at a sufficiently high price to cover not only the materials and labour costs of manufacture but also the costs of running the business. The money left over after these have been paid is profit. Profit may be disposed of however the organisation desires, perhaps as payments to shareholders, purchase of new assets for the company or bonuses to employees. It is usual for the owners to employ the profit in a way that benefits their organisation. The profit and loss account is the financial document which shows how much profit was made during the last financial period.

Jake has also prepared a profit and loss account for Year 5 of Jake Sounds, and this is shown in Figure 7.2. Unlike the balance sheet, the profit and loss account applies to a period of time and shows the cumulative effect of all the transactions over a period, usually a year. It records how much has been spent on, and how much has been received from, sales. Jake's account starts with turnover (all incomings) and subtracts net operating costs (all outgoings) to leave the profit. The rest of the account shows what happened to the profit ending with the amount retained by the organisation for re-investment. Dividends are payments made to shareholders. The profit retained by Jake Sounds in Year 5 can be seen to agree with the balance sheet figures and is equal to the difference between the profit figures for Years 5 and 4 ($53.5-42.4 = 11.1$).

The balance sheet and profit and loss account together form a meaningful summary of the organisation's financial activities. English law

Year	Sales	Cost of sales	Profit after tax	Loan repayment	Loan remaining
5	—	—	—	—	30
6	105	80	19	2	28
7	125	95	22	10	18
8	130	89	33	12	6
9	130	87	35	6	—
10	130	87	35	—	—

Figure 7.3 Cash-flow projection (New premises loan repayment — Jake Sounds)

requires that both documents are produced and contained in the annual directors' report.

7.3.3 Cash-flow projection

During his fifth year of trading Jake plans to buy new premises for the business. He has been operating out of his own house since the company began and now finds that the business is just too big to stay there.

Jake decides to borrow the money for this from the bank. They have asked him for some sort of justification for the loan and want to be convinced that Jake will be able to repay the loan together with the interest. To do this Jake provides copies of the balance sheet, profit and loss account, a report on how the new building will be used and a cash-flow projection. This last document is of greatest interest to the bank and it is shown in Figure 7.3.

In Figure 7.3, one can quickly see how many sales have to be made to allow Jake to repay the loan. The amount of profit consumed by the loan is clear and can be seen to be acceptably small. From a table such as this it is possible for the bank to cross-examine Jake and find out the likelihood of his being able to repay the loan through their thorough understanding of business.

In general, cash-flow projections are used to cover any situation where the distribution of cash is important. It is important to realise that it is perfectly possible for a healthy business to become insolvent simply by not making sure that sufficient cash is available where it is needed. In the same way that an army must not only have sufficient food but have it distributed appropriately amongst its men, an organisation must not only have enough cash, but have it distributed correctly to meet the expenditures for which it is required. If, for instance, there is not enough cash to meet the creditors who require imminent payment, insolvency may result even though the value of the whole company far exceeds the amount it owes.

There will often be intense debate about the numbers in the cash-flow projection, particularly the estimated number of sales. A bank making a loan will be keen to understand what evidence the organisation has for believing the predictions. A statement based on market surveys, previous experience, an analysis of competitors and customer research is much more likely to convince a lender to part with money than numbers made up by irrationally optimistic owners of new businesses.

7.3.4 Financial accounting ratios

We have seen how a balance sheet and profit and loss account can be prepared. We now have the ability to understand something of an organisation simply by looking at its published accounts. There are two parties who in general are interested in using the financial documents of a company to make decisions. The first is the organisation itself. The senior management of an organisation might well obtain copies of the financial documents of other organisations for comparison. They are concerned that their own organisation is going to fare well and seek to ensure that good decisions are taken by understanding the behaviour of their rivals. The second group of people are those interested in collaborating with the organisation in some way. For instance, these might be investors anxious that their investment will be profitable or they might be customers wanting to confirm the financial stability of a supplier.

The published financial documents of an organisation contain much information and the interpretation of them is a skilful process. To assist in analysing financial documents various ratios of the numbers contained within them have become commonly accepted as measures of important characteristics. The ratios are important in two ways: firstly, in absolute magnitude which betrays current performance; secondly, in their trend with time, which indicates likely future performance. Every section of industry has its own 'normal' values for these ratios and the relevance of comparing one organisation with another falls as the commercial differences between them increase.

There are a great many such ratios, too many to consider in this text. However, we will look at some of the most important, both as an introduction and to illustrate how ratio analysis in general operates.

Return on Capital Employed (ROCE): This ratio is defined as:

$$\frac{\text{pre-tax profit}}{\text{capital employed}}$$

Pre-tax profit can be read directly from the profit and loss account. It is common to use a profit figure before taxation and before interest in this ratio since this removes the investment performance of the organisation and assesses more directly the ROCE. The capital employed means the capital used in the creation of that profit — this is defined as total assets less current liabilities. The ratio is clearly of interest both to the organisation and to those who invest in it in any way. It indicates how much money was used in the generation of the profit. If the ratio is very large, the organisation makes a lot of money without using much in the process. If it is smaller than the prevailing rate of interest, investors would be better advised to place their money in a building society. If it is below the rate of inflation, the value of an investment in the organisation is actually decreasing.

Profit margin: This ratio is a basic performance indicator of the potential for the ROCE. It is defined as:

$$\frac{\text{trading profit}}{\text{value of sales}}$$

If the profit margin is high then each sale is generating much profit and if the rest of the company is efficient then the ROCE should be good. If it is low, then no matter how efficient the administration of the organisation, a high ROCE is unachievable. The ratio can be expressed in many ways and is improved by making it relate as closely as possible to the profit from sales without including any other activities of the business. Because of this the 'gross profit margin' is often used which is defined as:

$$\frac{\text{value of sales} - \text{cost of sales}}{\text{sales value}}$$

Stock turnover: Defined as:

$$\frac{\text{sales}}{\text{average inventory}}$$

This ratio indicates how much time stock is typically spending within the organisation (sometimes this is described as how fast the stock is 'being turned'). This factor has much importance in production control, financial efficiency and the market responsiveness of the organisation. In some industries it may be very fast, with the entire stock being purged every few weeks, in others it is much slower. In general the ratio is very dependent on the nature of the business but all organisations seek to minimise it since a large value implies that a lot of stock is being held which is financially less efficient.

Gearing: The ratio is defined as:

$$\frac{\text{fixed-term loans}}{\text{Net assets}}$$

This ratio is used to express how heavily an organisation is committed to repaying borrowed money. When the ratio is high, most of the profit is being diverted into repaying loans. If the ratio equals zero, the organisation has no repayment commitments and is entirely self-funded.

Liquidity ratio: Sometimes this is called the 'Acid Test'. It is defined as:

$$\frac{\text{current assets} - \text{stocks}}{\text{current liabilities}}$$

This ratio measures the organisation's ability to meet incoming debts. The top line indicates ready cash and the bottom incoming bills. In accounting, 'liquid' is used to mean 'readily converted into cash'. The current assets (also called ***liquid assets***) may be defined as either cash plus debtors plus short-term investments, or as current assets less stocks less prepayments. If the ratio is less than one, it is almost certain that the organisation will receive more incoming bills than the ready cash can pay. This can lead to instant insolvency. A value of about two is considered safe, much higher than this and the organisation is unnecessarily well covered and probably has capital tied up in its bank account not doing any useful work.

Output per employee: Many versions of this ratio are used, a common one is:

$$\frac{\text{Value of sales}}{\text{number of employees}}$$

In general all the ratios provide a crude overall indicator of the organisation's efficiency which allows comparison with competitors. The ratio values are very dependent on the nature of the business and all markets have their own minimum safe levels that organisations should exceed in order to be financially viable in the long term.

7.4 Budgeting

7.4.1 Master budgets

We have now seen how important the financial condition of an organisation is. In order to achieve its mission statement, it is crucial that the financial arrangements of the organisation serve its needs. Such arrangements will not just happen along and they must be planned for. The organisation must define the financial position it intends to achieve and then head towards it. It is not sufficient simply to aim to make each sale at a certain profit or perhaps to have a target output per employee. These are certainly sensible things to control but in themselves are not enough. To be successful the organisation must start with comprehensive plans that make clear the route to the desired value of all the important financial parameters. Of course, many things may go wrong along the way and simply having the plans by no means guarantees success. However, the organisation is much more likely to achieve its goals if good plans are prepared first. Financial plans of this kind are called ***budgets***.

A master budget is the starting-point and is the name given to a balance sheet, profit and loss account and a cash budget which have been prepared, as a group, to describe a desired future financial position. A cash budget is a projection of how much money will be in the bank account over time as a result of all the activities that the organisation is planning to undertake. From this budget, important issues such as the desired asset value of the organisation, the desired sales value, the desired ROCE, and all the other important values can be seen.

Once prepared, the master budget enables the objectives of each department to be quantified. A self-consistent group of objectives can be prepared that, when complete, will result in the desired financial situation. In this way the organisation's finances are directed at achieving the mission statement. This is clearly a superior approach over simply undertaking a lot of financially profitable ventures and seeing how much money can be generated.

7.4.2 Re-employing the profit

One of the figures within the master budget needs particularly careful consideration, the profit. Profit is usually re-employed in the organisation to assist in achieving the company's goals. The organisation will normally have many exciting possible enterprises into which resources could profitably be directed. It is not a question of finding something profitable to do with the profit, it is a question of choosing wisely a portfolio of the activities that will provide the greatest likelihood of achieving the mission statement. Once the master budget has been prepared and the budgeted profit is known, the organisation must choose exactly how this money will be re-employed as it is generated.

Case study: Tachatronics budgeting

Every year the Tachatronics Corporation have an intensive month of budgeting in which plans for the next year are agreed. Last year was good and prior to the 'budget month' the management met and expanded the mission statement to include the Japanese market from which they had previously excluded themselves. Next year's budgeted profit is £400k.

The company's five divisional managers put in preliminary budgets, each manoeuvring for the most cash, totalling £615k. The budgets are for money to spend on activities that meet the departmental objectives and do not include salaries. The senior management decide that to penetrate the new Japanese market a new 'Oriental Sales Division' employing four people should be created. The five existing divisions are Sales, Production, Research, Service and Administration. Money cannot be spared from the modest Service or Administration budgets and so priorities must be set between the remaining departments: firstly, to allow for the new Oriental Sales Division and, secondly, to bring the total budgeted expenditure of profit within the £400k available.

The senior management consider the provisional plans of each department. The Production Division has several projects in progress. First on the list is the capital-intensive acquisition of a new blanking machine tool whilst lower down is a value engineering programme. Sales forecasts are not as large as when the new machine was first considered. The Research Division has had four major programmes in progress and plans to continue all of them with increased expenditure, arguing that they will all produce products. Many such details are considered and tough decisions about which projects will stay and which will go are agreed. The decisions are not unanimous and the Board have to face the fact that they cannot afford to follow all the interesting avenues that have been presented. After several days, a package of objectives is chosen for the next twelve months, and the following draft divisional budgets then result.

> Sales £54k: Objective: Close contracts worth 10 per cent more than last year's contracts whilst using the same number of staff. This is to be achieved through the introduction of new technology to the division.
>
> Oriental Sales £57k: Objective: Close four contracts in Japan for existing products. Do this by identifying an agent already trading and enter the market by collaborating.
>
> Production £115k: Objective: Complete all deliveries within two months of receiving sales order and cut direct production costs by 7 per cent at the half-year by introduction of value engineering programme.
>
> Research £105k: Objective: Produce two complete products, including specifications, drawings, test data and any other documentation required to completely specify the products. The two are to be chosen from the five currently in progress. This selection is to be done in conjunction with Sales to identify best choices and complete reports should be written on the other three. The first product is to be completed by the end of month 4, the second by the end of month 9.
>
> Service £59k: Objective: Answer all calls within two hours. Have an engineer

on site within twenty-four hours for Europe and the USA. All breakdowns to be repaired in two days of service engineer's time.

Admin £10k: Objective: Complete all administrative tasks within one day of each request. Ensure pay-roll activities run error-free for the whole year.

TOTAL: £400k

The divisional managers discuss the draft corporate plans and after two weeks' intensive planning and organising with their teams they put forward some modifications, particularly to time-scales of introduction. The managers have each divided the objectives up into separate projects and have plans for each. The plans are presented to the Board in turn, at which some further modifications are made, but within the month plans for every division have been agreed in detail. Senior management are confident that the objectives serve their mission statement and the divisional managers are clear both about what has to be done, and what financial resources are at their disposal. ■

7.4.3 Individual budgets

Once divisional budgets have been prepared in the way described above, all the departmental managers have their own resources allocated in an indisputable way. Every manager knows what is expected of their own department and every manager knows how much money is available for use in achieving their objectives.

Chapter 8 deals with the planning of projects and describes some techniques for managing them. By using these project-planning techniques and estimating the cost of each undertaking, a departmental manager is able to produce an estimate of the total financial requirements of the projects. Such estimates are the basis of the budget allocated to the department. As the financial year progresses each department directs its resources at achieving its objectives. If all departments are successful the objectives of the organisation are met. Of course things never go exactly as predicted and it may become necessary to readjust budgets during the year. This does not mean the budgeting process was in some way wrong or is pointless; rather it shows that it is difficult to be accurate when planning the future and it is certainly true that it is better to know exactly what it is that one is changing and why.

7.5 Management accounting

In this section we will look at some of the financial data and information that is generated and used by engineers in the areas of costing, and investment appraisal; we will also look at what is meant by depreciation and how it is calculated. These are the most important topics in management accounting. The information that is generated through management accounting techniques is not normally divulged to people outside the company since it usually contains sensitive information. Knowledge of issues such as pricing structures or cost of manufacture could be used effectively by the organisation's competitors, or customers, to undermine profitability.

7.5.1 Costing

Costing is the process of calculating how much something costs to make. The most important use of this information is in determining a realistic selling price, usually with the aim of making a profit. In addition it allows you to determine which product areas are worth investing in and provides information on which to base management decisions relating to making or buying-in parts. It also allows you to decide which products need further development to give them the desired financial profile. The term 'costing' generally applies to existing products where manufacturing techniques are known and real data may be used. Before this time, costs are prepared by 'estimating'.

The role of engineers in costing and estimating can be very significant depending upon the type of company and products that are involved. Invariably, if you design a product you will have to calculate how much it is going to cost to make so that the company can decide whether or not to proceed with it. In a company that manufactures customised products an engineer may also be involved with estimating costs for customer quotations.

Costing is not as straightforward as might at first appear. Most people instantly think of some of the components of cost, materials and labour, for example, but these by no means explain how much something costs to make. What about necessary services that do not directly contribute to the product — lighting and heating for example? One might suggest that these should be totalled up and then distributed evenly amongst the products of the organisation but what if the products consume them unevenly or some not at all? Costing requires that you consider all the factors affecting the cost of the product: these are the amount of labour involved in its manufacture, the parts that go into it, any services that have to be bought in in order to manufacture, and how much it is costing to keep the factory operating in order to make that product. We will now explain two methods that are used for calculating costings. The first is based on direct and indirect costs. The second method is based on variable and fixed costs.

Direct and indirect costs

In this method the cost of a product is divided into two components, the direct costs and the indirect. The total price for the product is calculated by firstly producing a cost for each element. In the first case this is relatively straightforward but in the second is more complex since it is necessary to find a fair way to distribute the indirect costs, things such as lighting and heating bills, to the different products. The total cost of the product is produced by adding the two components together.

Direct costs: Direct costs are those that can be directly attributed to the product for which a cost is being prepared; they can be considered under the headings of *labour, materials* and *expense*.

Direct labour is that done usually by operatives or technicians who book their time to the job that they are working on. Costing of direct labour will involve determining the amount of time spent on that product and then deciding what labour rate applies. The time information can be gathered by reviewing time sheets or route cards. The labour rate is often more difficult because of the variability in the wages that different people will be paid. One way of dealing with this problem is to calculate a standard labour rate, based on the average wage bill, which will then be applicable for different processes or workshops. Some companies simplify this further by having one labour rate for all direct labour in the factory.

Direct materials are those which go only into that product; for example, if you imagine a china cup, the direct material will be so many grams of china clay. The direct materials will be detailed in the bill of materials prepared for the product or part. Direct material cost can be determined by considering the cost of making or buying the actual parts that were used in that product, although this may entail some rather onerous tracking of parts. Alternatively when you are using parts from stock which may have been purchased at different times and at different prices you can use standardised costs. Cost standardisation can be on the basis of the cost of the last delivery of parts, although this may lead to discrepancies where there are large fluctuations in price, or it could be on the average cost of stock held.

Direct expense relates to costs incurred by buying in services for that particular product and so would include subcontract machining costs, printing costs, finishing costs and so on.

Example

Calculate the direct costs involved in the manufacture of product A from the information below.

Material costs	£5.50
Standard labour rate	£7.50/hour
Finishing costs	£2.00
Total processing times	turning centre 14 minutes assembly 3 minutes

Answer:

Direct labour $= \dfrac{17}{60} \times 7.50 = £2.13$

Total direct costs = materials + labour + expense
= £5.50 + £2.13 + £2.00
= £9.63

Indirect costs: The indirect costs, also called 'overheads', are the costs that are associated with operating the business but that cannot be directly assigned to product or services being provided. As with direct costs there are three categories of overhead; these are *materials, labour* and *expense*.

For example, indirect materials might include things like solder flux, rags and cutting fluid. Indirect labour would include design engineers, supervisors and secretaries. Indirect expense would include heating, lighting and rent for premises. These are all costs to the business and the way in which funds will then be acquired to cover them is called *overhead recovery* and requires a two-stage process which ends up with a charge for overheads being made on products. The two stages involved are *apportionment to cost centres* and *absorption by product*.

Apportionment of overhead to cost centres: A cost centre is a part of the business that can be identified for the purpose of determining costs. A cost centre may be a whole factory, a particular department or a specific machine. Each cost centre should bear a fair share of the indirect costs, with the costs being apportioned on the most realistic and convenient basis that can be devised. Typical bases for apportionment might be floor area for apportioning rent, the number of employees for indirect labour and personnel costs, and the book value of assets for depreciation and insurance.

Example

Apportion the overheads for a company having the following departments:

	Department overhead	People	Floor area (m^2)	Direct materials
Personnel	£50,000	4		
Purchasing & Stores	£75,000	6		
Building Services	£20,000	3		

Light Machine Shop	£85,000	12	500	£200,000
Assembly Shop	£65,000	20	1000	£130,000
Press Shop	£90,000	12	1000	£150,000
Totals	£385,000	57	2500	£480,000

The only areas that would be directly involved with product are the three workshops and therefore the overheads for the other areas must be apportioned to these.

Considering Personnel:

This has an overhead cost of £50,000 and because it is dealing with people in the factory a fair way to apportion this would be on the basis of the number of staff in each area.

There are 53 people in the company, excluding Personnel, and the overhead is therefore distributed on the basis:

Purchasing & Stores: $\frac{6}{53} \times 50{,}000 =$ £5,660

Building Services: $\frac{3}{53} \times 50{,}000 =$ £2,830

Light Machine Shop: $\frac{12}{53} \times 50{,}000 =$ £11,321

Assembly Shop: $\frac{20}{53} \times 50{,}000 =$ £18,868

Press Shop: $\frac{12}{53} \times 50{,}000 =$ £11,321

This apportionment now gives the total overheads for the Purchasing & Stores as £80,660, and Building Services as £22,830. These will be apportioned to the three workshops.

The Purchasing & Stores overhead will be apportioned on the basis of direct material cost and the Building Services overhead on the basis of floor area. This gives:

	Department overhead	Purchasing overhead	Building overhead	Total overhead
Light Machine Shop	£96,321	£33,608	£4,566	£134,495
Assembly Shop	£83,868	£21,846	£9,132	£114,846
Press Shop	£101,321	£25,206	£9,132	£135,659
Total				£385,000

All the overheads are now apportioned to production departments from which they can be included as a cost of the product.

Absorption of overhead by product: In order to include the indirect costs in the product costing they are charged to individual products passing through the cost centre to which they have been apportioned. This is called overhead absorption because the costs are absorbed by the product cost. To calculate overhead absorption an absorption rate, also called *overhead recovery rate*, is calculated which is based on a standard or expected rate, or volume, of production. This recovery rate is then applied to the actual rate of production. Common bases for overhead absorption are direct labour (hours or cost), units of output, and machine hours. These would be used by calculating appropriate recovery rates as described below.

Material rate: In this method the overhead is divided by the actual or expected cost of the direct materials.

Example

For a light machine shop the total annual cost of direct materials is expected to be £200,000 and the annual overhead to be absorbed is £134,495, thus we can calculate a recovery rate as:

$$\frac{£134{,}495}{£200{,}000} = 0.672$$

Therefore for every job done in this shop we add an overhead charge of £0.672 for every £1.00 of direct materials. Thus, for job ABC:

Direct labour	=	£4.25
Direct materials	=	£7.25
Overhead (£7.25 × 0.672)	=	£4.87
Total cost	=	£16.37

Labour hour rate: In this method the overhead is divided by the total labour hours estimated.

Example

The annual overhead for an assembly shop is £114,846 and the estimated number of man-hours to be worked in a year is 35,520, therefore the labour hour overhead recovery rate will be

$$\frac{£114{,}846}{35{,}520} = £3.23 \text{ per labour hour}$$

Thus, for job ABC:

Direct materials	=	£7.25
Direct labour (30 minutes)	=	£4.25
Overhead $\frac{30}{60} \times 3.23$	=	£1.62
Total	=	£13.12

Machine hour rate: In this method the costs of owning a machine are to be absorbed by charging them per hour of operation expected.

Example

The running costs of a turning centre are as follows:

Power	= £0.25 per hour
Consumables	= £400 per year
Maintenance	= £850 per year
Tooling	= £3,500 per year

In addition the machine is depreciated at £4,000 per year. The expected run-time of the machine is 242 days per year with one eight-hour shift each day.

Machine costs per year	= 400 + 850 + 3500 + 4000
	= £8,750
Total hours per year	= 242 × 8 = 1936
Machine overhead recovery rate	= $\frac{8750}{1936} + 0.25$
	= £4.77 per hour

Thus for job ABC:

Direct materials	= £7.25
Direct labour	= £4.25
Machining time 30 minutes	
Overhead $\frac{30}{60} \times 4.77$	= £2.39
Total cost	= £13.89

Full costing

A full costing is the cost of a product based on both direct costs and overhead costs, as shown in the previous examples. It is an onerous and time-consuming task, particularly because of the overhead apportionment and absorption calculations and the subjective way in which bases for these calculations are chosen. However, it is only by following such a procedure that a rational answer can be given to the question, 'How much does the product cost to make?'

Variable and fixed costs

In the previous paragraphs, we have looked at how the real cost of a product may be calculated — cost that includes not only the direct elements of cost, but also the indirect ones. Another way of looking at the costs involved in manufacturing is to consider those that are variable and those that are fixed.

The variable costs are those that vary in proportion to the amount of output and so would include direct materials, use of temporary labour and overtime charges. The fixed costs are those which have to be paid irrespective of output and would include the cost of employing people and having the premises and machinery available.

The difference between the variable costs and the income from sales is called the *contribution* — it is the money available for paying fixed costs and providing profit. Some companies use a method of costing called *marginal costing* which only looks at the variable costs, the cost of producing one more unit of output.

Marginal costing is easy to understand and simplifies budgeting. In addition it lets you quickly identify low-contribution products; however, against this there are problems associated with people not understanding the difference between contribution and profit and it is possible to sell below full cost without realising it.

By considering the fixed and variable costs associated with manufacture you can determine the break-even point for the manufacture of any product. This is the point at which the income from sales is equal to the cost of manufacture, where no profit or loss is made, and the company breaks even. Break-even analysis is an extension of marginal costing as it allows you to determine the probable profit at any level of output. A break-even chart is shown in Figure 7.4.

The break-even point can be calculated using the formula:

$$B = \frac{F}{S - V}$$

where: B = break-even point
F = total fixed costs
S = selling price per unit
V = marginal cost per unit

The sales revenue at the break-even point $= F \times \dfrac{S}{S - V}$

7.5.2 Investment appraisal

Investment appraisal, which includes any capital expenditure, is a way of analysing the financial value or cost of investment decisions. It is a technique which looks at the financial return on an investment. This type of technique is not to be used on its own, as there will always

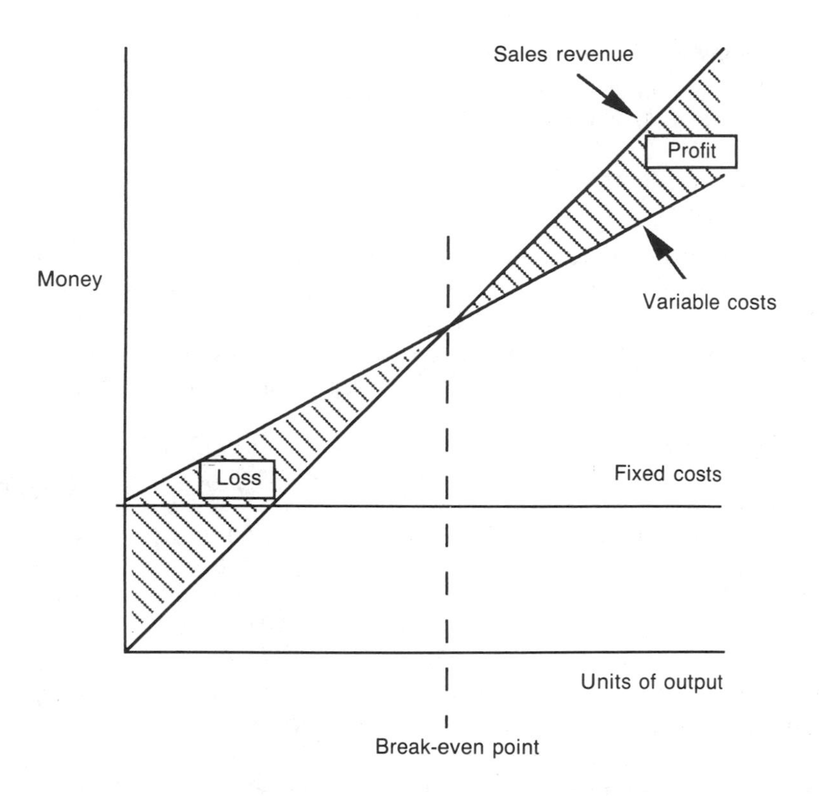

Figure 7.4 Break-even chart

be many factors that affect a decision; however, it does have value in that it is an objective assessment. Investment appraisal will help particularly in comparing investments and in forcing you to think about the cost of a particular decision, although appraisal is not a costing exercise in itself.

As an engineer you may be involved with considering such things as the cost and returns associated with purchase of a new piece of equipment or with a project for introducing a new system or product. We will look at two techniques that can be used for investment appraisal: the pay-back period and discounted cash flow.

Pay-back period

The pay-back period is the amount of time that a project must run for in order that the cash generated will repay the initial investment. Pay-back is calculated on the absolute values of the expected income after tax.

Example

Investment	=	£1000
Annual return	=	£200
Pay-back $\frac{1000}{200}$	=	5 years

This is a very simple technique and is therefore widely used. It shows that early returns are generally preferable to longer-term ones. However, it has major drawbacks in that it does not recognise that money devalues with time, it does not take into account any earnings after the pay-back date, and it ignores the timing of earnings prior to the pay-back date.

Example

		Project A	Project B
Capital investment		£7,000	£7,000
Returns:	Yr 1	£5,000	£2,000
	Yr 2	£2,000	£5,000
	Yr 3	£2,000	£6,000
	Yr 4	£8,000	£3,000
Total		£17,000	£16,000
Pay-back period:		2 years	2 years

In this example the pay-back periods are the same since after two years both projects have returned the original £7,000. Yet the total returns are different in both amount and profile. We need to know which one is best and by how much.

Discounted cash flow

Discounted cash flow looks at the earnings over the life of the project, the time at which the returns are received, and it allows for the fall in the value of money over a period of time. The fact is that £1.00 today is worth more than £1.00 would be in a few years' time. This comes about not because of inflation, which evens out over a period of time and also affects both costs and profit equally, but because £1.00 invested today will attract interest at some rate and therefore the investment will increase in value.

If the interest rate is 10 per cent we can invest £5.00 today and in five years' time the investment will be worth £8.05, assuming that the interest is immediately re-invested at the same rate and that the capital is left untouched.

In a similar way you can work backwards and say that if the interest rate is 10 per cent then £5.00 payable in five years' time will be worth

£3.10 today — this is called the *present value*. Tables are available which allow you to look up present values; however, they can be calculated using the formula

$$P = \frac{S}{(1+i)^n}$$

where: P = present value
S = sum in the future
i = interest rate (also called *discount rate*)
n = number of years

Example

An investor can get 8 per cent interest p.a. by investing money in the building society. The investor is offered an alternative investment by another company which will provide £388 at the end of each of the three following years. In order to calculate the value of this second investment you need to calculate what it is worth today and then compare this with the cost of the investment. As follows:

$$P = \frac{388}{1.08^1} + \frac{388}{1.08^2} + \frac{388}{1.08^3}$$

$$= £1,000$$

The present value of the investment is £1,000, therefore if the capital cost of the investment is less than £1,000 then this will make more money than if the capital was invested in the building society. The difference between the present value of the returns and the capital cost is called the Net Present Value (NPV).

Example

An investment of £13,000 yields returns of £4,000 per year for the first three years and £2,000 per year for the next two years. By calculating net present values determine whether this is a worthwhile investment, if the money could otherwise be invested at 5 per cent per annum, or at 10 per cent.

At 5 per cent:

$$P = \frac{4000}{1.05^1} + \frac{4000}{1.05^2} + \frac{4000}{1.05^3} + \frac{4000}{1.5^4} + \frac{4000}{1.05^5}$$

$$= £14,105$$

Therefore, at 5 per cent net present value:

$$= £14,105 - £13,000$$

$$= +£1,105$$

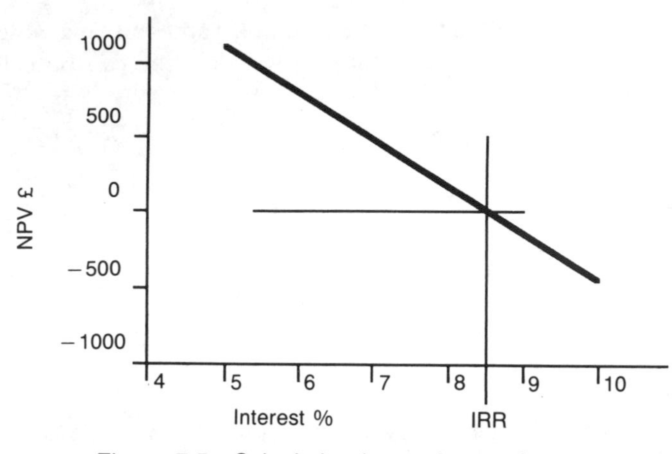

Figure 7.5 Calculating internal rate of return

If the money could only be invested elsewhere at 5 per cent then this is a worthwhile investment as it would bring a positive net present value, i.e. a profit.

At 10 per cent:

$$P = \frac{4000}{1.1^1} + \frac{4000}{1.1^2} + \frac{4000}{1.1^3} + \frac{4000}{1.1^4} + \frac{4000}{1.1^5}$$

$$= £12,555$$

Therefore, at 10 per cent net present value:

$$= £12,555 - £13,000$$

$$= -£445$$

If the money could be invested elsewhere at 10 per cent then this investment is not so profitable and the other investment would make more money.

Internal rate of return: This is the discount rate, or rate or interest, at which the net present value is zero. It is the point at which you break even. It can be calculated simply as shown below, which uses the previous example.

At 5 per cent	NPV = +£1,105
At 10 per cent	NPV = −£445

Using this information we can draw a graph as shown in Figure 7.5.
IRR = interest rate at which NPV = 0
By simple extrapolation:

$$IRR = 5 + 5 \times \frac{1105}{1105 + 445} = 8.56\%$$

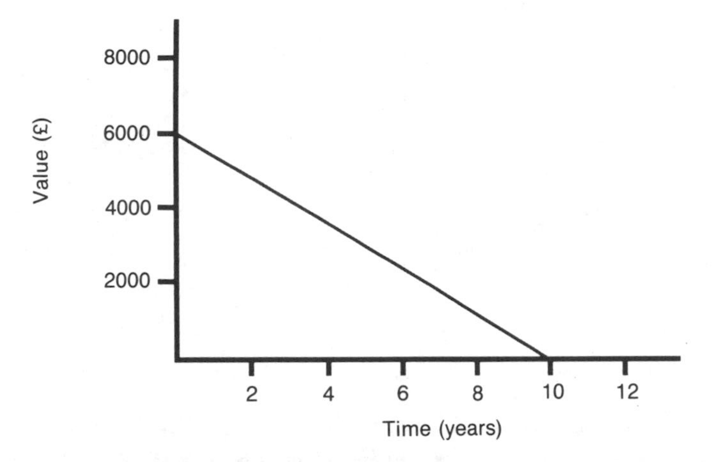

Figure 7.6 Straight line depreciation

The internal rate of return is 8.56 per cent, therefore if your money could earn more than 8.56 per cent elsewhere over the period of the investment, you would be better advised to take that alternative rather than the investment proposal described above. However, if you believe that the money would earn less than 8.56 per cent then the proposal is a good investment since you will earn less elsewhere.

7.5.3 Depreciation

When considering investments they are looked at in terms of the cost to buy and then in successive years they are considered as assets to the company which have a decreasing value. The term used to describe this reduction in value is *depreciation* and there are accounting techniques which are used to calculate it. Some factors that give rise to depreciation are wear and tear, shelf-life, and damage due to misuse. Depreciation must be incorporated in the balance sheet to show the reduction in asset value. Stock is not depreciated because it is valued annually using a stock-take.

There are many ways of calculating a charge for depreciation and companies can choose which one they like but they should be consistent across the company. We will look at three methods that are used.

Straight line method

In this method the asset is depreciated by equal annual charges spread over the asset's estimated life, as shown in Figure 7.6.

Example

Cost	=	£6,000
Scrap value	=	£400

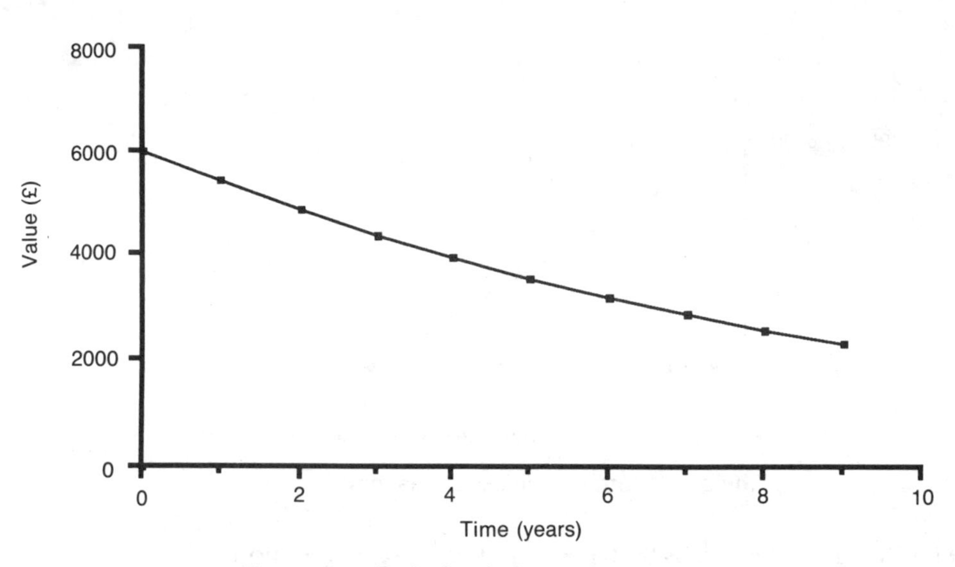

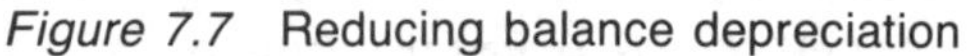
Figure 7.7 Reducing balance depreciation

Estimated life = 10 years
Annual depreciation:

$$\frac{6000 - 400}{10} = £560$$

Reducing balance method

In this method the depreciation charge is a constant proportion of the balance remaining at the end of each accounting period, as shown in Figure 7.7.

Example

Initial cost = £6,000
Depreciation rate = 10 per cent

Value	Year
£6,000	0
6000 − 10% = 5400	1
5400 − 10% = 4860	2

Production unit method

In this method the depreciation charge is based on the number of units of work produced, multiplied by a constant rate for each unit.

Example

Cost	=	£13,000
Scrap value	=	£500
Estimated production	=	250,000 units

Depreciation charge

$$\frac{13000 - 500}{250000} = 5\text{p/unit}$$

7.6 Summary of Chapter 7

In this chapter we began examining where the need for financial control comes from in the first place. It is clear that, even as individuals, we need some sort of financial control simply to ensure that we avoid living beyond our means. In the case study we found that a business like that of Jake Sounds has need for a far greater degree of control than we need for ourselves as individuals. We finished Section 7.2 by defining what an ideal financial system must provide.

In Section 7.3 we examined the area of the financial profession which provides information about the organisation to those in the outside world. In the section we explained the use and preparation of the three most important financial accounting documents. These were the balance sheet, the profit and loss account, and the cash-flow projection. By the end of the section we were able to prepare these documents for ourselves. The interpretation of such documents is a skill in its own right and in the last part of this section we met the concept of accounting ratios, which provide an indication of how particular areas of the business are performing.

In Section 7.4 we saw how an organisation can use the budgeting process to control its activities in a logical way. We found that there was common ground between the use of budgeting and management by objectives. In the case study we saw how a particular organisation set objectives for each department and budgets at the same time. At the end of the budgeting process each department had a very clear and unambiguous statement of what was to be achieved over the next year and what financial resources could be employed. This process was relatively rapid and by using budgeting in this way planning the next year consumed an insignificant portion of it.

In Section 7.5 we explained how financial data that is required for internal use may be prepared. The managers of the organisation must have a rational basis upon which to judge the value of future investments and with which to prepare costings of present products. This section explained the use of the management accounting techniques of costing, investment appraisal and depreciation.

7.7 Revision questions

1. Why is profit not an asset?

2. In the example of Jake Sounds a balance sheet was prepared for a very few financial transactions. How might preparing a balance sheet for a company that has conducted hundreds, or even thousands, of transactions differ from the way the one in the example was prepared? How might this process change further if the size of the organisation gets larger still?

3. When preparing a summary of an organisation's financial activities, what advantages and disadvantages of using a balance sheet are there?

4. Arrange the following data into a balance sheet.

Finance debt	325
Long-term mortgage	150
Permanent equipment	600
Stocks	1200
Profit and loss account	1617
Cash	850
Creditors	1500
Three-month bank loan	75
Called-up share capital	460
Depreciation on equipment	23
Buildings	500
Reserves	50
Debtors	1050

5. In an essay of less than 2,000 words: (i) list the sorts of records an engineering company must keep in order to make the preparation of a balance sheet possible; (ii) from your knowledge of the various departments within a company, discuss the ways in which the different departments might provide this data.

6. A balance sheet is often used by analysts to examine a company from a financial point of view. Describe the strengths and weaknesses of the document for this purpose.

7. What is the basic equation of a balance sheet? How does a balance sheet work? Why is it needed? What advantages does it bring? What disadvantages does it bring?

8. How is a profit and loss account arranged? What does it show? Why is it useful? How does it differ from a balance sheet? What is the difference between a profit and loss account as a financial document and the entry on a balance sheet called 'profit and loss'?

9. When might a cash-flow projection be useful to an engineer? How is it prepared? What limits the accuracy with which it might be prepared?

10. Why is the cost of a product not as straightforward to calculate as it appears at first? Describe two methods of preparing product costs.

11. What do the following terms mean:

Direct costs	Internal rate of return
Break-even point	Machine hour rate
Contribution	Marginal costing
Direct labour	Overhead recovery rate
Discount rate	Production unit method
Indirect costs	

12. Why are budgets prepared? Describe a process by which they may be prepared. What advantages do they bring? How do budgets affect senior management? How do they affect engineers in the organisation? How do they affect junior staff?

13. Eruco Ltd manufacture luxury cat toys. Their electric mouse department is particularly profitable, producing 150,000 mice per year at 30 per cent profit.
The bill of materials for an electric mouse gives the following information.

Part	Quantity	Cost/unit
Fur fabric	$0.0225m^2$	6.50
Motor	1	0.20
Revolving eye	2	0.03
On/off key	1	0.03
Plastic tail	1	0.01
Frame	1	0.02

The mice are made in batches of 500, each taking 43 hours. Labour is charged at £5 per hour.
Overheads for the mouse department are £73,000 per annum and are absorbed on a direct materials basis.
Calculate the manufacture cost and the selling price of an electric mouse.

14. The overheads for a subcontract machinist are:

Cost centre	No. of people	Direct material	Total overhead
Admin & Finance	3	0	75,000
Sales	1	0	25,000
Capstan Shop	5	80,000	90,000
Press Shop	5	60,000	120,000

Apportion the overheads to the cost centres and calculate the cost of job XN01 on the basis of the following information.

Job XN01:	Direct materials		£15.13
	Direct labour	Capstan Shop	30 mins @ £6/hr
		Press Shop	15 mins @ £7/hr

Overheads are to be absorbed on the basis of direct labour. The factory works forty-seven weeks a year and operates an eight-hour day.

15. A company is offered two investments: the first requires a capital outlay of £40,000 and will bring a return of £12,000 per annum for the following five years. The second investment requires a capital outlay of £60,000 and will bring returns of £15,000 per annum for the first two years but this will fall to £10,000 for the next six years. Assuming the rate of interest is 10 per cent calculate the net present value of each of these investments.

On the basis of net present value, which would you suggest is the best investment for the company? Would your advice change if it was likely that the interest rate would increase to 15 per cent during the life of the investments?

7.8 Further reading

Droms, W.G. (1990), *Finance and Accounting for Non-financial Managers*, 3rd ed., Addison-Wesley (ISBN 0-201-55037-7).

A readable 250-page textbook introducing finance and accounting and requiring no specialist knowledge. The book is in six sections. The first introduces financial management and tax. The second covers the fundamentals of financial accounting, the third financial analysis and control. The fourth section covers decisions regarding working capital, the fifth long-term investment decisions and the last long-term financing decisions.

Chadwick, L. (1991) *The Essence of Management Accounting*, Hemel Hempstead: Prentice Hall International (ISBN 0-13-284811-2).

A 170-page distillation of management accountancy ideal as preparatory reading for a short course or for reference. Very compact and synoptic in its approach.

Chapter 8
Project planning and management

Overview

Projects are the contemporary solutions to resolving the conflicting needs of meeting requirements with limited resources and time constraints. All engineers are involved in projects at some stage of their career or undergraduate programme. In this chapter we look at projects, their definition, and at techniques that can be used to plan them. Through the media of a case study we also look at the way in which projects are controlled and executed in industry.

8.1 Introduction

The marketing department of Vammu Tins Ltd has identified a niche in the market place which could be filled by a new product. The product identified is a self-evacuating tin that could store biscuits indefinitely. This product would be new to Vammu Tins as previously they have only produced conventional tins and boxes *and* it would require some considerable investment on the part of the company. We will look at how Vammu Tins can make a decision about whether or not to proceed.

Firstly the Board of Directors will need information so that they can decide if the project is feasible; they need a project proposal. It will have to identify what the aim of the project is, what it is going to cost the company, and what benefits will accrue; it will also have to show that the project is achievable.

A project specification should be used to identify the aim of the project. This is the document that sets the limits on the project and that can be used by others to evaluate the implications of the project. In the case of Vammu Tins the project specification would be drawn up by personnel from both the marketing and engineering departments. The engineers will be able to make an assessment of the broad activities that will have to be carried out in order to execute the project, such as designing, drawing, prototyping, and finally manufacture. They will be able to make an assessment of the resources required for each of these activities, and they will be able to give an estimate of the cost of the engineering aspects of the project. The marketing personnel will be able to identify the benefits that can be obtained from successful introduction of the new product. They will also identify any constraints, such as deadlines, as well as any marketing activities that have to be carried out and the costs associated with these.

All of this information can be put together in the form of an outline project plan which will show whether or not it is feasible to meet the deadlines with the resources available.

The specification, the cost–benefit analysis, the resource requirements, and the outline project plan form the project proposal. The Board can then use this to decide whether or not to proceed, knowing the implications of their decision for the company.

If the Board decide to go ahead then the project team can get to work. However, the deadlines set in the proposal have to be met, specific tasks have to be identified, resources and work have to be planned. In order to do this the project team has to carry out detailed project planning and resource scheduling. This is particularly important when, as in this case, a number of people will be involved since they all need to know what they are required to do and when.

The level of planning done at the proposal stage depends upon the type of project to be considered, but in general if it is very speculative, a high risk for the company, or requires a lot of investment then more detail will be required.

Planning after the proposal has been accepted is much more detailed but the level of detail will vary with the time to the project completion. For example, if the Vammu Tins project is estimated to take six months it will be possible at the start of the project to schedule tasks for the first month on a weekly, or daily, basis; beyond the first month the schedules may be given on a monthly basis. However, the project plan should be reviewed throughout the project and can be revised to include more and more detail. So, after the first week the schedule for the next month is given on a daily basis but for the remaining four months and three weeks a monthly schedule is outlined.

The alternative to planning would lead to a great deal of uncertainty for the company. The Board of Vammu Tins would not know if the new product is feasible and might decide against it, missing a major opportunity, or they might decide to design the product and find that it consumes more money than it could ever make. Worse still, they might realise that it is feasible and work on the design only to be beaten by the competition because the engineering and marketing departments each thought the other was working on a particular aspect of the project when neither of them were.

Projects are a major part of an engineer's job, whether a product development project, like the Vammu Tins example, or a more specialist one such as the installation of a new piece of machinery. Projects vary in size in terms of the time that they take, the resources that they require, and the impact that they have upon the business. However, whatever the subject matter of the project it should be planned to ensure that the resources are used in the most effective way to achieve the desired result.

Most projects, such as the design of a new product, will have very well-defined goals; however, some are more speculative such as researching into the ways of manufacturing high-temperature superconductors for the electronics industry. An industry-based project will always require justification in terms of there being some form of reward for the company. Often this will be in terms of financial reward; however, some projects are done because of more intangible benefits such as improving teamworking skills by providing interpersonal skills training. The main reason for the project may be defined for you, from the corporate objectives; however, you will be constrained by resources, including time, and of course you will want to do the job well. Doing a good job starts with good planning. This chapter will help you to plan well.

The first step of project planning relates to what is to be done and why, and usually results in some form of project proposal. We will look at this step in Section 8.2.

In Section 8.3 we will look at detailed planning and at two project planning techniques that can be used. In Section 8.4 we will look at how a project can be controlled using a project plan and we will also look at a case study of a project carried out by Edwards High Vacuum.

The learning objectives of this chapter are as follows:

1. To understand how project management techniques are used.
2. To consider some examples of project management as it relates to engineering.
3. To consider the constraints that can be imposed on projects and the effects that these can have.
4. To learn how to plan a project using Gantt charts and critical path analysis.
5. To learn how to allocate resources.
6. To consider the problems of project control and how they might be resolved.

8.2 Defining the project

In this section we will examine how a project may be defined, using a project specification. We will also introduce ways to consider the implications of the project, and examine some of the things that will constrain what happens during a project.

8.2.1 Specifying the project

The project specification is a description of the project so that all interested parties know what is planned and what the outcome should be. In industry a project specification might form part of a proposal which would be used to sell the project to the people who control the resources. In an undergraduate programme the project specification would be used by the course tutors to ensure that the work planned is of an appropriate standard for the course and that it is achievable; it may also be used by external examiners and the engineering institutions as part of their assessment of the student.

The project specification is similar to a design specification in that it provides the terms of reference for the people employed on the project. From this initial specification the project team can analyse the project in more detail and they can prepare a project plan. Finally, a project proposal can be produced which can be used as a working document during the execution of the project.

A project specification should include the following:

1. The title of the project.
2. The scope.
3. The objectives.
4. Any conditions under which the project is to be carried out.
5. Priority in relation to other projects.
6. Authority.

Title: Giving the project a title ensures that everyone knows what the project is about and avoids confusion when there is more than one project.

Scope: The scope defines the scale of the project. For example, a design modification project may be limited to the products supplied to one particular customer, or a computer system implementation project may be limited to the purchasing department.

Objectives: The objectives are what are to be achieved by successful completion of the project. Objective-setting and use of corporate objectives have already been discussed in Chapters 6 and 2 respectively; the same rules apply to setting project objectives. The objectives must be clear and understood by all the people involved in the project; they should also be consistent with company strategy. Project objectives should be realistic. In a design environment the objectives of the project may be fully defined in the design specification. Before finalising objectives it is always worthwhile discussing them more widely to get views and ideas from other people who may have interest and expertise in the field, to check congruency with other projects, and to ensure that your views are consistent with company policy.

Conditions: The conditions under which the project is carried out need to be specified. This is particularly important when the limits of the project extend into other areas; for example, if a design project needs to have some manufacturing element one of the conditions may be that daily production schedules must not be compromised by the project. Conditions will also include deadlines to be met, any special conditions relating to funding or personnel, and points at which decisions have to be made by other people.

Priority: The level of priority the project should have within the organisation has to be established and clarified in order to make sure that it has the right level of support and commitment. In order to do this the project's compatibility with other projects in the organisation needs to be assessed.

Authority: The authorisation for the project should be stated and the authority of people working on the project should be defined. For many projects one person will have sole responsibility for planning and executing the work, but for large projects there might be a whole team of people and it is imperative that the authority for particular parts of the project, and the authority for making decisions at different stages, is defined at the start.

Figure 8.1 shows the project specification that the authors used for writing this book.

8.2.2 The implications of the project

The implications of the project relate to the benefits you expect to achieve and what these will cost you. They also relate to the resources that will be required to complete the project successfully. In order to consider these implications the tasks that have to be carried out must be defined. The level of detail with which the tasks are analysed will depend

Title: *Management in Engineering* textbook

Scope: This book is intended to provide an introduction to the management topics and techniques that an engineer will come across in the normal daily activities associated with working in a small or medium-sized electromechanical engineering company. It is intended that the book will provide a standard text for management courses that are run in the first and second years of many BEng courses and will also sell to qualified scientists and engineers already practising engineering.

The book is not intended to be exhaustive, and many other books will be referenced. It does not intend to cover the law as it affects management and engineers as this would dilute the material which it aims to cover, although it will mention the areas in which the law applies so that the reader can be guided towards further reading. Similarly it does not cover the European Market as it is felt, partly because of the amount of material, that this would be better dealt with by a subject specialist in its own course.

Objectives: The main objectives is to produce a book, of approximately 450 pages, that meets the market requirements for an undergraduate mainstream text that is applicable to UK, European, American and Australian markets.

Each chapter in the book will contain an overview of the topics to be discussed and will show how this relates to the other sections of the book. Each chapter will be able to stand alone, thus providing a reference text, or can be used as a whole, thus providing a core course text.

The chapters and sections will be illustrated with case studies and examples: there will also be typical exam questions and some worked exam questions, appropriate to first- and second-year degree students.

Each section and chapter will be summarised, including a list of the main points made. References for each chapter will also be given, enabling the student to carry out further reading in order to develop each topic area further.

Conditions: The first draft of the book will be available to the publisher by the end of January 1992 in order to be available for the 1993/94 academic year.

The book is to be written as a joint project with each author contributing 50 per cent of the text. It will then be edited with each author having specific responsibility for a number of chapters.

Priority: The book will take priority over other research work but will take second priority to teaching commitments.

Authority: The copy will be shared out equally and the individual chapters will also be shared out, with each author having total authority for editing and amending copy in their chapters in preparation for the final draft.

Figure 8.1 Project specification

upon what stage you have reached in planning; for instance, if an assessment is being done for a project proposal then the tasks might be very broadly defined, while if a detailed work plan is being prepared then the tasks will be much more specific.

In order to cost the project we have to be able to identify the resources that the project will consume. The resources may take the form of money, people, machines, materials. For some tasks there will be no flexibility in the way in which the resources can be used, but for others there may be some flexibility. However, at this stage we are concerned with gross requirements and not schedules. For example, if laying a cable will take one person three days then the resource requirement is three person-days — it does not matter at this stage if the project will eventually use three people for one day.

In determining the cost of a project you should always consider the costs in terms of labour, materials and expense, as discussed in Chapter 7. The labour cost would be the cost for the time of the people who are required to work on the project. Both direct and indirect labour must be costed, and the cost if people are required to work overtime, shifts and so on must also be counted. If new staff are required for a project then the costs escalate as you have to account for the costs of recruitment and allow for the fact that initially they may not be as productive as you might expect. Material costs will include costs associated with any new parts that have to be purchased specifically for the project. Expense will include the direct monetary costs associated with capital investment and the use of specialist services, as well as machinery and equipment costs that are calculated in terms of the time for which you need to use them. It is particularly important to identify this cost if it means that equipment is not available for manufacturing when it normally would be. Expense will also relate to the cost of scrapping equipment and parts that are made obsolete by the project. The costs of any project should be considered in terms of those that are one-off and those that are recurrent such as annual maintenance charges.

Example

Consider the costs associated with implementing a new computer-aided drawing (CAD) system and conversion from a manual drawing system.

The cost of the system will include hardware, software, training and documentation. you should include an allowance for consumables as there will have to be a first-off purchase, and there will be costs for installation and probably for delivery.

After considering about the cost of the system in terms of what is going to be paid to the supplier for it, consideration must be given to where and how it is going to be used, who is going to use it, and how the

> organisation will move from the current situation to the planned one. Planning how and where the system is going to be used might mean that you have to consider converting office space, you might need new furniture and services, such as air-conditioning or stabilised power supplies. You will also have to consider the cost of administering the system, particularly if this means a new post or someone being given extra responsibility.
>
> When you consider who is going to use the system you have to plan for training, and for perhaps redefining duties and responsibilities. At a minimum you will have to cost the time for which that training will take people out of the workplace, but it is also likely that the costs of travel and expenses during training will be significant. You may also have to cost the time for someone to prepare in-house documentation for using the system.
>
> You will need labour to input data, such as drawing frames and parts libraries, or to transfer data from the manual system, and you will have to charge the time for the project manager and the person who is actually carrying out the installation of the system and writing the procedures for using it, and determining how the change from the manual system is to take place.
>
> Finally, you should not forget the first-year running costs which will include maintenance and consumables.

This all seems very detailed but it is important to consider all factors so that a decision can be made on the value of the project. The problems of not properly costing a project are numerous, not least of which is the fact that you might look silly (at best) or incompetent. The worst consequence of not having a correct costing is that of getting part-way through the project and the budget running out — this may mean that you have to reach some sort of compromise on a scaled-down project, or even scrap everything that has been done. If the project cost is over-estimated then the decision may be not to proceed because it appears not to be economically viable.

If you are asking someone to 'buy' your project you need to give them a good reason to invest, you need to justify the project in terms of the benefits that it will bring. Generally it is much easier to reach agreement on a project if the justification is done in terms of financial benefit as the decision is then based on what you get back, in monetary terms, for what you spent, in monetary terms. However, there will be times when it will not be possible to give a financial equivalent as benefits can be both tangible and intangible. Tangible benefits are those that you can see clearly and quantify, such as the amount of money you expect to be able to save from doing the project. Intangible benefits cannot be quantified in financial terms. Consider a value engineering project that will result in a product having less individual parts. This might lead to a cost-saving in materials and a cost-saving in labour — these are the tangible benefits.

However, a project which involves training to improve teamworking will have intangible benefits. It is very difficult to cost improvements in morale.

When calculating the cost and identifying the benefits of a project it is important to realise that the costs will mostly be one-off, while the benefits will accrue over a long period of time.

Example

Let us consider the case of the CAD system and look at how this might be justified.

1. The speed of producing drawings and parts lists will be increased. This will be labour-saving and therefore the cost-saving can be estimated.

2. The speed of amending drawings will be increased. Again this is a saving in labour cost which can be calculated by estimating the number of amendments made each year.

3. Storage and retrieval of drawings will be more effective. This is more difficult to quantify in terms of cost but you may consider factors such as the time-saving, the reduction in space requirements, or a cost-saving by not having to make archive copies on microfiche.

4. It will be possible in the future to expand into a linked computer-aided manufacturing (CAM) system. This benefit may be very difficult to quantify in financial terms but that does not detract from its importance if computerised manufacturing is a company objective.

5. Use of the system should improve the quality of designs. This benefit may arise because you employ a higher calibre of staff to use the system or because they are able to use a more systematic approach to design by having recourse to a library of standard parts. It may be because the system offers 3-D graphics which allows the designer to consider the part in detail before it is manufactured. It is not likely that you would be able to quantify this benefit, but if it is likely to be a real benefit it must be highlighted.

6. Finally, another intangible benefit is that having an up-to-date system would attract more, and better, recruits to the department.

This list is not exhaustive but it provides an insight into the sort of areas that really have to be thought about when you are putting forward a project.

8.2.3 Constraints

Constraints are the factors that restrict what can be done in the project. Typical constraints might come from deadlines that have to be met for an external body — for example, in a design project there may be meetings with the customer. Other constraints that might be imposed could relate to the level of funding or other forms of resource available, for example people, equipment or a requirement to use certain suppliers.

Constraints are also imposed when you require something to be done that is outside your sphere of influence, e.g. a director has to make a decision before the project can continue to the next stage.

It is very important to identify the constraints because they do affect the success of the project.

Example

Constraints on the CAD implementation project might include the following:

1. The need to avoid making anyone redundant.
2. The requirement not to employ any new people.
3. The system must be compatible with existing company hardware.
4. The software must be able to link with a CAM system.
5. Financial restrictions.
6. Deadlines for various stages of the implementation.

8.2.4 The project proposal

The project proposal is the document that brings together all the information about the project so that it can be accessed by other people. If it is your project then the proposal is your sales brochure — if accepted it should also be your working document. The proposal must include the project specification, the costs, the justification, any constraints, and a project plan. It might also include some background to the project. It should be well written and appropriate to its audience, and there should be an executive summary so that anyone can see at a glance what you want to do, why, and how much it is going to cost.

In the next section we will look at how you prepare the project plan.

8.3 Planning the project

In the previous section we looked at how a project is defined and a project proposal prepared; we will now examine how one determines what needs to be done in detail to plan the activities and

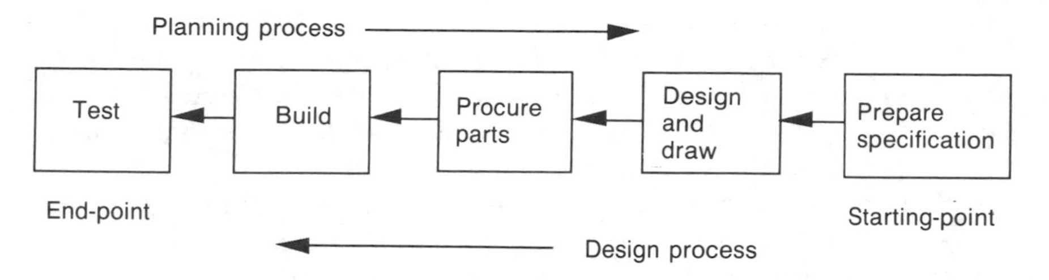

Figure 8.2 Steps in producing a piece of equipment

resources in order to achieve the project objectives. We will look at two planning techniques: Gantt charts and critical path analysis (CPA).

In order to plan activities we have to determine what those activities are and how long each activity is going to take. We also have to use the information that we had collected previously defining deadlines, as these set the limits of the project plan.

8.3.1 Project activities

Activities

The activities are the specific tasks that have to be followed through in order to meet the project objectives. Determining what tasks have to be done can be very difficult and it is important to consult widely to ensure that all activities are considered. One way of trying to determine the tasks required is to work back from the planned project end-point.

Example

A product development project requires that you design a product and produce a prototype. If we look at the production of a piece of equipment we can see that this will involve a number of steps, as shown in Figure 8.2.

From the full list of activities we have to establish the sequence in which they should be done and which activities are dependent on other activities.

Example

The activities the authors identified for preparing this book were as follows:

1. Research and prepare copy for each chapter.
2. Divide chapters and topic-writing responsibilities.
3. Proof-read each other's work and verify material.
4. Assemble book and proof-read.
5. Prepare index and table of contents.
6. Send to publisher.

Duration of activities

In order to determine the time it will take to complete the whole project and to plan the different activities you need to have an idea of how long each activity will take. The more realistic the estimates used for the plan the more accurate it will be.

Estimating time can be very difficult initially, but it should get easier as you build up more expertise through the life of the project and as you get more experience of other work of a similar nature. One way of estimating is to forecast on the basis of historical data, for instance if a similar task has been done before. It is easier to estimate time for small pieces of work and therefore another way of estimating is to break up the tasks into smaller and smaller pieces until it is possible to identify some particular activities and put a time to them.

It is impossible for anyone to be perfectly accurate at estimating time but by concentrating on the things that affect the time required for each activity it should be possible to keep errors small. For example, in the preparation of this book we had to consider that we had other jobs to do and we had to consider the time that we could really spend writing, allowing for teaching commitments, holidays and so on. Unless you are realistic and accept that other things still have to be done you will end up always producing plans that can never be achieved. This is not only demoralising personally but can reduce the confidence of a supervisor in your ability to get things done. Contrary to this, care must be taken to make sure that times are not greatly over-estimated or else people will start to wonder what is being done and may assume either that you are lazy or that you are incompetent. When estimating activity times never under-estimate the value of asking for advice from people who have been doing that type of work for some time. Chapter 6 provides further information on time management which will be useful when planning a project.

In the project proposal resources for the project will have been identified, and these now have to be tagged to specific activities as they will affect the duration of the activities. For example, if is known that two people are available to carry out some activities then this should be acknowledged when the duration for those activities is being calculated. In Section 8.3.5 we will examine how projects can be scheduled on the basis of meeting particular resource requirements.

8.3.2 Milestones and targets

Milestones are natural key points in the project, defining the end of one set of activities which must be completed before the next set can start. They divide large projects into a series of smaller ones.

Targets in a project are forced key points, they define certain tasks that have to be completed by a certain time. Targets must relate to the project objectives. They must be set because they are decision points, they are motivators, or they can be used to check progress in order that remedial action can be taken before the project has advanced too far.

Milestones and targets are particularly important as motivators in long projects, as they provide goals that can be worked towards and achieved in a reasonable time. In an undergraduate project most of the project milestones would be set by the course tutor and they would relate to handing in reports, giving presentations, and having drawings available for the workshop. For this book we set some of our own milestones, for instance when we would have copy ready to swap with each other, when we would have decided on the case studies we were going to use and so on. In addition to these, we had external targets to meet which were set by the publishers relating to preparing the first draft and responding to reviews.

Having determined the activities and milestones the next step in the project management process is to plan, and as mentioned previously we will look at two planning techniques — Gantt charts and critical path analysis.

8.3.3 Gantt charts

Gantt charts are named after Henry L. Gantt who was a pioneer of scientific management working in the early part of the twentieth century. Gantt charts take the form of a bar chart which provides a graphical picture of a schedule. On a Gantt chart you use the vertical axis to indicate the activities to be carried out and the horizontal axis indicates the time. An example of a Gantt chart is shown in Figure 8.3.

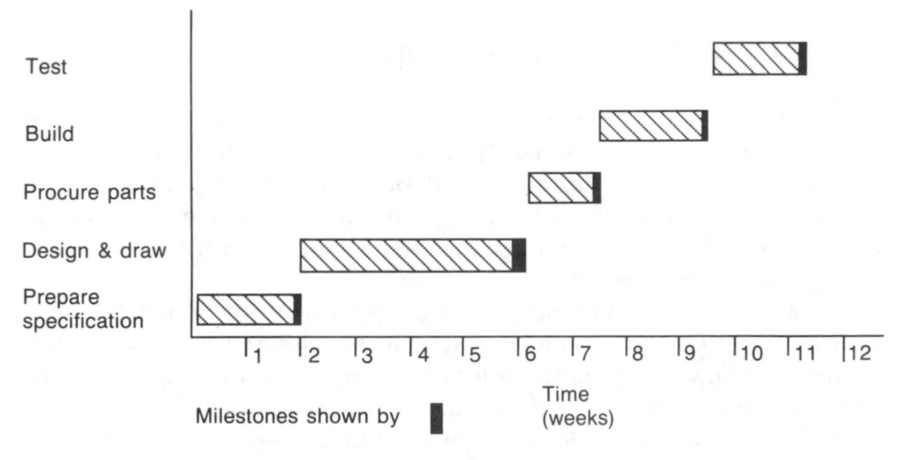

Figure 8.3 Gantt chart

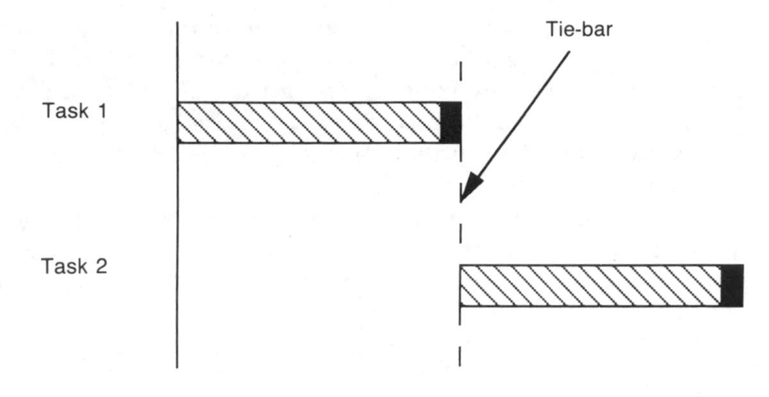

Figure 8.4 Tie-bars on Gantt charts

As you can see from Figure 8.3 it is very easy to see what needs to be done from the chart. For this reason Gantt charts are widely used for planning in industry, particularly for production planning. They can be produced on a computer or written on a piece of paper but more commonly they take the form of a wall chart or planning board.

Gantt charts are limited in their use because it is not possible to show easily how activities depend upon one another. For example, if you are planning to manufacture a new product that requires special tooling you would have to show that the manufacture activity was dependent upon the procurement of the tooling. You can do this by using tie-bars or arrows, as shown in Figure 8.4; however, this rapidly makes the chart confusing to look at when you have many dependent activities.

It is also possible to group tasks together vertically so that they form blocks of similar tasks with a common output, thus reducing the complexity of the chart.

8.3.4 Critical path analysis

Critical path analysis is a more sophisticated planning technique that allows the representation of all the activities in the form of a diagram showing their sequence and any interdependence. It also allows analysis of each activity to determine when it should start, when it should be finished and how much room for manoeuvre there is if the project is to finish in a certain time.

We will look at a method of critical path analysis which involves drawing a network of activities in which the activities are represented by arrows. We will look at the rules that apply for drawing the network and then look at how it is analysed.

Critical path analysis does have disadvantages: it can be harder to see at what stage the project is, unless the network is drawn to scale, almost like a Gantt chart. It can also seem quite complex to people who are not trained in its use.

Drawing rules and conventions

A network diagram is made up of two elements — activities and events. An activity is a time-consuming task, and is represented by an arrow or line. An event is an instantaneous point and it may be the completion of one activity or the start of another; it is shown in the network as a circle or box.

A sequence of events is called a *path* and it is usual to number the events sequentially as they are viewed reading from left to right in the network. It is also conventional to draw the network with time increasing from left to right. A path of activities and events is shown in Figure 8.5.

The network is constructed by putting all the activities in a logical sequence, with no activity starting until all the activities upon which it depends have finished.

The path in Figure 8.5 shows dependent activities, that is where one activity has to be completed before the next can start. However, there may also be independent activities where two activities can go on at the same time, as shown in Figure 8.6.

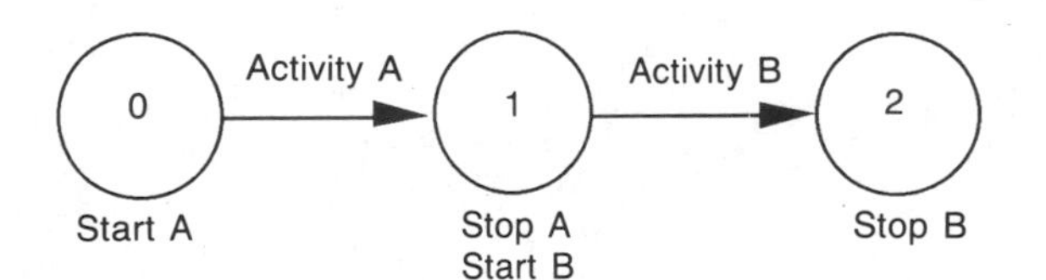

Figure 8.5 A path of activities and events

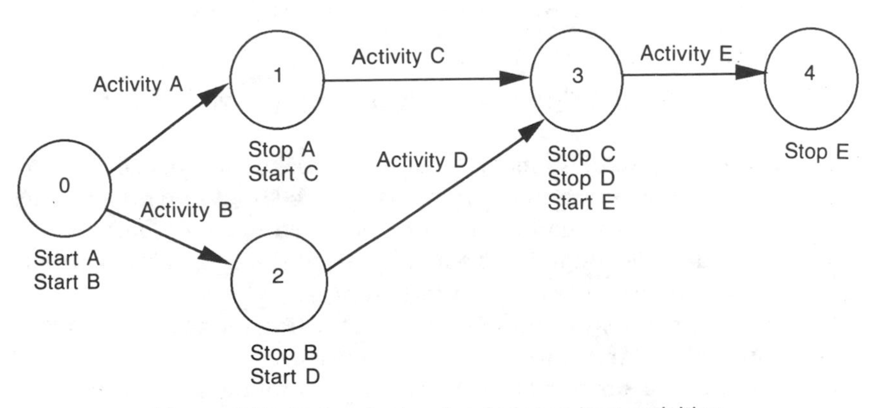

Figure 8.6 Network showing independent activities

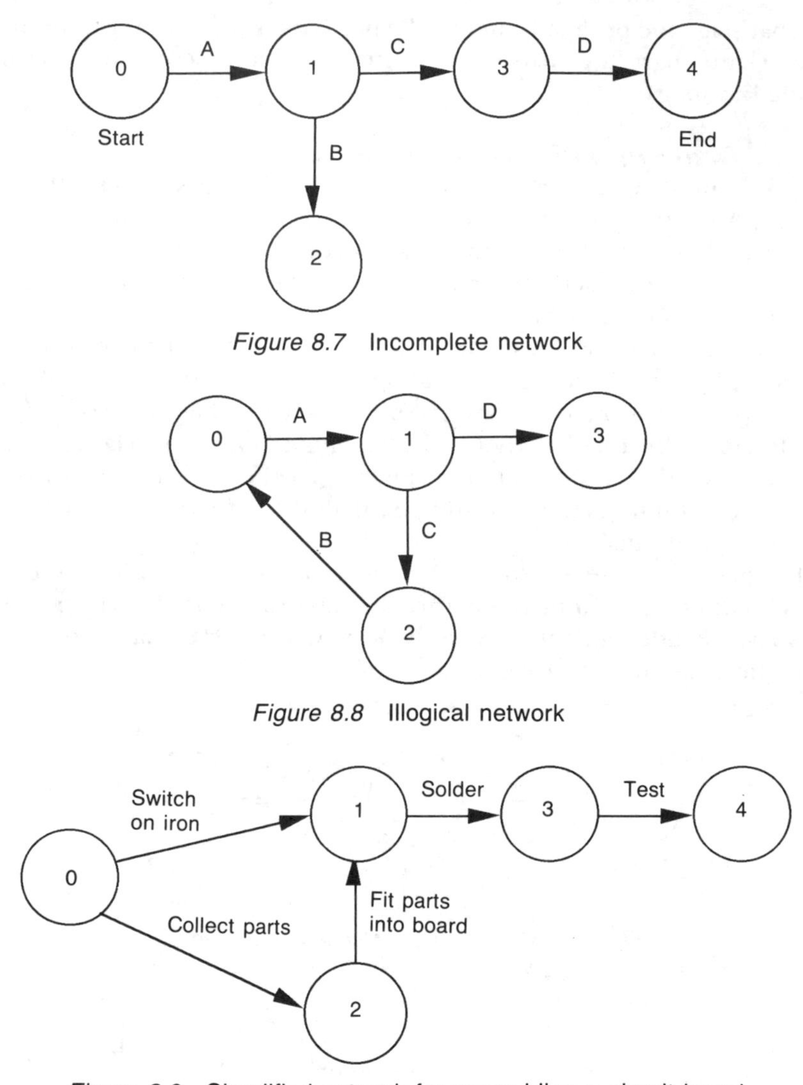

Figure 8.7 Incomplete network

Figure 8.8 Illogical network

Figure 8.9 Simplified network for assembling a circuit board

The logic of a network should be checked to make sure that all activities have an appropriate start- and end-point. There should be no activities that do not lead to the end of the project, as shown in Figure 8.7.

Similarly, paths should follow from the start to the end and they should not go round in circles, as shown in Figure 8.8.

It is vital to start and end a network with a single event since only a single state defines the start or end of a project.

An example of a simplified network for assembling a circuit board is shown in Figure 8.9.

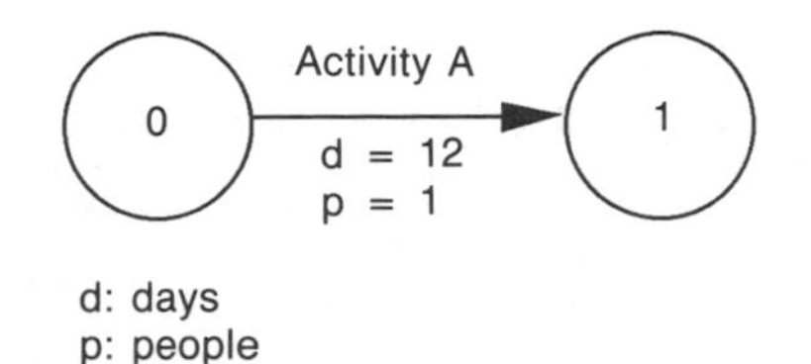

Figure 8.10 Labelling of activities

In Figure 8.9 you see that the activities are labelled which makes the network easier to use, if you can make the labels clear descriptions of what the activities are. The convention is to label above the activity arrow for horizontal arrows, or to the right of the arrow for vertical arrows.

Activity times should be shown on the network below the activity label. You may also want to indicate how many resources are involved in each activity, and this is also shown below the activity label, as shown in Figure 8.10.

When you initially draw a network you may find that it takes several attempts to get it right; however, once finalised it is always worth redrawing the network neatly. It will certainly be easier to follow and therefore easier to work with.

Summary of rules and conventions

1. Activities are drawn as arrows, events as circles or boxes.
2. Events are numbered sequentially, activities are labelled either above or to the right of the arrow.
3. Time increases from left to right.
4. There should be no activities dangling and no loops.
5. Activity times and resource requirements can be shown below the activity label.
6. Draw the network neatly.

Example

Draw the network for the following list of activities. The preceding dependent activities are given.

Activity	Predecessor
A	—
B	—
C	A
D	B
E	D
F	C, E
G	F

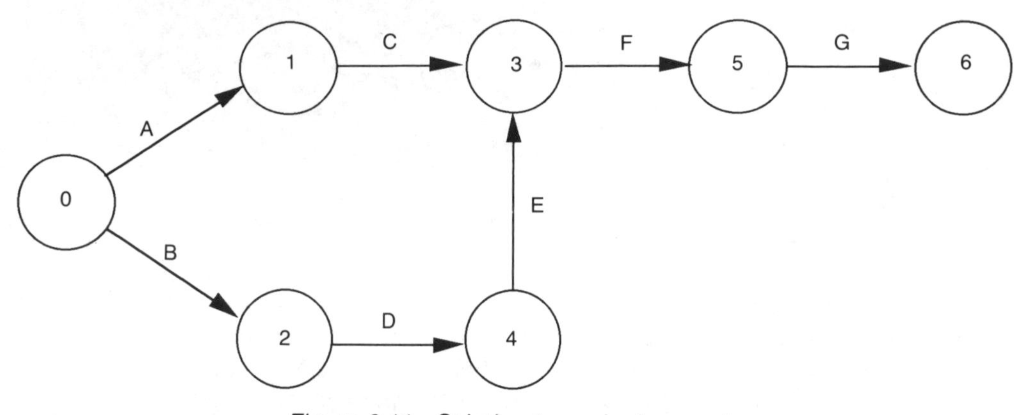

Figure 8.11 Solution to worked example

This gives the network shown in Figure 8.11.

You see that the project starts with a single event, labelled 0, and ends with one event, number 6.

Dummy activities

Dummy activities are often used when drawing a network, although they have no time associated with them. They are represented by dotted lines and can be used to ensure that the logic of the network is correct, to prevent activities having the same start- and end-points, or if their use makes the network easier to read.

Example

Draw the network for the following list of activities. The preceding dependent activities are given.

Activity	Predecessor
A	—
B	A
C	A
D	B
E	D
F	B
G	C
H	B, G
J	H
K	E, F, J

This gives the network shown in Figure 8.12.

The dummy activity translates the end of activity B to event 5 indicating

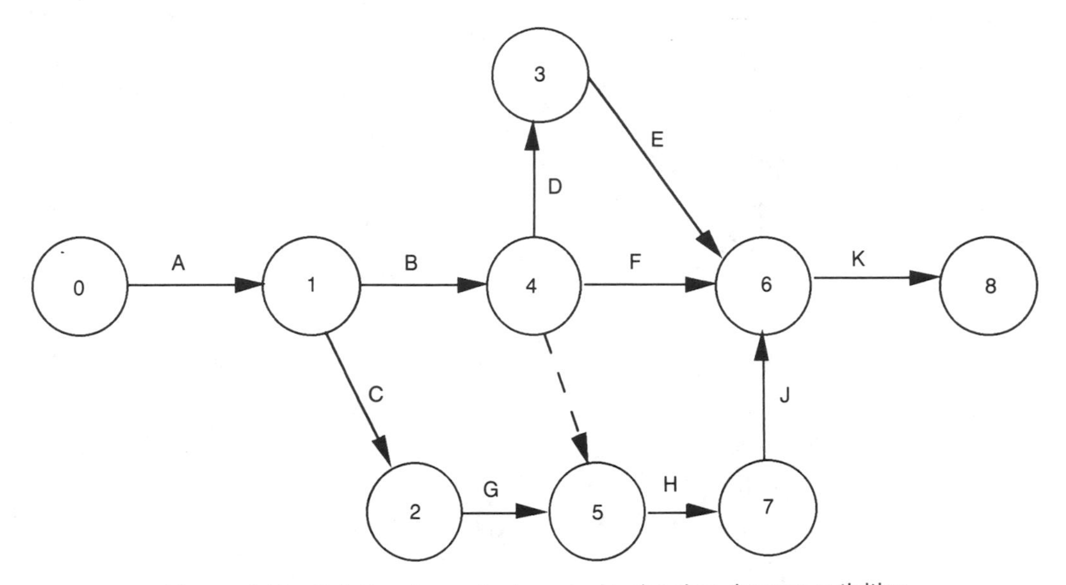

Figure 8.12 Solution to worked example showing dummy activities

clearly that both B and G must be finished before H can start. To have drawn B between events 1 and 5 would have been incorrect as this would show that B and G both have to be completed before D and F. This would not be true, D and F are dependent only upon B.

Analysing the network

Analysis of the network will provide the overall time that it will take to complete the project, assuming that you have all the resources that you need. You will also be able to calculate the earliest start and latest finish dates for each activity. You can calculate the float on each path, and you will be able to identify the critical path.

When analysing a network it is usual to start at time 0.

Earliest start date (ESD): The earliest start date can be indicated above each event, or it may be written within the event circle. Both methods are shown in Figure 8.13.

The ESD for any activity is the earliest that it can start due to the activities that precede it having to be completed. The ESDs are calculated by working through from the beginning of the network totalling the durations of all the previous activities — this is called doing a *forward pass*. When two or more events have to be completed before an activity can start then the start date of the activity will be determined by the longest duration preceding the activity.

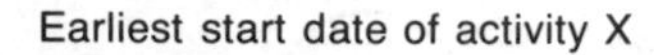

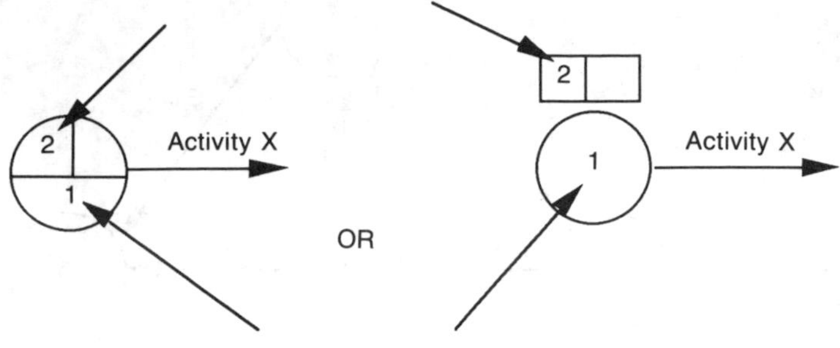

Figure 8.13 Showing earliest start dates in the network

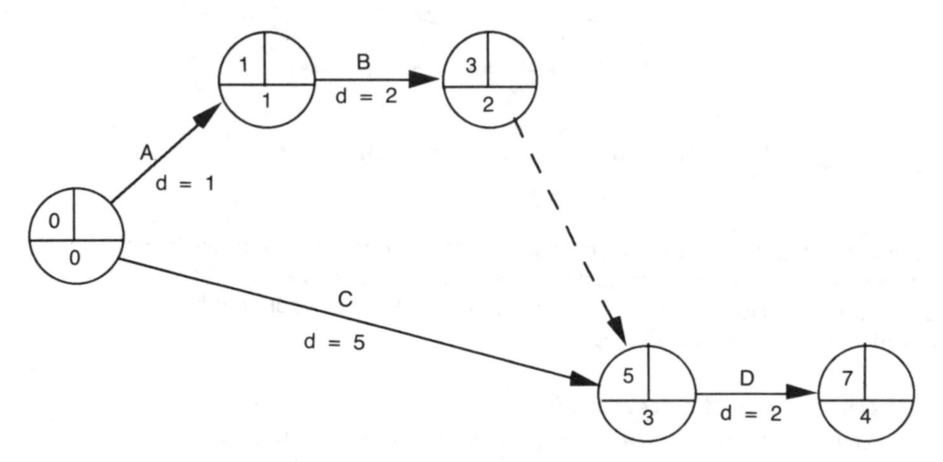

Figure 8.14 Calculating earliest start dates

Example

Activity:	A	B	C	D
Predecessor:	—	A	—	C, B
Duration (days):	1	2	5	2

The network, with earliest start dates, is shown in Figure 8.14. You can see that in order to start D both path AB and path C have to be completed. AB will take three days and C will take five days, therefore the earliest that D can start is day 5.

When you reach the last event in the network the ESD is calculated for an activity subsequent to the project completion, thus the ESD on the last event is the minimum overall project duration.

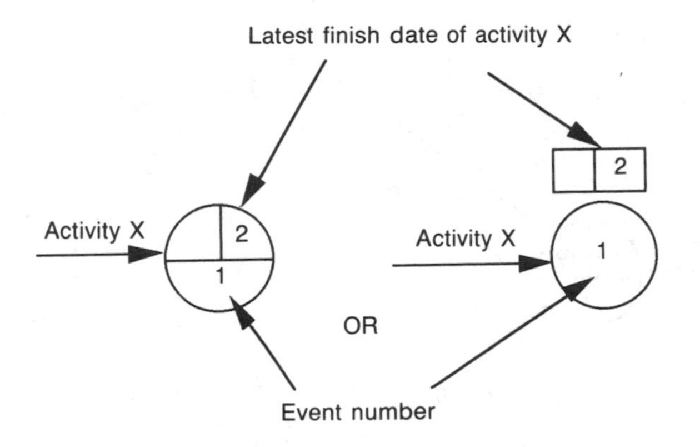

Figure 8.15 Showing latest finish dates in the network

Latest finish date (LFD): The latest finish date can be indicated above each event, or it may be written within the event circle. Both methods are shown in Figure 8.15.

The LFD is the latest time that any activity can finish if the project is to be completed in the earliest possible time.

The LFDs are calculated by working from the end of the network successively subtracting durations from the project completion date — this is called a *backward pass*. When two or more activities precede one event the earliest date will determine the latest finish date for the activities.

Example

Activity:	F	G	H	K
Predecessor:	E	E	G	F
Duration (days):	6	10	5	2

Gives the network with latest start dates as shown in Figure 8.16. Note that the completion time for the project is thirty days.

The LFD for the last activity will be the minimum overall project duration.

Note that for both the first and the last events in the network the earliest start date should be the same as the latest finish date.

Float: Float is the difference in time available for the series of activities in a path and the time required for the path — it is the amount of *slack* in the project. Knowing the float allows the project manager to schedule and optimise the use of resources whilst having full knowledge of the effect on the overall project duration.

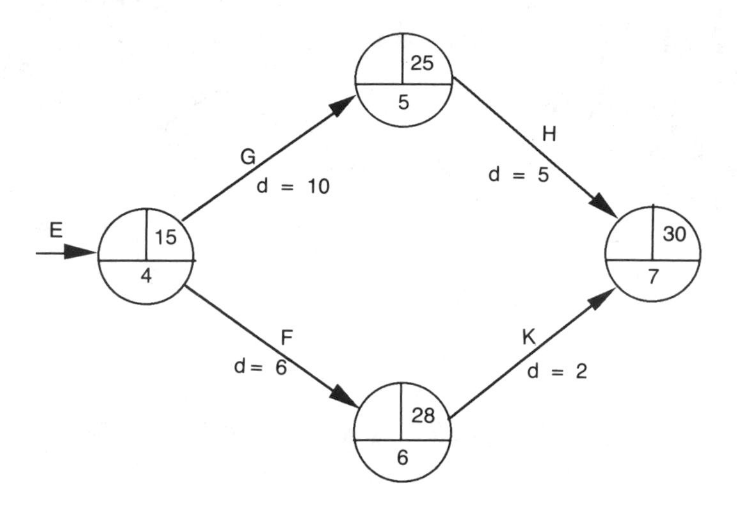

Figure 8.16 Calculating latest finish dates

Time available = Latest finish date for activity X (LFD) − Earliest start date for activity X (ESD)

Time required = Duration (d) of activity X

Float on activity X = LFD − ESD − d

Float is a characteristic of a path and not a single activity. For example, if the float on a path is five days, if any activity takes five days longer than expected the float is reduced to zero and none of the other activities can be allowed to overrun if the project is to be completed in the minimum time. Section 8.3.5 shows how float is used.

Critical path: The critical path is the longest path through the network, and can usually be identified by the fact that the float on the path is zero. The critical path activities can have a float, although it will be the lowest in the network, if a project duration is imposed rather than if it has been calculated from the activity times. Consequently a critical path can also have negative float which highlights the fact that too little time has been allowed for the project and therefore steps need to be taken to reduce activity times if the project is not to be delayed.

Activies on the critical path must be finished in the time given or else the finish date of the project will be affected. The only way in which you can bring a project forward is to reduce the duration of an activity on the critical path, such that a new critical path applies with a shorter overall duration. Determination of the new critical path is achieved by re-analysis of the network.

Summary of analysis

1. Calculate earliest start dates, using a forward pass.
2. Calculate latest finish dates, using a backward pass.
3. Calculate float on paths.
4. Identify critical path.

Example

Draw and analyse the network for the following:

Activity	Predecessor	Duration (days)
A	—	1
B	A	2
C	A	1
D	A	1
E	A	3
F	B	2
G	D	2
H	E	5
J	F	1
K	CFG	1
L	H	1
M	J	2

This gives the network shown in Figure 8.17.

The project overall completion date is day 10. The critical path is AEHL, the float on the other path is as follows:

Path	Float (days)
BFJM	2
C	7
DG	5
K	4

This indicates that each of the four paths, shown above, can be scheduled within a time slot that is longer than the duration estimated originally. They can therefore be delayed to the extent of the float without affecting the overall project duration.

8.3.5 Resource allocation and levelling

We have seen that by calculating float it is possible to determine the time that can be used for a particular path of activities without affecting the overall project duration. It follows that if the changes cause the float on a path to be exceeded then this will affect the project duration. We have also seen that any changes on the critical path will affect the

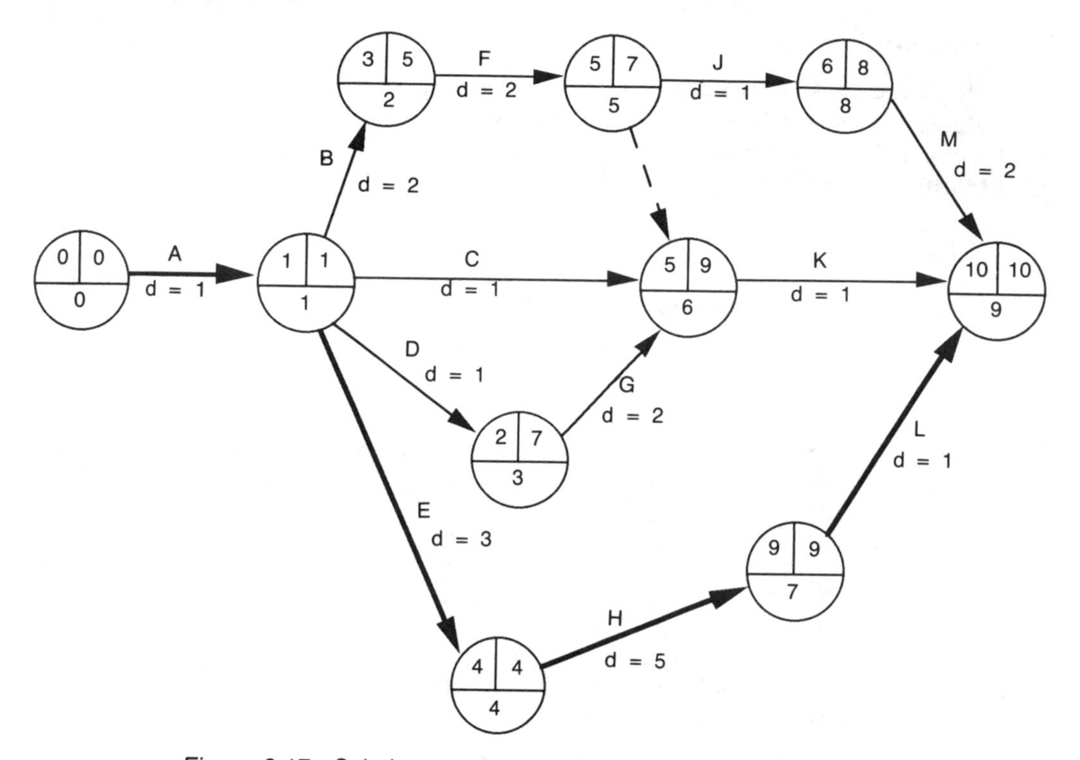

Figure 8.17 Solution to example showing full network analysis

project duration. We can use this information to schedule all the resources that are required for the project. The scheduling of the resources is the most important aspect of project planning — for example, you may need to know exactly when certain monies are required in order to get them from the bank, or you may need to book a particular piece of equipment.

All projects are resource-limited in some way: for instance, a project may be restricted by a certain number of people being available or by certain materials not being available until a particular date. In addition, there are always upper limits on the amount of money available for a project. The project manager needs to know how these restrictions are going to affect the project.

Example

Let us consider the following project:

Activity	Predecessor	Duration (days)	People required
A	—	2	2
B	—	4	2

C	—	5	2
D	A	10	4
E	B	1	2
F	C	3	1
G	D	4	2
H	EFG	3	5

The network is shown in Figure 8.18.
The project completion date is nineteen days, assuming that all the resources are available as required. The critical path is ADGH, the float on the other paths is:

Path	Float (days)
BE	11
CF	8

This shows that we have an extra eleven days within which we can schedule activities B and E, and an extra eight days for activities C and F. Rescheduling up to these limits will not affect the overall project duration.

We can now draw a resource histogram which allows us to see when the resources are required and in what numbers (Figure 8.19). At this stage it is assumed that all the activities start at their earliest start date.

In real life it is likely that you would have a fixed number of people available for the duration of the project and therefore you do not really want a situation where some of them will be idle one day and overworked the next — you will want to *level* the requirements. There are complex

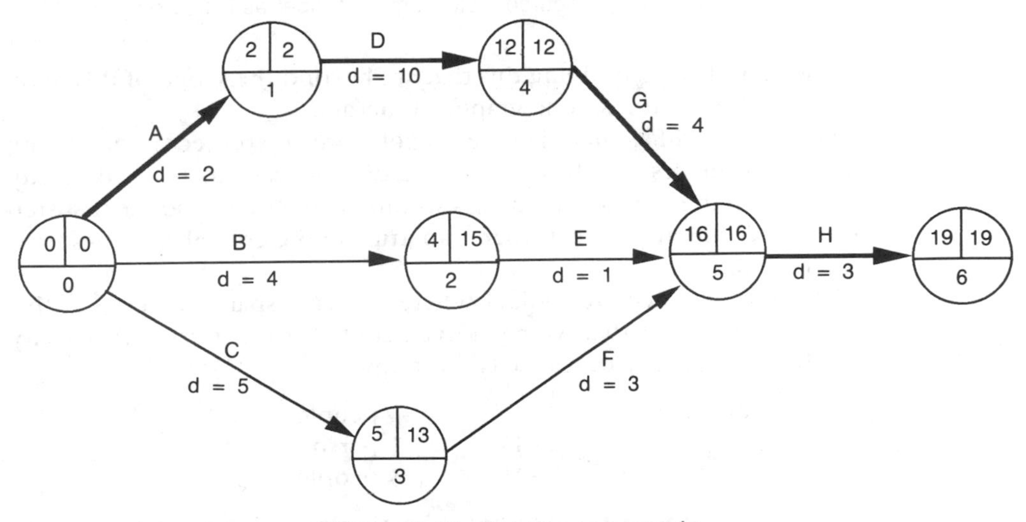

Figure 8.18 Solution to example

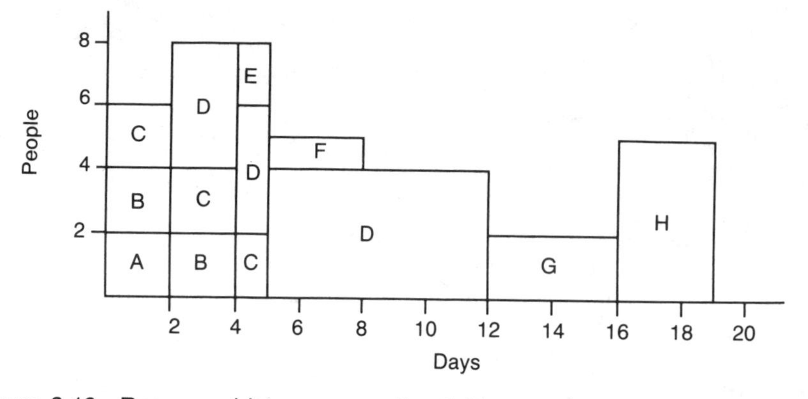

Figure 8.19 Resource histogram — all activities starting at their earliest start date

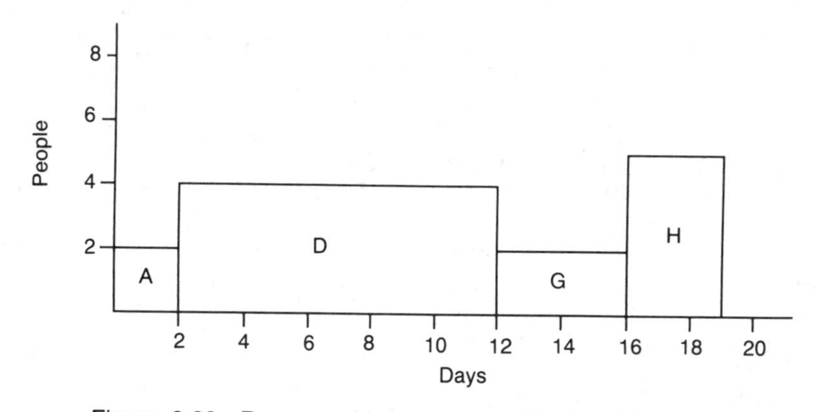

Figure 8.20 Resource histogram — critical activities only

mathematical ways of doing this that are beyond the scope of this book so we will look at a simple graphical method.

Firstly, we will assume that we do not want the project to go on any longer than necessary. In this case we do not want to reschedule any activities on the critical path unless absolutely necessary, so the first step is to redraw the resource histogram putting in the critical activities first, as Figure 8.20.

Secondly, we need to look at where we have spare capacity; in this case if we consider that we need five people for a critical path activity (H) then we have spare capacity as follows:

From day	0–1	3 people
	2–11	1 person
	12–15	3 people

We can now look at the non-critical activities and see what resources

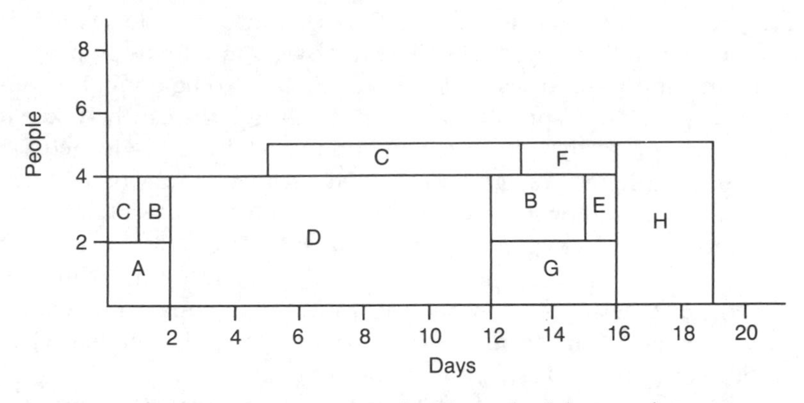

Figure 8.21 Resource histogram showing schedule to reduce personnel requirements

are required and when. From the network we know the float available and we know the time slots that can be used for each activity. What we do not know from the information given is whether a particular activity, say C, has to be done by two people for five days, or if it is possible to use up some of the float extending the duration and using only one person for some of the time. This would obviously be known if it were a project with which you were involved.

In Figure 8.21 a solution is given which assumes that activity C can be done by one person for a longer period of time, and also assumes that activity B can be split into two sections with a delay between the two sections. You will see that in this way it has been possible to reduce the maximum personnel required from eight to five and the demand is constant at four people for the first five days and five for the remainder of the project.

This rather simple approach to resource allocation is equally applicable to other resources such as cash and materials. In this way we can schedule a project to make optimum use of resources available, or to take account of resource constraints.

8.4 Project control

The project plan is a working document. If used correctly it will enable control of a set of activities and resources to achieve the desired end-point.

In the same way that design work should be reviewed at regular intervals to ensure that the specification is being met, so project plans should be reviewed. Review dates should be incorporated into any plan and they should be appropriate to both the time-scale and the complexity of the project. Reviews should be carried out by management in conjunction with the project team. The review should focus on progress made to date and look forward at the work that is to be done. It should highlight any potential problems in order that action can be taken to prevent them, or so that the consequences can be acknowledged.

At the initial planning stage estimates of activity durations will have been made, so as the project progresses it should be possible to refine the estimates based on experience. It will also be practical to incorporate more detail about the activities to be carried out in the near future.

The project review process should be formal, reflecting its importance for project management. In addition the process should result in revision of the project plan so that it reflects the real situation.

Project management is no different to the management of any other engineering activity. It requires effective use of resources, in particular personnel. It is likely to involve financial control, which we have discussed in depth in Chapter 7. Most importantly, good project management depends upon strong leadership and teamworking. The team effort is not restricted to the project team but to all the people who will be affected by the project; this is clearly seen in the case described below.

8.4.1 Case study: Edwards High Vacuum

Edwards High Vacuum is a division of BOC Limited. It manufactures a large range of vacuum equipment on three UK sites. Its site at Crawley in the UK houses the company headquarters and is also a manufacturing base for the company's range of diffusion pumps.

One of the pumps manufactured at Crawley is the DIFFSTAK MK2 diffusion pump. In October 1989 the factory at Crawley was organised by function and the manufacturing system had to cope with eight families of products, each with six variants, and production of between 1 and 200 per month for each variant. The DIFFSTAK MK2 pump was manufactured in batches to a monthly demand forecast indicated by the Master Production Schedule (MPS). Buffer stock of finished pumps was held to cover any fluctuation in demand. The lead-time for the pumps was three months.

Following a review by a team of consultants the management team at Edwards decided that the method of producing the DIFFSTAK MK2 could be changed to reduce lead-time and they therefore established a project team to achieve this. The Project Manager chosen to lead the team was the Materials Manager. His team consisted of two professional consultants, one full-time and the other as an adviser, and two Edwards' production engineers. The decision to include the consultants was made because they had previous experience of other production systems and in implementing modifications to existing systems.

The brief to the project team was that they should work towards a make to order (Just-in-Time, or JIT) production system for DIFFSTAK MK2 pumps, starting with the fabrication and assembly processes. As this would inevitably entail significant changes on the shop-floor, communication with the workforce and the unions was of paramount importance. The project team realised that a major part of the project would be a method study to determine what methods were actually being used in the existing manufacturing system and that this data-gathering exercise would have to be undertaken before any decisions could be taken about the way forward. Edwards' General Manager therefore consulted with the union shop stewards and got their agreement to the exercise. Following this the Project Manager made a presentation to the workforce, which included an introductory address by the General Manager. The workforce were also trained by the consultants in JIT techniques by means of practical exercise.

The first phase of the project, the method study, took two months. It was carried out by the two production engineers from the project team. During the method study they employed a part-time toolmaker and draftsman so that any problems identified by the study could be quickly rectified. This showed the workforce that the project really was intended to have an impact. The method study confirmed that the suggested JIT system could be used successfully and a more detailed project plan was prepared and issued by the Project Manager. The preparation of the plan followed discussions with the project team. The plan covered the implementation of the JIT system and was presented in the form of a Gantt chart, allowing it to be understood quickly and easily.

Prior to the implementation phase the project team presented the proposed solution, including a new plant layout, to the senior managers and then to the workforce. The changes to the plant layout were scheduled by the plant engineer to take place during the Christmas holidays.

Project monitoring and control was achieved by daily meetings and monthly reviews. The daily meetings were held throughout the project and were to report on progress. They lasted usually for about thirty minutes and included the project team, any operators who wanted to go along, the Production Manager, and sometimes either the General Manager or the Works Manager. The daily meetings also served to engender team spirit and to foster good communications between the shop-floor and the project team. When the project entered the implementation phase the meeting also involved a manufacturing systems engineer and the plant engineer who had responsibility for plant layout. More formal project reviews were held monthly with a presentation being made to the senior site managers by the project team.

In addition to the planned daily and monthly meetings *ad hoc* brainstorming meetings were held when problems were experienced.

The Production Manager took over ownership of the line from 2 January 1990 and the new production system, operating as a single-unit flow line, started on time on 7 January 1990. The lead-time for the DIFFSTAK MK2 pump was ten days. The project team was disbanded after the line had been running for a week. As well as the reduction in lead-time the project achieved a 30 per cent saving in floor-space and considerable reduction in work-in-progress and finished goods stock. Finally, continuous improvement groups were established to encourage worker participation in the continuing development of both the product and the production system. ■

8.5 Summary of Chapter 8

In this chapter we have examined the role of projects in an industrial environment and the importance of planning and costing. We have considered the way in which a project can be fully defined in order to enable the costing and planning process to be carried out effectively. The role of the project proposals as a document defining the project, the implications of it, and the constraints imposed upon it, was also described.

We have described two project planning techniques that are widely used. These techniques are Gantt charts and critical path analysis. We saw that there were many advantages to using critical path analysis, particularly because of the way in which float can be calculated. Knowledge of the float on any path of activities allows optimum resource allocation and scheduling to be carried out. We showed a simple technique that can be used for this.

In Section 8.4 we gave a case study of a large industrial project and considered the way in which it had been planned and the techniques used for execution and control.

8.6 Revision questions

1. (a) For the information given below:

    (i) Draw an activity network.
    (ii) Determine the earliest finish for the project assuming that it starts at day 0.
    (iii) Determine the float on each path.
    (iv) Determine the critical path.

Activity	Predecessor	Duration (days)
A	—	5
B	—	1
C	A	5
D	B	7
E	C	2
F	B	6
G	C, D, F	10
H	E	8
J	G	2

   (b) (i) If activity A can be reduced by two days how does this affect the overall project duration?
    (ii) If activity E is increased by three days how does this affect the overall project duration?

2. (a) Briefly explain the following terms:

   Gantt chart
   network
   milestones

   (b) Give two advantages of using Gantt charts rather than networks.
   (c) Draw and analyse the following network:

Activity	Predecessor	Duration (days)
A	—	2
B	—	1
C	—	3
D	A	1
E	B	3
F	C	4
G	D	2
H	E	1
J	B, F	1
K	G, H, J	3
L	J	2
M	L	4

Indicate the overall project duration, the latest finish and earliest start dates for each activity, and the critical path.

3. Draw and analyse the following network:

Activity	Predecessor	Duration (days)	Workers req'd
A	—	2	3
B	A	3	2
C	A	3	1
D	B	4	2
E	C	1	4
F	C	2	3
G	B, E	1	1
H	D	1	1
J	F	2	2

   (i) Draw a resource histogram showing the numbers of workers required each day, assuming that each activity starts at its earliest possible start date. What is the maximum number of workers required?
   (ii) Reschedule the activities so that the maximum number of workers does not exceed six at any time. Redraw the resource histogram and comment on the practical implications of this project if workers are to be hired specially.

4. A product is made up of five parts, W, V, X, Y and Z. Parts W, V and X form

a subassembly, and parts Y and Z form a subassembly. The two subassemblies are assembled to make the product. The supplier lead-times and assembly times for each part are shown below. Show this information on a Gantt chart and determine the lead-time for the product and the latest dates on which each part should be ordered if product is to be delivered at the end of week 6.

Part	Supplier lead-time (days)	Assembly (days)	Assembly (days)
W	2		
V	10	2	
X	4		
			3
Y	3		
Z	4	2	

5. What information should be included in a project proposal? Give reasons for the inclusions.

6. Discuss the personnel management issues raised by the case study in Section 8.4.1.

7. Prepare an analysis of the communication techniques used in the case study in Section 8.4.1 — what alternative methods could have been used? How did the communication methods adopted contribute to the project? What would have been the likely effects of the alternative methods?

8.7 Further reading

Lock, Dennis (ed.) (1987), *Project Management Handbook*, Aldershot: Gower.

Lockyer, K.G. (1984), *Critical Path Analysis and Other Project Network Techniques*, London: Pitman.

Chapter 9
Communication skills

Overview

In the previous chapters we have seen the various important tools of management described as individual subjects. In the first two chapters we explained the environment in which engineering businesses operate. In the subsequent chapters we saw how the needs that result from this operating environment may be met. We have seen how the personnel needs of an organisation may be served. We have seen how teams operate, how objectives are set, how money and information are controlled, and how projects are managed. However, there is one skill that underpins all of these, a skill without which none of these things would be possible. This is the skill of communication.

The case study in Chapter 10 illustrates all of the managerial skills we have described in the previous chapters. Teamworking issues arise, financial issues are involved. Sometimes information control is foremost, at other times it is appraisal. Communication is there all the time. So all-pervading is communication that every single paragraph of the 200 paragraph case study involves a communication of some sort.

9.1 Introduction

Communiction is a vital skill for any professional. No one can be effective in an important job without being a good communicator. It has even been said that there are no management problems in the world, only communication problems. In this chapter we will see the importance of information and consider examples showing how crucially important particular pieces of information can be. Communication is the process by which we acquire all information and it is therefore worth studying in order that we may become good at it. Communication is important, it is fundamental to engineering, as it is to so many other things in life. Unless an engineer can effectively communicate designs, ideas and instructions the requirements of the job will not be met. Engineers who are ineffective communicators may also endanger life, by not adequately defining critical processes or giving warnings. They are unlikely to be able to achieve satisfactory positions in their companies because they not only fail to 'sell' themselves but they give a bad impression to customers, and they may cost the company money by not correctly determining the customer's needs.

All management tasks involve communication and it is not possible to be good at managing without first being able to communicate well. In this chapter we will first examine the role of communications in the workplace. In Section 9.3 we will look at wider communications and at some of the techniques that can be used to gather the information which will allow us to do our jobs effectively. In Section 9.4 we will present techniques for ensuring effective communication in the common forms of writing. Section 9.5 deals with oral communications in a similar way. Finally, in Section 9.6, we will examine some of the ways in which meetings may be managed to ensure that they are effective.

The learning objectives of this chapter are as follows:

1. To understand the need for effective communication in the workplace.
2. To describe some of the methods that are commonly used to aid communication in the workplace.
3. To provide guidance for achieving good oral communications.
4. To show the accepted forms of written communications that will be used by engineers.
5. To explain the role of meetings and to consider the way in which they can be run efficiently.

9.2 Communications in the workplace

The workplace of even a modestly sized organisation is a microcosm of the outside world. The complexities that exist in the outside world usually exist in some form within the organisation. Where communication is concerned this is also true. To understand how communications operate in an organisation we need two things. The first is to understand what the organisation needs from its communication system and the second is to look at the various methods that can be used to aid the communication process. If we understand communications properly we should be able to answer questions like: 'How can an organisation tell if the communication system in use is satisfactory?'. 'What factors affect its efficiency?', or 'How may the existing methods be altered to improve communications?' In Section 9.2.1 we will look at the needs of an organisation that result in a communication system. in Section 9.2.2 we will consider the methods that are used in the communication system and the aids that assist the process.

9.2.1 The purpose of a communication system

Only by understanding what the organisation hopes to achieve from its communication system can we expect to understand how to improve it or to design it well. Understanding a communication system means firstly knowing what purpose the system serves. We can then see how it might achieve that purpose and understand the limitations that a system might bring. We therefore turn our attention to why the comunication system exists.

To understand what purpose a communication system serves we must, as with so many other aspects of engineering management, look at the organisation's corporate objectives. In Section 2.4 we examined the nature of corporate objectives. We saw that an organisation of any size needs somehow to make sure that the efforts of the people that compose it are all directed in some beneficial direction. We saw that using corporate objectives is the way in which this is achieved. An example of a corporate objective is given in Figure 9.1. In order to achieve this, or any other, objective the company needs a communication system.

The ideal communication system for an organisation is the most cost-effective one that makes it possible to achieve the corporate objectives. Let us consider the communication implications for the objective in Figure 9.1.

The first and most important implied communication need is for the directors who set the objective to communicate it to the managers of the

To return a profit of £75,000 by supplying centrifugal water pumps to European customers over the next calendar year.

Figure 9.1 A corporate objective

company. With any objective there will be wide-ranging consequences and it is imperative that the managers not only understand what they are required to do but they know how the rest of the company is going to act. So we immediately have a requirement for downward communication, the directors to the managers. We also have a requirement for horizontal communication, the managers must communicate with each other. At this level of operation the communication must be accurate and formal, there is no room for misinterpretation or lack of detail. The consequences of inaccurate or complete information are too great.

If the objective has been set so that departmental managers can develop their own strategies for meeting it, this again implies cross-department communications. However, it also implies that there must be a route back up the organisation to the directors, particularly if there are likely to be any problems with meeting the objective.

If we now look at the other communication requirements for meeting this objective we can see that there is a need for customer contact. To trade effecively one must understand and know the world of one's customers. One cannot depend on their complaints to guide strategic commercial decisions, one must actively gather information and to do this one must communicate with them. In doing this the organisation, and the individuals who communicate on its behalf, may use many different techniques but they are all employed to find out what actions will lead to satisfied customers. In the case of this objective the organisation must know what the customers expect from a technical, commercial and service point of view.

Different customers may have different expectations and one type of pump and method of sale may suit one customer and not another. If the organisation is to be successful in making both satisfied with a purchase from the organisation, different technical and commercial methods may have to be used for each customer. The market for the pumps may change with time, developments in technology may alter customer's expectations or changes in exchange rates may affect the price of competitive products. To be effective in selling to its customers the organisation must be knowledgeable of all such developments in the customers' situation and this means communicating to find out.

The next communication need implied by the objective is for knowledge of the financial performance of the organisation. The organisation cannot achieve this objective without knowing such things as how much each product costs to manufacture, how much the overheads are costing or what the profit margin on each sale is. It is only by ensuring that this important information is communicated to the managers who need it for decision-making that the information can be effectively used. Similarly measurement of the achievement of this objective requires adequate financial data. Many communication techniques are used in this process and there will certainly be a requirement for communication in all directions, with cost data being communicated across departments but budgets being set at a higher level. The directors will also want feedback on the costs of implementing the strategies for meeting the objective. As the information is communicated upwards in the organisation it becomes more synoptic and describes the performance of larger and larger portions of the organisation. Computers often play a major role in producing financial data for decision-making and these certainly make the data to be communicated more widely acceptable and understandable. Ultimately the communication of this financial information to the Board of Directors is the responsibility of the financial director — such a post can even be defined in this way.

A further communication implication of the objective is the communication between the individuals of the organisation. If the objective is to lead to actions for every person in the organisation they will somehow have to be made aware of what actions are required of them. The whole process of implementing strategies to meet objectives is one of delegation and the managers at every level have to be capable of taking responsibility for their own particular area and then sensibly deploying the resources under their control to achieve their own objectives. The communication skills involved in managing a group of professional engineers are considerable. Communicating with those for whom they are responsible can absorb a large proportion of a manager's time, and for some it is the only task they do. The need to be good at communicating with people at a personal level cannot be overlooked, while the need to have an organised approach to handling departmental communications is equally important. Nothing is so frustrating for the employees of a department as not knowing what is going on.

Finally, the last communication issue implied by the objective is with the ouside world. There will clearly be a need to identify and trade successfully with the organisation's suppliers. There will also be a need to keep up to date with any legal requirements of the organisation's trading position. Most organisations have some sort of relationship with at least the local community and often with a much wider public. Large organisations may employ a public relations department whose sole purpose is to ensure that communications with the outside world are handled effectively and the public image of the organisation is managed.

In all of these communications there will be protocols that define the way in which the comunications take place, and which set limits on the methods of communication. For example, it may be agreed that all communication with customers is done via the sales department and that everything is confirmed in writing. One of the most commonly used methods of communication is the meeting and most companies will have rules that define when meetings should be called and how they will operate. Meetings are so important in the workplace that we have included a separate section that explains how they work, Section 9.6.

Having seen the need for communication in the workplace we will now look at some of the methods available.

9.2.2 Communiction methods and aids

In most orgainsations many different methods of communication are employed and we shall describe some of them below. We will also consider some of the aids to communication that exist. The order is alphabetical and is not intended to attach importance to any particular method.

Appraisal interviews

These are used as a communication device between employee and manager. They concentrate on a review of the performance and possible development of the employee. They can also be used to discuss specific problems but they are not very effective for this as they usually take place at fixed times, annually or six-monthly. One of the main advantages of appraisal interviews is that they provide a forum for looking forward and agreeing personal objectives for the next period. There may be a two-tier appraisal system in which the appraisee also has an interview with the manager's superior. This interview serves as a check that things are well between manager and employee as well as providing the opportunity for the employee to discuss wider issues such as company development. Provided the manager is someone whose judgement is respected and is the sort of person that will discuss issues objectively, appraisal interviews can be very useful. Appraisals are a very important part of personnel management and are dealt with in detail in Section 4.4.

Computers

There is wide use of computers in the workplace. The communication advantages they bring are speed, transferability without misinterpretation, and the ability to store a vast amount of information and retrieve it quickly. Management information systems, computer-aided drafting systems, and computer numerical control machining programs are all examples of the way in which computers are now used. In addition to these specialist packages there is the ubiquitous word processor which allows even computer-illiterate people to prepare documents that can be read by others. A single floppy disk can contain thousands of documents and yet can be transmitted down a telephone line in a few seconds, and it can even be posted if necessary. The continued reduction in price and improved software makes computers feasible for the smallest firm. Information is often transferred directly from computer to computer. This offers advantages of speed and cost as there are no pieces of paper involved or laborious manual processes. This is not even limited to technical information as many large networks now have electronic mail systems where messages can be written directly to a user's terminal or file. Personal computers can be used for facsimile transmissions, word processing, communication with other users on a network, reading floppy disks sent by others, even operating your diary and daily planner. There is practically no communication task for which software has not been written and communication has replaced computation as the primary role of a computer. Being competent in the use of a personal computer is now an expected skill for professional people.

Conversation

The majority of workplace communications result from ordinary conversation. It is one of the most useful forms of communication because it is so flexible. In a world of memos, meetings, reports and procedures the skill of holding a successful conversation is often neglected. There is more to having a conversation than simply walking up to a person and starting to talk. The way a conversation proceeds depends very much on the sort of conversation it is and the expectations of the parties concerned. A chat with a friend about the state of the country is likely to proceed along different lines to a more formal discussion with a superior about some aspect of work. In all cases a good conversationalist will be able to measure how well the information being sent is received: they use their interpersonal sensitivity to see the conversation from the other person's point of view and adjust it to produce the most effective result. They control the direction of the conversation so that the main point is not lost. They normally know what they want from a conversation and during its course they examine the possibility of achieving their goals and constantly aim to achieve the best they can from the situation. They can readily turn their speech to convince, persuade, alienate, sympathise, praise, chastise, caution, commend or to any other message they want to impart. Above all they always leave the conversation with everyone involved clear about what has been discussed and who is responsible for exactly what actions.

Drawings and signs

Most would immediately consider technical drawings as being the main use of these in the workplace. There are well-established standards and codes of practice for the representation of physical objects as drawings — for example British Standard BS308. A good example of the use of signs is the standard safety logos for various hazards: hard hats, safety glasses, and corrosives. Their main advantages are that they are very descriptive, overcome many language barriers, can be designed to attract attention and are interpreted very quickly.

Facsimile (fax)

Faxing involves generating a copy of an original document at another location by sending the information via the standard telephone network. Faxes have all the advantages of the telephone for speed and distance without its limitation of being unable to see a picture or have a written record. The person you are faxing does not have to be in attendance in order to receive the communication, it is transmitted and is then ready for collection at the receiver's convenience.

Grapevine

This is the unofficial communication system of any workplace. It involves the sending of information from person to person in an ongoing chain. Most of the information conveyed by the grapevine concerns individuals in the workplace. It may be slanderous, malicious, untrue, or any combination of these. Provided the information is from a reliable source and has not been through too many 'hands' it can provide an invaluable insight into what is going on, especially in judging matters that are otherwise difficult to quantify, like the mood of another department. Clearly, the grapevine is an intrinsically unreliable communication system, depending as it does upon the vagaries of human conversation, and for this reason it can never be considered as a formal communication technique. For this reason the grapevine ends up filled only with scandal, rumour and gossip and, exciting though these may be, their truth will always be questionable.

Meetings

These are convened to discuss all sorts of subjects. A potential problem is that as more people attend, it becomes more difficult to achieve anything. They are very useful for getting people to air their views and criticisms before a relevant audience or to review progress on a project. It is important not to consider them as a luxury for higher management as they give an opportunity for individuals to talk at length on a specific subject, allowing a rapport or sense of co-operation to be achieved.

Memoranda

These are either communications with people outside the organisation that are less formal than a business letter, or they are used as a method of communicating within the organisation which can be as formal or informal as required. They are characterised by being short communications that contain a few important pieces of information.

Sometimes memoranda (memos) are copied to more than one person to indicate to the others on the circulation list that the information has been passed on. For this reason, it is important that memos are dated. There is no defined format for a memorandum but many establishments have pre-printed memo pads or sheets which are convenient to use and which can be folded to save having to use envelopes.

Noticeboards

Noticeboards have a variety of uses. They can be used to display general information of interest, or to inform personnel of specific rules relating to work. It is important when using a noticeboard to put any information on display well in advance as very few people read a noticeboard more than once a week. Their main advantage is that they are cheap because only one copy is needed; however, unless they are properly managed they can become cluttered and will go unnoticed.

Pre-printed forms

These are used when something is required repeatedly: sickness, leave, expenses, stores issue. They give a standard format for anyone to use and it is easy to check that all the required information and authorisations have been included. Their great advantage is that they fix a protocol for communication in a specific case and provide a record for reference should any problems arise. Often filling in a form once produces several copies so each person involved in its implementation has their own copy in case the original is lost.

Suggestions box

Often workers on the shop-floor have very good ideas about improving the way their tasks are done. As this is not usually considered part of their job there has to be a method of getting these ideas to the people who can change the system. Rather than appeal to individuals' better nature, sometimes a financial incentive is used. Suggestions may be reviewed by a manager or a committee and a decision whether or not to implement will be made. Although relatively informal suggestions boxes can be quite motivational, as for shop-floor workers they may see the box as their only means of communicating with the 'management'.

Telephone

The telephone is the most important aid to communication that has been invented. It has revolutionised communication over the entire planet. The use of the telephone is a special skill and is so important to engineers and other professionals that we shall take a more detailed look at its use in Section 9.5.2 as part of our examination of oral communications.

9.3 Information-gathering

Information is often the most valuable asset we own. By knowing where the enemy will strike, a defending commander can win a battle before it has even begun. A company that accurately knows the internal affairs of a rival is at a great competitive advantage. Even something as trivial as a purchase is affected by information — one cannot know that a purchase is good value for money without first having information about the options available. Information can therefore be of immense value, but not all information is valuable. In the example the information has three main parameters that give it its value. The first is quality. The quality is clearly important: where something important is at stake, the truth of the information must be questioned. Before undertaking a major commitment one must be sure that the information upon which it is based is sound. The relevance of a piece of information also affects its value. If two people are bargaining over a sale, knowing the upper price acceptable to the purchaser is of great value to the seller, but of no interest to a third party who has nothing to sell. Lastly, the timing of the information is also important: no one is interested in what their commercial rivals were doing this time last year, it is the next year that is of value. Likewise, knowing where the enemy will attack is priceless beyond measure, while knowing where the enemy actually did attack is history.

Good engineers are good users of information. Like great statesmen or military heroes, they make well-informed decisions based on a rational treatment of the relevant facts. For engineers there are many types of information — technical details, personnel or commercial, for example. There are also many places from which the information can be gained. In this section we will look at some of the many sources of information engineers have at their disposal and discuss the best ways to use them. We will consider the customers of the organisations, a special group of people with information of particular importance to engineers. We will also consider suppliers, the boss, the use of libraries, personal filing, reading skills and note-taking.

In our everyday lives we are constantly in contact with potential sources of information. Every day we are exposed to far more information than we could possibly use. The important thing to do is to use one's objectives to define the information required and then efficiently obtain it. With the invisible commodity of information, it is quality not quantity that counts.

9.3.1 Customers

Just as car engines run on gas, so trading organisations run on customer satisfaction. People only spend money on things that bring them benefits. Yet many organisations hardly make any effort to find out what their customers really want. A common technique is to listen to complaints and react to minimise them. While this is better than nothing it is not nearly enough, and can never provide a complete understanding of the organisation's customers. A fact that is often quoted is that whilst only one in twenty-five customers complain of their dissatisfaction, on average they will each tell ten other people about a bad experience. The organisation is not only the last to hear about its errors, its poor performance is widely advertised.

Complaints are therefore a very poor indicator of customer satisfaction. If you want to know about your customers, you are going to have to find out actively. This can be done in two ways: by listening to them and by visiting them.

Listen to customers

This means much more than sending out a few questionnaires, listing the causes of complaints or looking at service department records. The need is for active listening, actually putting effort into getting the required information. One cannot aim to meet the needs of one's customer without actually knowing what those needs are, and one cannot expect customers simply to come along and describe them to you. Customers are often unsure of their needs themselves. An organisation that assists them in accurately identifying their real needs will stand a far greater chance of generating customer satisfaction than a less attentive rival.

Visit customers

The easiest way to get a clear understanding of your customers' situation is to go and spend time with them, doing what they do and seeing things from their point of view. The stereotyped image of good product development coming only from the application of research and development in a well-equipped laboratory, peopled with excellent academic brains, could hardly be further from the truth. Success in developing new products means meeting customers' needs and an organisation cannot expect to know what these needs are without being in good contact with the world of their customers. Visit them often.

Customers not only have information about their own needs that is of value but often are in possession of information that may be difficult to obtain elsewhere. They often know much of the market-place and products of rival organisations. Customers often freely volunteer this information as they try to play one supplier off against another.

9.3.2 Suppliers

Just as customers are useful in the provision of information so are suppliers. Organisations that supply products or services put effort into selling their products. Like any other organisation, they attempt to produce happy customers. A great service that suppliers offer is technical advice. There are so many processes that are used and so many factors that limit these processes that no engineer can be acquainted with all of them. However, if the supplier is offering a specific process then there will also be technical advice available on the process. When engineers face problems with which their organisation is unfamiliar, it is far better to get information from a knowledgeable supplier and to collaborate with them than it is to do it all yourself.

9.3.3 Boss and colleagues

One's boss and colleagues should be an invaluable source of information. A competent boss will be able to direct subordinates to information and guide them around the organisation but will often only do so if asked. The information held by the boss is especially useful to new recruits. Whilst it is certainly appropriate to use the boss to gain information about the upper levels of the organisation which cannot easily be found out in other ways, it is certainly not appropriate to expect a boss to provide general information available from other sources. The most important information a boss owns is a knowledge of which objectives are most important and why. In addition, a competent boss will be able to sort out an employee who gets stuck, either by helping directly or by directing to a source of information that will help.

One's colleagues are also sources of information and can be used to assist in learning the procedures of the organisation, the technical issues or the history of the work currently under way. This should be a two-way process with the individual providing something in return. Colleagues soon tire of repeated requests for help without getting anything in return; learning by helping is both infomative and fair. Colleagues that need to exchange information more and more soon develop into teams. Teams can be especially efficient at information handling. Chapter 5 describes this subject in detail.

9.3.4 Library

A library is a vast information storage and retrieval system. Libraries vary in size considerably but even the smallest usually contain far too many books to allow users to find their way around unaided. Many systems of referencing have been developed to assist the users of libraries in their search for information. At most libraries it is possible to search the whole collection on a computer index by author, by subject, or by 'keyword'. Good information-gatherers are expert users of their library systems and regard their library as an extension of their own filing system. Good engineering libraries contain not just books, but journals, trade magazines, abstracts, standards and newspapers. Often there are 'CD-ROM' machines with a database of the articles not contained in books that may be searched to provide listings of subject-related references.

9.3.5 Personal filing

We all have our own personal filing system. The function of the system is to permit rapid retrieval of previously gathered information. This extends from the way we categorise and store our own books, magazines and collections of other printed matter, through to how we arrange to store information such as when the local swimming pool is open. Most people have reasonably effective storage systems for their books and notes, but few people have a comprehensive system that deals with all the information they collect and might want. Personal filing is a part of personal management and a simple system is described in Section 6.2.2.

9.3.6 Reading

Practically everything that is worth knowing is written down in some way, somewhere. Reading is therefore a particularly important part of information-gathering and it deserves special attention. Printed matter has many advantages over other forms of information. It is permanently recorded and may be left and returned to later, unlike a conversation. It does not fade or change as verbal accounts may. By using the conventions of good writing such as contents list, structure and cross-references the author makes it possible for the reader to search quickly entire volumes for specific pieces of information. No speech or meeting can do this. Printed matter, or rather the characters, syntax and grammatical rules that compose it, is designed to assist the information-gathering process. There is more to using it well and people like engineers, who need good information skills, need to be effective readers.

Reading is a skill. We normally learn to read very early in our lives and many people cannot even remember being unable to read. As we grow up we improve our reading and use it to serve many needs. In our adult life we read at many different levels. At one end of the spectrum is the acquisition of factual, detailed — even mathematical — material which is very condensed and contains a great deal of meaning. At the other end of the scale might be a collection of articles passed to an executive for glancing at to see if anything of interest is contained within them. In this case the reading is characterised by great speed and even skipping over complete sections. The reader here uses the structure of the material to decide what level of attention is appropriate. No one should assess the relevance of an article by reading the whole thing. On coming across a new scientific paper, for instance, the reader will probably read the title, rejecting the work if it is not of interest. If the title is not recognised or sounds as if it might be interesting then the paper might be opened and the abstract or introduction read. The decision to proceed is constantly reviewed. This process emphasises how important it is to structure and prepare written material well. A reader will conclude that poorly structured material is a worthless product of a muddled mind. In the vast majority of cases the assumption is correct. After all, if someone cannot even marshal their thoughts into a sensible and logical structure what hope is there that the material itself will be of any use?

Even when the reader finds a section of particular interest, the material will not necessarily all be read in depth — far from it. The skilled reader will use a speed and depth of acquisition appropriate for the particular reading need at any given instant. The following techniques show the sort of range of reading techniques any competent reader can use. Such a person will often use all the techniques during the reading of just one page.

High resolution: In this technique reading is very slow, often going back to earlier sections and re-reading sections in detail. When finished, the structure of the whole section can be remembered. The whole structure and the line of reasoning are memorised, allowing the reader to be able to reproduce the material and explain it to someone else in a different way. Perceiving the structure of a body of knowledge is the gateway to deep understanding. Such understanding, not surprisingly, takes time. The ability to re-explain material in a different way is characteristic of deep learning and a good test of it. In general, material once deeply learnt will be retained for a long time.

Normal speed: This is the speed at which the reader naturally takes printed matter. It is the speed usually used for recreational reading. The normal reading speed varies greatly from person to person but it does change as the reading ability changes. It can be used to measure reading development in children and, in general, a slow natural reading speed goes with poor general reading skills and a lack of ability to vary the reading style. A high reading speed is associated with the opposite. This is due to the fact that good reading skills are a matter of practice.

Speed reading: In this mode the reader does not acquire detail and after reading the section the reader may well not remember large portions of the text. The purpose of the technique is to let the reader get quickly to areas of interest. Once interesting material is found the speed goes down and the detail acquired goes up. In this mode the reader often does not settle the eye on every word. Often only a few words per sentence are actively 'read'. Sometimes groups of words or whole phrases are acquired at once. This technique takes advantage of the intrinsic structure of the language being used to allow the reader to gain understanding of the material in an overall way. Practice can greatly improve this technique and it is often the case that people under-use this mode.

Scanning: This is an extension of the above. More material is left out. Sometimes only the first and last sentence in each paragraph are read. Whatever is done the important thing is that the reader gains an overview of the document, knows what sort of thing is in it. As with speed reading the purpose is to allow the reader to make a decision about whether more or less detail is required.

9.3.7 Note-taking

Note-taking is a process of synoptically recording information whose delivery speed or situation does not permit the receiver to control the acquisition process. The notes may then be used for detailed learning at the individual's own pace. Good notes reflect the structure, key elements and unique facts gained during the receipt of the communication. The best notes are made by collating rough notes, made at the time, into a complete and well-presented set. Usually this involves researching the information from other sources such as books to augment what was presented and make clear any areas of uncertainty.

Notes are used in may ways. They might be used to make a temporary record of direction and be thrown away afterwards. Alternatively, an engineer visiting a customer might take notes during the conversation and later they would be written up in detail, being careful to record every last item that might be of interest. Notes are most commonly used for recording the contents of a lecture of presentation.

9.4 Written communications

We all need to communicate but some of us have a greater need than others. Engineers and technologists are amongst those whose requirement to communicate, as part of their job, is absolutely essential. There is little point being a creative desgner if you can't sell your idea to someone, can't explain how to make it, or can't explain how it works or what it does. Other professionals will need to read your explanations, your manager may need to read your reports, the financial staff of your organisation may need to have the financial implications of your work explained to them. How will you achieve these aims without being effective at written communication? Similarly we need to be able to understand other people's communications: as engineers it is often necessary to understand many different forms of communication such as design reports, technical specifications, test reports, and market reviews.

Outside the workplace we still need to communicate; for instance, applying for a job involves communicating ideas that suggest that you are better suited to a job than anyone else. More simply, you might just want to make a complaint about a product that you have purchased. In all of these activities written communication is of the utmost importance.

9.4.1 Factors affecting written communication

Some of the information that humans communicate to each other is of transitory value only. Knowing that the departure of a particular flight has been delayed by two hours is only of interest to passengers waiting to catch it. However, other blocks of information will always be useful. The works of Isaac Newton, for example, are constantly referred to and the texts that are published are simply contemporary accounts of the original principles. Written communications have evolved to serve the communication needs associated with information that has a significant lifetime. When speech is in use it is possible for feedback to occur — if the receiver does not understand the communication, it can be questioned and the meaning explained by the sender in a different way until understanding has been confirmed. Written matter, however, does not answer back. For this reason any system of written communication must have firm rules of construction upon which it is based to enable understanding. It does not matter what these rules are, only that they are known to the communicators, and humans have developed many very different languages that can all cope effectively. Some languages even have significant differences between the written and spoken usage. In China, for example, some regional differences in dialect and interpretation of the language are so great that people from different regions can easily communicate in writing though they cannot understand a word of each other's speech.

The rules of language exist to make communication easy. The only things that set limits upon these rules are the limitations and needs of humans themselves. Our voice-box and acoustic interpretation skills have a limited speed of operation, our physical size and hand–eye co-ordination dictate ideal dimensions for characters upon a page.

In essence, written communication is only a permanent record of spoken communication and so the rules of language apply equally to both. However, it is within written communication, where the very structure of the language is open and visible on a page, that the rules of language are most easily scrutinised.

Many things impinge upon the ease with which a section of writing can be understood. In the following six sections we shall explain the role of some of the most important.

The parts of speech

Language is a code. Each word has its own unique meaning but some groups of words have similar functions. The function that a group of words serve in language is called a 'part of speech'. The parts of speech exist to make the language clear and complete. Examples of commonly known parts of speech in English are the noun, the pronoun, the verb, adjective, adverb, conjunction and preposition. By studying the functions of these parts of speech we can see how the language operates. For instance, consider the verb 'to run'. Phrases such as *he ran* or *they were running* or *don't run*, all have clear, unambiguous meaning because of our knowledge of how the verb 'to run' should be modified in each case. In other languages there are different rules. These rules are particularly important in written communications because we have no opportunity to say something again if we make a mistake. Written text must be correct. The purpose of dividing the language up into these parts of speech is that with the description of the language that they provide, we can use them to define the absolutely correct usage of our language.

Punctuation

Blocks of words that do not have punctuation are impossible to understand easily. Punctuation is a written substitute for the pauses and non-verbal communication that accompany spoken language. However, punctuation can be more than just this, and if used properly it greatly assists the rapidity with which a piece of writing can be understood. Correct use of punctuation is not trivial and people usually underestimate its importance — even the humble comma can be used to alter completely the meaning of a sentence. For example, '*I pressed the button, not meaning to endanger life*', has a very different meaning to, '*I pressed the button not, meaning to endager life*.'

Spelling

To a good reader, poor spelling is an offence against the language. Good spelling exists not only to avoid ambiguity in meaning, as between '*too*' and '*two*', but also to allow speed in reading. Good readers go too fast to stop and construct each word from its letters, and the whole word is often acquired at once, especially for small words. If a particular word is misspelt the experienced reader's eye stops on the word because it is noted not to be the word at first imagined. This interrupts the flow of reading and efficiency of communication and must therefore be avoided. Some people find spelling mistakes annoying simply because they indicate a lack of care and attention, others feel the language is being weakened by colloquial acceptance of alternative spellings and jump on the errors as proof. Whatever the reasons, you are unlikely to impress your reader by spelling badly. The implication will always be that someone who cannot get such simple things right cannot have spent much time on it and therefore the content is also suspect.

Sentence construction

The sentence is the basic unit of the language. It is a collection of words that has a complete meaning in its own right. Any sentence abstracted from any piece of good writing will have meaning. This does not mean that one can always know what the author meant by the sentence; for example, '*They are always doing that*' is a sentence but we have no idea what '*they*', actually are, nor what they are '*doing*'. Nevertheless, the sentence is complete on its own. Many issues impinge upon correct sentence construction: 'numerical agreement' between subject and verb, for example — '*two bolts are steel*', not '*two bolts is steel*'. The 'tense' of the verbs within the sentence must agree. The 'case' of a sentence must be correct: '*they were the ones*' is correct, not '*them were the ones*'. Most rules of sentence construction come naturally through using the language — we instinctively know which version is correct and many people don't even know the rule (in this case 'that the pronoun should be in the nominative case if it is the subject of a verb') that they are applying. At its most sophisticated the science of sentence construction is extremely complicated; once again, however, this complex system of rules, conventions and exceptions exists to make the language easy to follow and being skilful in its use is likely to aid your ability to communicate effectively through the written word.

Style

Everyone has their own style of writing, language is such a versatile thing and people so different that it is possible to identify authors simply by the style of their work. The most important issue about style is that it, like the factors above, exists to improve the reader's ability to understand the material. Even the distinctive style of a great novelist or playwright exists for this purpose and the enjoyment that well-written text brings is an important ingredient of style. Good advice about style stems from the factors that improve the effectiveness of written communiction. Below are some points for consideration.

Avoid using many words when few will do: For example, '*the audibility of oral delivery should be received by the listener with sufficient acoustic energy for reception*' is an unnecessarily complex way os saying, '*speak clearly*'. No one is fool enough to think over-complication is a sign of intellect.

Avoid ambiguity: Many simple ways of expressing things allow for misinterpretation. Ambiguity can creep into text in the most unexpected ways. '*Nothing is more effective in relieving stress concentration than increasing the radius of the concentrating feature.*' Does this mean that increasing the radius *is* the most effective thing, or is '*nothing*' the most effective thing, meaning doing nothing is better than changing the radius?

Avoid saying nothing: '*An appropriate value of ballasting resistor is selected by using a technique which permits the determination of an ideal resistance.*' The sentence really says only that '*to choose a resistor, you use a resistor-choosing technique*'. No informed reader will be impressed with such an insult.

Avoid accidental untruths: It is quite easy to make untrue statements in writing whilst trying to illustrate a point. For example, '*No one who has simply written a business letter and posted it immediately can have prepared it properly.*' This may well be true for an individual case but the statement says that it has been true throughout all time and this is clearly not something the author could know even in the unlikely event of it being true.

Avoid jargon: Science and commerce are pervaded with jargon words that restrict the audience of the text. It is very annoying for the reader to find a TLA slipped into the text without explaining what it is! The correct use of jargon is an efficient way to avoid unnecessarily repeating a long word many times over. Deoxyribonucleic acid is commonly replaced by DNA — this one simple abbreviation alone has probably saved acres of paper. Everyone knows what it means (unlike TLA above which stands for 'three-letter acronym') and the effectiveness of a written communication is not compromised by its use. If a particular text involves repetition of a particular group of words for which there is an abbreviation then use it, but define it the first time it is used, and either include it in a glossary or index the first entry where it is explained.

Handwriting

Most written communications are not handwritten but even the wide availability of word processors has not meant that handwriting is dying out, far from it. Handwriting is a very versatile technique for making marks on a page and those who can do it quickly and legibly will always be at an advantage. The way the characters that compose the English language have evolved is by no means arbitrary. The characters have conflicting design constraints impinging upon them. On the one hand they need to be very different in shape from each other so that they are distinctive on the page and the eye can easily distinguish one from another. On the other hand they need to be very similar to one another in order that the hand can easily produce them in a flowing, easy-to-control way that permits speed. The arrangement we have today represents the culmination of much evolution. The character shapes are all based around the same simple shapes and the curves used to join one character to the next are all of the same form, making them easy to reproduce rapidly. The differences between words come from surprisingly small differences in the characters and some of these differences are added after the word has been written. In writing the word '*section*' most people write out the whole word and then come back to cross the '*t*' and dot the '*i*', allowing them to produce the word at greater speed. It is also true that the current design of the characters allows for a great variation between individuals, which is a disadvantage since it makes the character set open and the more an individual leaves the basic font the harder it becomes for the reader to understand the text. Just as with the other important factors in written communication there are books describing handwriting in detail and contrary to popular belief it is quite possible for adults to make big improvements in their handwriting abilities if they so desire. For those with bad handwriting this is certainly advisable: being unable to write something that another person can read easily is a great life disadvantage.

In this section we have considered various factors that affect written communication. Some of them, grammar and parts of speech for example, often seem to be tiresome rules that serve to make written communication difficult. In fact these rules and protocols exist for the opposite reason and are very effective at their task. As professional engineers we must stop thinking of them as a source of loathing for generations of school children, and start thinking of them as the science of communications — a science at which we can excel.

9.4.2 Preparation for creative writing

A written communication must be effective. You must be able to state what you want in a way that your reader will be able to understand. With oral communication testing this understanding is easy, you can see by the body language of your listener, you can ask questions. Because you are dealing directly with the receiver you can react to feedback and change the delivery to make it appropriate. You cannot do these things with written communication and therefore it is particularly important that you prepare well beforehand. When your reader uses your material you will not be on hand to explain it, so the text must be self-explanatory. Something written by you is your envoy, it goes before you, people form opinions about you based upon it. Because of these issues most people want their writings to be looked upon favourably and so they employ some effort in making sure that the communication if effective. In this section we shall look at the seven most important issues that should be addressed before pen is put to paper, or finger to keyboard. A careful consideration of these issues should be made before embarking on any type of written communication. After we have introduced these important general issues of writing, Section 9.4.3 will look at some specific types of communication that are important to engineers.

Your reader

It is important to think about who your intended reader is. You need to have a picture of what the reader will want, or need, from your communication. You must also consider the knowledge level of your reader, particularly when using technical terms and jargon. You do not want to alienate your reader by detailing every basic point, but at the same time you want to ensure that the terms you use and the things you explain are understood. Your reader is your customer and, as with any business, your success depends on happy customers.

Your objectives

You need to be absolutely clear about what you are trying to convey. What are the limits of your subject matter, what topics should be included and which things left out? Are you trying to inform your reader or explain something, do you wish to persuade or to influence your reader? You have to decide what your motives are so that you can choose the most appropriate style and language. The reader will look at the text and formulate an opinion on what the work is about. If your writing is unstructured and often leaves the central aim, drifting into things without explaining to the reader the need for them, the reader will soon tire of the text and consider it too poorly written to deserve attention. You would not read something so poorly written, why should your reader?

Authority

The authority behind the communication should be made clear. A discerning reader will want to know why the communication should be believed. It might take the form of a quotation list showing the sources of information or it might be a statement of the author's position. It might even involve a description of the author's life, pervious publications and contemporary reports. Not all communications, however, have such a presitigious validation. A personal letter from a friend might contain a recommendation we choose to follow; in this case our knowledge of the individual gives the letter its authority. The authority or power behind a communication is so important that almost whenever we receive an important text we ask, 'Who is it from?'

Readability

The language you use must be appropriate to your intended audience. You should not write a report for your boss in the same way that you write a letter to a friend and no one would write to their bank manager in the same way that they would write instructions for a child. The general issues covered in Section 9.4.1 should be considered from the specific point of view of each communication you make. The issues of spelling and punctuation need always to be beyond reproach but the complexity of the grammar, the vocabulary, the style, and the structure you use should be varied to suit the reader.

Media

You need to decide what medium is most appropriate for your communication. You need to be sure that writing is the most effective way of conveying the message but then you need to decide whether it should be handwritten, word-processed, or produced on a desktop publisher. How large should the communication be? An advertising leaflet and a specification leaflet may be describing the same thing but they usually have a very different size, number of words and approach to the communication. Will a photocopied sheet of text be acceptable to your reader or is a glossy pamphlet expected?

Pictures, diagrams and tables

You need to decide whether these things will enhance or detract from your message. 'A picture paints a thousand words' but if you use a poorly produced diagram or inappropriate picture you will confuse the reader and make understanding difficult. As with language your use of pictures and tables must be thought through to ensure that they are appropriate to your reader and that they convey your message effectively.

You must always ensure that when you are using these to show points or describe something you make reference to them in the text and clarify what you want your reader to gain from them. Diagrams can take a great deal of effort to get right and , just as with text, the reader will conclude that if the diagrams have not received attention it is likely that the content did not either and the quality of the whole communication may be questioned.

Checking understanding

It is difficult to check understanding in a written communication since there is usually no immediate feedback between author and reader, but it is not impossible. Redundancy is the term used when points made previously are re-explained using different language. The intention is that if the reader did not understand the first explanation they will understand one of the others. The danger with redundancy is that in excess it will bore the reader, but if too little is used understanding may be compromised.

Other techniques for checking understanding include making statements that explain what the reader should have understood by particular points in the text. If you come across '*We have therefore demonstrated that only the principal stresses play a role in failure prediction*', it is clear what should have been gained from the text. If the reader does not understand then a return to earlier sections is needed.

Questions are an excellent way to test understanding and it is common for textbooks to contain progressive questions that test successively deeper and deeper sections of the material. A very good test of deep understanding is whether the reader can express the information in a different way. Of course the author is not there to check the answers and so either the answers are reproduced, or the answer must be available within the body of the text. In this way the readers can monitor progress for themselves.

It is common for text to be written for a range of abilities and this makes it difficult to ensure that all the material is appropriate to everybody. If it is difficult to do this it is better not to try: rather, include sections for people with little understanding or knowledge and other sections for those with previous experience. If the sections are clearly identified readers will easily be able to use the contents, cross-referencing and index to find the way to the sections they need. If, however, these issues are poorly treated it will be impossible for the readers to find their way around and the text will be practically useless.

If there are large amounts of prerequisite understanding, too much to reproduce in the text of the communication, it may be appropriate to define the knowledge required. This may be done either as a start to the text or within the body of the text by guiding the reader to other helpful texts at the point where they are likely to run into difficulty.

9.4.3 Specific writing techniques

We will now consider some specific forms of writing that are widely employed in engineering. These are reports, specifications, instructions and business letters.

Report writing

In the world of engineering you may be required to produce many types of reports. Technical specifications, progress reports, feasibility studies, laboratory reports, project management reports, personnel reports, or customer visit reports. A report is usually a fairly formal document used for consideration by others in the formulation of a decision. Reports may have much information to present, they may need to highlight many issues raised by the information and they may need to draw conclusions. All reports are self-consistent rational bodies of text that permit the reader to understand the issues at stake. Reports are characterised by introducing data, discussing its meaning and limitations, and then drawing logical and relevant conclusions.

Below are some rules for preparing management and project reports. We will consider two issues: firstly, the format of the report and, secondly, the contents of a report.

Format of a report: The format that you use for your report defines how your report will look. It is important to present a consistently formatted document since it assists the reader in following the structure of the report. The format of this book, for example,has been very carefully maintained throughout and is the same from cover to cover. References are always made in the same way and the chapters all follow a similar structure. These things make it easy for the reader to follow. The sorts of issues that need careful standardisation are chapter and section labels, how many lines are left between each section, whether there will be more to separate major sections than minor ones, how lists are given and how references are made.

It is particularly important to think about these and decide what you want to do before you start writing, since you can then write in the correct format and reduce the time that would have been given over to 'tidying up'. If you are writing on a word processor, it is just as important as you can often set up many of the format parameters within the software and so save time.

Tables and figures also need careful format attention. Clearly you will have to decide what they are going to look like and what typeface you will use for labelling. You will also have to give them numbers, or references, and titles. You will have to decide where they are going to go and how the text will reference them. Diagrams are usually included in the body of the report as the text references them but sometimes they are all left to the end. If you want to put them in the text you have to decide how much space to leave above and below, i.e. how you break the text. For figures and tables less than a half a page long it is possible to include them within a page of text as near as possible to the reference; if they are much longer it is usually best to put them on their own page as soon as possible after the reference. If you put all the tables and figures at the end of the report then they are normally put after the references and before the appendices. Often all the tables are put together first and then all the diagrams.

Finally, always proof-read your work to ensure that you have eliminated spelling and grammatical errors, as these will only detract from your work.

The contents of a report: In its longest format a report may contain all the sections listed below:

Title page
Abstract
Contents
Nomenclature
Introduction
Text
Discussion
Conclusion
Recommendations
Acknowledgements
References
Appendices

Although it is not normal for a report to include all of these sections most reports contain many of them. We will now examine the typical contents of each one.

Title page: The title page should be the first page that you see when you open the cover of the report. It should give clearly the title of the report, the name of the author, the organisation for which it was commissioned, the date, and other information appropriate to the situation. The title page may also include the abstract depending upon the size of the report and the size of the abstract.

Abstract: The abstract is a summary of what the report is about. It should be short but must provide enough information for readers to decide whether or not there is anything of interest to them in the report.

Contents: The contents list should give chapter numbers and titles with the number of the page on which the chapter starts. In a long report, or even a short but complex report, it may also be appropriate to list sections and subsections. A contents list for figures and tables should be given, after that for text, containing a list of the items to be found in the report and the appropriate page numbers. The way in which titles are written in the contents list should be the same as the way in which they appear in the text.

Nomenclature: The nomenclature section defines terms, symbols and abbreviations that are used within the text and which you want to be sure that the reader understands. It saves you having to repeat definitions throughout the text and prevents misunderstanding, particularly where common terms or units are used.

Introduction: The introduction is the background to the report. It may be an introduction to a specific problem that your report attempts to solve, or it may be a more general introduction giving some of the background explaining why the report was written.

Text: The text is what your report is all about. It should be divided into logical chapters that follow through the subject matter. If appropriate, chapters should be subdivided into smaller sections and subsections. The titles that you use for each chapter and section should give an indication of their content. It is normal to include data in the text although if there is a large amount it may be summarised and then included in full in an appendix. If the data is too large even for an appendix it should be referenced as a separate entity.

Discussion: If your report gives results or work that should be compared with other similar work then you will need a discussion section. You may give it a different title but within this chapter you will be discussing the relevant and pertinent points of your data and arguments. Discussions can be very lengthy and if so they should be divided up with their own structure to make reading them easy.

Conclusion: The conclusion to any report should not be very long. It should provide an end to your report, summarising your findings and your arguments. The analysis and background will all have been done elsewhere in the report and there is no point in duplicating it.

Recommendations: If you have done work that requires further work which you believe could be logically carried forward, or if you expect some action following the submission of your report, this is where you make it clear. The recommendations of very large reports are often very short; for example, at the end of a 200-page marketing report you may find, '*In order to achieve its strategic goals Haldone Freionics should open a US subsidiary company, introduce a new model of medium-power fridge for the home market, discontinue its off-site manufacturing*

and secure a loan to finance the redevelopment of the main factory.' As with the conclusion, all the argument for this should be contained within the body of the text.

Acknowlegements: You do not have to include acknowledgements although it is usual if anyone has provided you with specific help that has allowed you to do the work upon which your report was based.

References: Throughout any project report it is likely that you will have drawn upon data, ideas and material from other people. Your report should clearly indicate when this happens by making reference to the source material. The reference section of the report should provide a list of all the material used and it should be possible from the way in which the text is referenced to find which particular document is being referred to. If the number of references is fairly low you may choose to distribute them throughout the text rather than collect them all up into one section. A typical method of referencing is to insert a bracketed number which then points the reader to the reference that has been used: '*The necessary understanding of hydraulic pumps may be gained from Sayers (14), his explanation of pump options makes clear the need for* . . .'. Later in the report, possibly at the bottom of the page or in a separate reference section at the end of the chapter or report, details of the reference will be given. There are many standard formats in use and a common one is shown below — the important thing is that the reader has enough information to obtain a copy of the referenced text.

(14) *Hydraulic and Compressible Flow Turbomachines*, A.T. Sayers, pub. McGraw-Hill Book Company (UK) Limited, Berkshire, England, 1990 (ISBN 0-07-707219-7).

Appendices: Appendices should be listed at the end of the contents section and included at the back of the report. If you have more than one appendix it is usual to label each appendix as Appendix A, Appendix B, or I, II etc. It is not usual to number the pages in the appendices, although if you do this each appendix is numbered as a document in its own right, and the report page numbers do not carry on through. An appendix should be used whenever you have some background material that some readers may wish to look at but which might be known, or not required, by other readers. For example, if you were reporting on a project that had required a lot of statistical analysis you might include statistical tables, or relevant theory, in an appendix. This would mean that your text only contained your results and the results of the analysis.

Design specification:

1) Height must not exceed 1.2m
2) Manufacturing cost to be less than £5.00 per unit
3) Weight, fully loaded, not to exceed 50kg

Product specification:

1) The height is 1.15m
2) Manufacturing cost is £4.00 per unit
3) Weight, fully loaded, is 43kg

Figure 9.2 Specification

Specifications

A specification is a detailed description of all the important aspects of the subject it describes. Sometimes the specification will describe technical issues and will be a complete technical expression of a product's performance; other specifications include marketing specifications, in which the commercial profile of a product, often imaginary, is defined. Specifications are prepared by engineers to describe designs, commercial performance, purchases, or assembly procedures. At any time when important parameters need definition specifications may be prepared.

Engineers are perhaps most frequently involved with specifications for design work. This is a two-stage process which includes, in the first instance, preparation of a design specification. In this case the specification will define what is actually required of the design but will not suggest how the criteria are to be met. This allows other people to make an input to the specification but ensures that the designer can apply maximum creativity in meeting the requirements. After the design work has been finalised the designer will then produce either an equipment, or process, specification which defines the criteria that the equipment or process actually meets. It will usually be much more detailed than the original design specification. For example, the design specification might give a maximum size and weight for the equipment whereas the equipment specification will say what the actual size and weight are.

An example showing the difference between a design specification and a product specification is shown in Figure 9.2.

Section 3.6 contains a description of how a technical specification may be used in the product development process and should now be read since it is only by understanding how such a document is to be used that it is possible to write one effectively.

Some general guidance for the preparation of specifications is given below.

1. Specifications must not be ambiguous; it may be difficult to achieve this if you do not use technical language and terms.
2. Aim to use the minimum number of words and clauses; for example, some requirements may be presented in tabular form and contain many numerical descriptions.
3. Language should be simplified and care taken to avoid subjective statements such as 'of adequate strength'.
4. When preparing the specification, aim to define optimum quality for the job rather than highest quality. Over-specifying can increase the design time and cost to a point at which the equipment ceases to be economiclly viable.
5. The specification should define essential features and therefore should not contain phrases such as 'if possible' and 'preferred'. The obvious question is 'at what cost' are such things wanted? It must be possible to decide whether these things are needed and the specification should define the answer, not pose the question.

Instructions

Instructions in an engineering environment can be given in a written or pictorial format, or a mixture of both. Instructions are the means by which you get someone to produce exactly the result you want by following a defined procedure. This does not mean the instruction has to define every last elemental step the operator must follow. This would make the instructions far too lengthy to be practical and would limit their use to a very specific task. Good instructions make good use of the skill level of the person who will use them. An engine assembly manual might contain the instruction '*tighten the head bolts to* . . .'. This instruction is not enough to describe the whole process to the layman but a skilled mechanic will know that implied in the instruction is the tightening of each nut in turn, the use of a torque wrench, the correct 'pinching up' of the gasket and checking of both faces for burrs.

Some companies will have standard forms for various types of instructions, for example test, manufacture, inspection. Although the forms may be different the guide-lines for their production are the same.

1. Break the task down into a series of small, logical steps.
2. Describe each step in understandable language.
3. Refer to the equipment, or draw suitable analogies, so that there is no misunderstanding.
4. Use diagrams and models to illustrate what is required.
5. Ensure that the reader knows what the aim of the task is, when the task is complete, and what is supposed to happen next, if anything.
6. Include tests for correctness at important stages: statements like '*You should now be able to see the flange inner-rim bearing on the swashplate*' help the reader to confirm that things are going as planned.

INSTRUCTIONS FOR ASSEMBLY OF CIRCUIT BOARD W23

CARE: You must use a static protected workstation for this assembly

1. Locate components as shown in drawing W231
2. Solder all joints in accordance with soldering standards
3. Clean all joints and carry out continuity checks
4. Attach connector strip to board as shown in W231
5. Wrap board in bubble-wrap and place in Stores

Figure 9.3 Example of an instruction

It is particularly important to be clear for whom you are writing the instructions, so that you use appropriate vocabulary and appropriately detailed steps. You should also check the instructions carefully to make sure that you haven't omitted anything important, and that you have given any necessary warnings.

Figure 9.3 shows an example of an instruction. Notice that it makes clear reference to more detailed documentation.

Business letters

Business letters are constantly in use by professional people. You may wish to apply for a job, you may need to contact a supplier, you may need to complain to a customer for non-payment of an invoice or you may need to complain about service received from a supplier. All these situations need well-written letters. It is a mistake to think that composing a good business letter is quick or easy — some of the most important ones can take hours to deliberate over and get exactly right. If you have simply written the letter and posted it, the chances are that it has not been properly thought out. Often a well-written letter will produce a response when a poorly written one will not. If you were on the receiving end of a poorly written complaints letter in which all the problems were your fault and there was no acceptance of responsibility on the part of the letter writer you would, like most people, probably feel less inclined to act in favour of the complainant. People often mistakenly think that because business letters are about business they should be cold, formal and cannot offend. Businesses are operated by people and it is still a person who will read your letter; therefore be courteous, fair, clear, straight to the point and give all the information required but don't be long-winded.

Figure 9.4 shows a widely accepted format for a business letter.

Your
address

1st January 1993 Date of writing

Name and
address of
correspondent

Ref: LB.dbf

File reference

Dear Sir Depending on how well you know the person, could be Charlie/Sir/Mr Person

Text

Text

Text

Yours faithfully If you address the person by name use sincerely instead of faithfully

Your name signed; appropriate to the
way you addressed the letter

Ann Other
Senior Engineer Your name and job title printed

ENCS If anything enclosed with the letter

Note: if someone else signs on your behalf it is shown:

Yours sincerely

pp Ann Other

Figure 9.4 Business letter format

9.5 Oral communications

Oral communication allows an exchange of information between people able to see or hear each other. It probably developed in the social groups of our primeval ancestors and was used initially to alert members of the group to the arrival of danger, or the location of water and food, and to impose the hierarchy of the group. However, before long a communication system had developed and the exchange of more general information became possible. The tool developed for this purpose was the question. Early questions might have been sounds to ask 'Which direction is the danger?', or 'How far is the water?' The greater the ability to communicate such things the better the groups were likely to fare. consequently there is an evolutionary effect which favours good communicators. From these simple, early oral communications, and the evolutionary pressure that favours those who are good at them, language developed.

Oral communications have limitations, however, and other types of communication have developed to overcome them. The greatest weakness is that the whole communication has to be remembered, which limits the complexity of the communication to the memory capacity of those involved. Thus to communicate lengthy messages or ideas that would be required in the future, an alternative to the transitory memory was needed. The permanent marks of mud or charcoal on cave walls were probably the first use of a permanent record and their modern counterpart is the printed word. The pressure to succeed by communicating still exists today, communications are still evolving and within organisations there is no doubt that success comes preferentially to those who communicate well.

Oral communications are used in many ways, from public speaking to private scheming, bringing a board of directors to a consensus, dealing with an argument or conducting a job interview. We all communicate orally, every day. Nevertheless, some people are much better at communicating than others.

Apart from the memory problem another major difficulty with oral communication is that the quality of a communication relies on the communication skills of two people, a receiver and a sender. The receiver is the person receiving the communication, listening, and the sender is the one talking. As a sender one needs to exert some influence over the receiver to ensure that the message is being both understood and given the weight that it deserves. This is a skilful matter; however, there are some ground rules which will help, such as structuring the delivery, checking understanding, summarising periodically and delivering in a way that is likely to interest the receiver.

In the workplace we use many different methods of talking to each other. There is the primarily one-way communication, as exemplified by the oral presentation. There is the two-way, information-gathering type of communication, as exemplified by the job interview and the more discursive negotiation process. Finally, there is the group communication that takes place in meetings.

In this section we will start by looking at the factors that affect oral communication, the sending and the receiving. We will then look at specific techniques for achieving effective communication in presentations, interviews and negotiations. Finally, we will consider some ways that can ensure effective use of that much abused instrument, the telephone.

9.5.1 Factors that affect oral communications

In any oral communication there will be at least two parties, the sender and the receiver. The success of the communication depends on the sender's ability to make the message clear, and on the receiver's ability to listen to the message and understand it.

As a sender there are a number of areas that will affect the clarity of the message. The first point to consider is that the message should be audible. Oral communication means using the mouth to communicate. Usually this means making a noise with it, although there are many non-audible communications such as mouthing, kissing, pulling faces, and sticking your tongue out. When the mouth is used to make a noise it is clearly important that it makes enough noise to be received and understood by the target of the communication. Words should not be mumbled, they should rise above the background noise. Adjusting the volume of your voice is difficult, particularly when dealing with unfamiliar surroundings and you do not know how well the sound will travel. It is almost as important not to shout at people, as generally this is perceived as aggressive and may deter people from listening.

Different audiences have different speeds at which communication is best. An engineering course which extends earlier knowledge might well open with a review lecture, which would cover the material only once and at quite high speed. On the other hand, a message delivered using a public address system at a rail station announcing a change of stations for a connecting service would be given very slowly and repeated often. Determining the speed at which the message is conveyed can be done by knowing something of the audience; however, a sender should also be able to vary the speed according to audience reaction, either speeding up or slowing down.

Making the message audible and controlling the speed with which it is delivered obviously improves the ease with which the message can be understood. However, on their own they do not ensure that the message is likely to be received and understood. To meet these needs we need to hold the attention of our receiver, and we need to make sure that the receiver understands what is being said.

To hold attention we need to make the communication interesting. Interest in a verbal delivery can be stimulated in a variety of ways:

humour, anecdotes and enthusiasm are powerful techniques. In some cases the interest may come from circumstances, as in the case of the station announcement above, in other cases it is important to engender interest. For instance, a group of communication students will not respond well to a communications course unless it is delivered by staff who fully appreciate the importance of the subject and construct a course that includes practical seminar experience together with theory and relevant examples. The language used and the style of the delivery must be appropriate to the target audience if you want to maintain interest.

Making sure that the receiver understands what is being said is a continuous process. Firstly, we need to use devices that will help understanding, such as using diagrams, repeating information, and using analogies. Secondly, we need to check understanding. The outcomes of conversations are often very different from the outcomes imagined by the participants. The development of misunderstanding is a constant risk during conversation. For this reason, good oral communicators always check understanding at key points during the communication regardless of what variety of communication it is. This is most effectively accomplished by asking probing questions that require the receiver to process the information in some way in order to prepare the right answers. In ordinary conversation this may be done simply by asking if the other person understands or is happy with the conversation; however, the sender should also look for signs that indicate incomprhension — the glazed look, doodling, doing something other than listening attentively.

Summarising so far; to send a message effectively we need to make sure that it is audible, that the speed and style of delivery is appropriate to the audience, that it is interesting, and that the receiver understands what is being said.

In addition to considering the content of the communiction and general style of the delivery there will be other factors that affect the way in which the message is received. Consider the effect of things like your behaviour, dress, appearance, actions. If you are communicating with someone shy and introverted it is not appropriate to be forceful and dominating, as might be appropriate towards a loud extrovert. By being sensitive to the personality of others you can greatly increase the chances of effective communication.

Often it is necessary for communications to involve criticisms. The other people in the conversation are just as entitled to their points of view as you are. If you respond to their criticism with a retaliation, either open oral warfare or silence will result. Both of these are barriers to effective communication and so must be avoided. As emotive subjects such as criticism enter the conversation the sender should respond by increasing objectivity and further reducing the margin for ambiguity. Depersonalising issues is an excellent way to avoid confrontation. If someone feels they are under attack on a particular ussue they will be very defensive; if, however, they feel they are being offered constructive ideas aimed at solving difficult issues they are more likely to respond positively.

Active listening

Active listening is an oral communication skill in its own right. In essence it is a way of ensuring that errors of oral communication are eliminated, but it requires the active participation of the receiver, not just the sender.

When conversations are under way the participants often have a list of things that they 'want' to hear. Imagine an employee who really does not want to do a certain task and is in a conversation with the boss. It may be that the boss finds it hard to ask the employee to do this task, perhaps because even though the employee should be doing it the boss didn't make this explicitly clear at the start and so it reflects badly. Subconsciously, neither party wants the discussion to turn to the sensitive area. More often than not, in such circumstances, the participants will avoid the subject and the two people will have different recollections of what was said because their hidden 'wants list' has coloured their objectivity.

When active listening is to be employed the following points should be considered. Remember that this is a technique that is used by the listener to ensure that the message is received accurately.

Use open questions: Attentively acquire all the information on offer, and where there seems to be a shortfall use open questions to fill in the gaps. For example, instead of saying 'Was the unit delivered on time?' use 'How do you feel now that you have the unit?' Open questions lead the sender to provide full information.

Make sure you get all the points out first: If there is a problem to be discussed or an issue debated then the active listener can take some control by ensuring that all the points relevant to the issues are agreed. If this is done prior to the wider debate then it will be clear to both parties what has to be addressed.

Handle criticisms effectively: Often it is necessary to make comments that may be percieved as criticisms. All people are entitled to their points of view. If you respond to criticism with open oral warfare you are unlikely to be effective in the communication. Your arguments will not beat another person into submission, rather your 'opponent' will perhaps take up the challenge and the discussion becomes a heated argument. The alternative to this may be total withdrawal and the other person does not then divulge important information that might be of use to you. If you argue points with someone you may make them give in on some factual points but you will not alter their own opinions. Again actively listening and watching for signs that show people's sensitivities can ensure that these situations do not arise.

The customer is not always right: People in conversations are not always right about things, even when it is their specialist area. Handling this area of a conversation is especially important if you don't want the other person to feel threatened and so stop communicating. Don't confront the other person and so force him/her into a position from which they cannot win something. Use gentle questioning to examine ideas using his/her own knowledge to find the solution. Saying things like 'Yes, I see that approach, how would we then cope with . . . ?' will be helpful in this situation. Calmly make firm statements which do not attack the other person's conversational territory. In this way the conversation will carry along a lot further and more information will be exchanged as the other person learns to trust you. When a difficult part of a conversation is reached it often helps to de-personalise the issue. For instance, if you want a customer to explain the details of a rather complex design they are suggesting, don't say 'Using your system is bound to introduce errors', say 'Yes, that is always the most difficult area to get right, how important are the errors of that technique in your situation?'

9.5.2 Specific oral communication techniques

All of the points discussed so far are relevant to any oral communication. We will now look at how they apply in specific circumstances and consider some techniques that can be used to improve effectiveness for different types of communication. We will start by considering the oral presentation.

Oral presentations

Presentations are a particularly important form of oral communication which engineers, and other professionals, use to brief others on particular topics. Engineers may present their work to colleagues, personnel managers may make presentations about recruitment plans, sales engineers may present solutions to customers.

There are two aspects to making a good presentation: preparing it and giving it. We will look first at the preparation.

Preparation of the presentation: Preparation is essential to a good presentation, although if you are a member of the audience the preparation may not be obvious. What is obvious is the person who has not prepared and overruns the time, finishes too early, cannot find the right slide, or starts to wonder what topic should come next. The time to prepare for even a short presentation can be significant and should always be allowed for in the planning process. There are five main areas to consider in the preparation of a presentation: these are timing, objectives of the presentation, the structure, visual aids, and questions.

Timing causes more problems with presentations than anything else — if you overrun it looks like you have not prepared; if you are too quick then it gives the impression either that you raced through the material or that you did not have enough material. To get some idea about time start by taking the time slot you have been allocated, then take away a time for questions at the end, a time for getting everyone settled at the start and time for operating any visual aids or demonstrations that you may be using. You can then prepare the delivery to fit the remainder. Estimating how long it will take to deliver a given piece is not easy and even experienced presenters can get it wrong. The best thing to do is to give a dummy presentation to an empty room speaking the words aloud, just as you would for the real thing, and time it. Talking aloud is important because you always read faster than you speak. You may feel silly giving the presentation to any empty room but you will feel far worse if the timing goes wrong during your presentation because you did not rehearse it well enough.

When preparing your material consider what your *objectives* are for the presentation and then ensure that the objectives can be met with the material. For example, if you are trying to sell a design to a board of directors consider what aspects of the design they will want to know about and make sure that these are covered in your presentation. Also consider your audience and their level of understanding so that you prepare material at an appropriate level, particularly in terms of using language that will be understood by the audience. Finally, you must ensure that your material has a coherent structure: plan it like you would plan a report, with an overview (abstract), an introduction, the text, and finally a conclusion.

You should make a clear plan of the presentation showing its *structure* with timing markers. The best thing to use is a single side of big text that is easy to read during the presentation and which contains the headings of each section and the points that are to be described within each. You can then easily pick up your place by glancing at the sheet, you will be sure not to miss anything and if you are not in the margin of time by which each section should be completed it is easy to correct as necessary. Your plan should clearly have on it the one or two critical messages you wish to convey in the presentation; you must be absolutely clear what the message of the presentation is and resist the temptation to include material that is interesting but not really relevant.

Visual aids can be a very effective way of reinforcing points, or demonstrating what is meant. They should help the audience understand your message. Unfortunately too few people give as much weight to the preparation of their visual aids as to the extent to which they are used. Visual aids can only be effective if they can be seen by the audience, understood by the audience, and if they form part of the material. Do not use overheads or other visual aids that you have not tested out yourself to ensure they can be read from the back of the room. Visual aids take just as much effort to perfect as the body of the presentation. Diagrams should be simple and easy to understand, yet they may convey a great deal of information. Clear diagrams and visual aids can make a presentation, they offer the audience new interest and stimulation and if well chosen can explain points with great clarity. On the other hand, they can be a real embarrassment and when poorly prepared can ruin an otherwise good presentation. The golden rule which will ensure your visual aids are of acceptable quality is to try them out yourself first by putting them up and having a look from the back of the room in which you will give the presentation.

Overhead or slide projectors are the most common form of visual aids; certainly overheads are the easiest to prepare. To guide you in your preparation there are a few simple rules:

Keep the number of words per slide low.
Keep the number of ideas per slide low.
Characters should be large enough to be seen clearly from the back of the room.
Keep within the bounds of the screen.
If you are handwriting a slide, then print.
Label diagrams clearly and use the same drawing standard that you would if it were part of a report.
Use colour and highlight key words and phrases.

Questions are usually taken at the end of a presentation, although this can vary depending upon the style of the presenter, the topic of the presentation, and the length of it. If the presentation is long and many ideas are being covered it is unreasonable to expect the audience to have to sit through and remember their questions at the end. Wherever you decide to take questions it is always worth preparing for them so that you are not thrown 'off guard'. Similarly, if no questions are forthcoming you may wish to proffer one yourself to prompt the discussion. When you have completed your material preparation consider a few questions that might be asked and prepare answers to them.

Making the presentation: When making a presentation there are five areas that will have to be considered and addressed. These areas are timing, presentation of material, using visual aids, reacting to the audience, and helping the audience.

Timing: Throughout the presentation you should monitor progress and make sure you can see a clock or watch without constantly looking at your wrist. If questions are unexpectedly long adjust your talk to ensure that the main points are still delivered within the time. The hallmark of a skilful presenter is the ability to be flexible and change the details of the presentation as it progresses without losing the flow or dropping material. This is difficult to accomplish but will come with practice and experience.

Presentation of material: When presenting the material the aims should be to present it clearly and in a way that is interesting to the audience. First of all, begin with an 'icebreaker', rather than leaping straight into the body of the presentation. Introduce yourself and explain what you will be talking about; if appropriate check attendance so that missing people can be accounted for. When you start talking you are in control and authority is with you — it is up to you to take control, shut doors, start things going, settle the audience and so on. The icebreaker is the time to do all this and it serves many functions at once. Primarily it establishes you as being in control, it settles the audience and brings up the concentration level. It confirms what the audience are about to hear and it gets everyone focused before any of the important material is transmitted. A presenter should make a time allowance for the icebreaker which increases with the size of the group and what is to be said. For a group of ten colleagues it may only last a minute, while with a mixed group of a hundred or so in an unfamiliar place it may take several minutes.

When making the presentation talk clearly and talk slowly. Some presenters talk to individuals in the audience, changing person from time to time. It is certainly good to make eye contact with your audience since this will help to keep you in contact with them. Making a presentation is not like talking to someone standing next to you. To be effective in a presentation you must be heard and understood by everyone you wish to address. This means you must go at the pace of the slowest and be loud enough for the most distant. Your audience cannot control the speed of your delivery and so you must do it for them. It is good practice to ask your audience if they can hear you, although this must be done at the start during the icebreaker to avoid having to repeat material if some cannot hear.

Unless it is really necessary avoid reading from a script. Nothing makes presentations more boring than having them read like this. The implication is that the presenter is so bored or unfamiliar with the content that they are likely to forget what they have to say. The audience are left wondering why they weren't just given a copy of the script and left to read it in their own time. Giving a presentation involves adjusting the speed of delivery to suit the audience, going over particular things as that audience needs to, taking feedback and making changes as you go along. Rigid scripts afford none of these and indeed if this is all a presentation becomes then it gives no more than the audience would receive by reading the script for themselves.

There are occasions in life when a script is called for, at times when every last word of a delivery will be subsequently dissected and its meaning closely scrutinised. Perhaps a presidential speech, a statement of the company position at a shareholders' meeting, or a very technical lecture attended by world authorities where the ambient level of understanding is very high. These are suitable subjects for the prepared speech and it should be reserved for them.

Visual aids: When using visual aids ensure that they complement the material and that they add to, not detract from, the overall presentation. In particular use them when required and remove them when they are finished with. If a slide is kept in view long after it has been discussed the audience will be distracted by it and won't give the weight to the rest of the presentation. Always refer to diagrams and slides, incorporating them into the presentation — the audience should not have to wonder what the point is. You should also make sure that you do not obscure any visual aid, and that it is clearly visible to the audience, checking focus, that a slide is the right way up and so on.

Reacting to the audience: During the presentation the good presenter breaks down the barrier with the audience by reacting to them rather than remaining aloof and untouched. The presenter should not stand still and deliver a monotonic slab of material but walk around, address individual people in the audience, establishing eye contact with them. In this way the audience will feel that they are being spoken to directly as individuals; this will hold attention and make the presentation more interesting for them. It is also important not to alienate your audience: dress like them, speak like them, talk like them and be like them. People can be very prejudiced against differences and take the opinion that people unlike themselves cannot have something useful to say. If such an atmosphere prevails during a presentation your audience will be less receptive to the message and more damning of the material. Therefore to minimise such effects and give your presentation the best chance your audience should be able to identify with you.

Helping the audience: To help your audience you should summarise at the end of each section and especially at the end summarise what happened, what was done, what it shows and what the point of it all was. The summary is a very powerful teaching tool: people remember things in a structured way and reinforcing the structure by restating it synoptically at the end of a section is a most effective technique for assisting learning. Summarising should not be confused with repetition. Repetition can be used to ensure key concepts or buzz-words are remembered, whilst summary is used to ensure the structure of the knowledge is remembered. A common piece of good advice for making a presentation is '(i) tell them what you are going to tell them, (ii) tell them, (iii) tell them what you have told them'.

Finally, do enjoy it — enthusiasm is contagious. If you are interested and dynamic about your subject your audience will be too. If, on the other hand, you — as the champion of the subject — appear bored or indifferent your audience will conclude that the material itself must be boring and dull and they will not pay attention. Be confident.

Interviews

Interviews cause more apprehension and stress than practically any other type of oral communication, which is surprising really since they are simply a discussion between two parties about whether they can make a deal which both parties find fulfilling. Usually interviews are associated with employment, with an employer attempting to ascertain if an applicant is acceptable to the organisation and the applicant trying to assess what the job will entail, what the prospects are and what it would be like to work for such an organisation. Interviews also take place in many other circumstances when one person, or group of people, needs to obtain information from another person or group, such as trying to establish a customer's requirements.

The success of interviews depends chiefly on those who run them. When they are run well interviews are exceptionally good at communicating the position of both sides and identifying critical information. Sadly the converse applies to badly run interviews. Interviewing is a skill and there are many books and training courses that exist to enhance the effectiveness of both interviewer and interviewee. In this text we will review the most important aspects of the interview, looking at the interview process and at the role of the interviewer.

The interview process: The process must begin with a definition of the aims of the interview and where appropriate these should be agreed by both parties to the interview. From the aims the interviewer can draw up an agenda to ensure that these are met. For example, if the aim is to choose a suitable person to join a team of designers then the agenda should cover areas such as previous experience, technical ability, what the person is like as a team member, and what the person expects from the job. It might also include some informal meeting with the dssign team.

Establishing an agenda will also require determining who should attend, whether technical people or marketing, accounts and so on. The more people attending the greater the problems of co-ordination and communication and therefore the need to make sure that everyone is clear about the aims of the interview and the procedure that will be followed. The procedure will define the times of the interview, where it is to take place, if any demonstrations, presentations or tours are to form part of the process, and the way in which decisions will be agreed and ratified.

The next stage requires that the two parties are brought together in a suitable environment so that the interview can take place.

Bringing the parties together means arranging facilities, agreeing dates and times, and ensuring that travel arrangements are practical. The greater the distance between the two parties and the more important the interview the more difficult this arrangement becomes; certainly, when trying to get many people together, it is often necessary to plan months in advance.

If a number of people are to be involved in carrying out the interview then they should prepare beforehand, ensuring that they agree and understand each other's role. In particular agreement should be reached about who is to run the interview, and who will cover which aspects of the interview. Most importantly all should agree on the criteria that will be used as a basis for any subsequent decisions. One thing that should never occur in this situation is that the interviewers start their own conversation, excluding the interviewee, or argue about issues in front of the interviewee.

A suitable environment is very important: it has to be able to cater for the parties and has to be conducive to the free flow of information. In the recruitment situation this may mean having informal surroundings that are screened from external interruption. If dealing with a major customer this may mean having good communication links and catering facilities. These domestic arrangements are an important aspect of the interview — no one is going to relax if they are in desperate need of a toilet, or perhaps if they haven't had a cup of coffee in the last four hours, or have just finished a long car journey. Certainly when dealing with sensitive issues there should be no disturbance, the telephone should be switched off and people notified that the room is unavailable.

As the final part of the process the interviewer and interviewee should agree on the follow-on procedure, how the interview will be recorded, or how the interviewee will be notified of actions and decisions.

The role of the interviewer: The interviewer will be responsible for ensuring that the process is defined and understood.

However, in addition to this the interviewer has the responsibility for ensuring that the requisite information is extracted from the interviewee. There are a few simple techniques for ensuring that this happens. Initially the interviewee should be welcomed and put at ease. If it is a job interview then the candidate could be engaged in conversation about some of the interests given on the application form, or there could be a more general conversation about the weather or the journey. Following the general introduction the interviewer should 'set the scene' by discussing the aims of the interview, the procedure for the interview and either agreeing the agenda for the interview or indicating the structure that the interview will take.

During the interview the interviewer is trying to gain information upon which a decision can be based; therefore it is necessary to get the interviewee talking. In fact in any interview the interviewee should do most of the talking. Obtaining information is done by asking questions and there are two rules that should be followed:

1. Ask open questions.
2. Ask single questions.

Open questions are ones that will allow the interviewee some opportunity to give detail; they cannot be answered in one word. For example, if a manager was trying to find out if a subordinate was settling in to a new job the question could be asked 'Are you settling in?' If the answer is yes then the manager may feel his/her job has been done but they may have no evidence to substantiate this. If the answer is no then more questions have to be asked and these have to be developed from zero input. However, if the original question was 'How do you feel about the new job?' then a more detailed answer is likely to be forthcoming. More questions may still be required but at least the interviewer has a direction to follow.

Asking single questions is also important: asking double, triple, or worse, questions can mean that both interviewer and interviewee forget the original question, and it becomes very difficult to answer any question. An example of a triple question is 'What is your experience with that programming language and don't you think it's very hard to learn, I used it a long time ago but I think it's just been upgraded, am I right?' — where would the interviewee start?

The interviewer can have a very hard job to do, particularly if the interviewee is unwilling to contribute. Unfortunately as with so many management skills the only way to increase effectiveness is to practice.

Negotiations

Negotiation is the process of reaching agreement between two parties that do not share the same view. Negotiations may be very large and protracted as is the case with nations negotiating defence or economic settlements, or they may be small and rapid as with two people bargaining a price for a good. In either case the parties are negotiating and this involves various stages. The purpose of negotiation is to arrive at some agreement which is acceptable to both parties. Anyone can reach an agreement with someone else when a mutually beneficial possibility exists; only experienced negotiators, however, can be sure that they have gained the most and given the least that the particular situation afforded.

There are four stages to effective negotiation: these are preparation, examination, confrontation, and resolution.

Preparation: Experienced negotiators know that this is the most important stage. Without taking time to consider what is actually wanted from the negotiation one cannot be certain of heading towards it. On a simple level an engineer might prepare for a meeting listing the desired outcomes and actions together with responsibilities. Alternatively a union negotiator may well discuss with colleagues all positions to be taken, set upper and lower limits on issues, decide how the cases will be argued, even consider how the other side will respond and prepare counter-measures. If aggressive tactics such as sudden mood changes, unannounced adjournments or decoy concessions are to be used it is clear that good preparation will be required.

Examination: The early part of negotiation is concerned with an examination of the position taken by the other party. In particular the negotiator is looking for indicators to the upper and lower acceptable limits on issues. At the same time negotiators will try to keep their own position secret. A lot can be deduced in the examination phase about the ranges of acceptability and various tactics are available for examining them further. Suggesting values indicates an expectation and so tends not to be used. This phase also includes a 'sizing-up' of the opposition which can be useful later, particularly where the choice of tactics can have a critical effect.

Confrontation: Inevitably during the negotiation confrontation will occur. Good negoiators never let it escalate into an open battle and control it long before such a point is reached, avoiding an impasse. An impasse means dialogue has stopped and unless one side gives way negotiations cannot continue. This is a very unpredictable and potentially ruinous position and is therefore avoided. Nevertheless if two parties have incompatible views then they must be confronted and negotiated away. Of course compromise is the way forward and during the confrontation phase the disagreements have to be dismantled one by one. If a large impasse seems likely it may be best to tackle simpler, less contentious issues first and build good relations.

Resolution: As the negotiations progress compromise is established and principles upon which concessions can be made are discovered. The negotiation becomes the resolution of these differences once the confrontation has been passed. In the end all the contentious issues are either resolved and both sides have a way forward or they cannot be resolved and the impasse is insurmountable.

The success people have in these different stages has a lot to do with their experience and negotiation is a skilful process. Experienced negotiators nearly always out-manoeuvre their opposite numbers and where there is much negotiation to be done in a job special training is given. Sales personnel, for instance, are usually trained by their organisations to be effective in commercial negotiation. There are many tactics to use in the negotiating game: bluffs, making much of small concessions, negotiating with the most favourable person opposite, seeding dissension, undermining power, adjournments, tactical behaviour, stating one's position early, stating it late, opening much too high. The list is very long.

Negotiations are always about important things and so it is not surprising to find that they take some time. People new to negotiating often feel that the process is unnecessarily lengthy and if all sides simply stated their case a solution could be found much more simply, however, important issues need to be examined from all angles to make sure nothing is missed and especially to ensure that an agreement will endure. Agreements will collapse if, subsequently, important issues prove to have been inadequately covered.

Telephone

The mass communication device invented by Alexander Graham Bell is now totally integrated into life on earth. One can now phone practically anywhere in the world, and a vast network of satellites, cables and fibre optics makes conversations with people on other parts of the planet a normal part of life at work. The power of the telephone is more than its long-distance capability. Communications over short distances, within single companies, even within single departments, are very different as a result of its existence.

Using the telephone is not as trivial as it is often considered. It is common to train people in its use if their job requires them to use it a great deal. Because the telephone permits verbal communication but prohibits non-verbal communication, successful use of it cannot be achieved by having a conversation in the normal way. During telephone conversations it is necessary to provide verbally the information normally provided non-verbally — often this is achieved through an exaggerated range of voices and verbalisations that are not speech.

People often mistakenly think that because the telephone has been answered the person at the other end is ready for a conversation. People usually answer a ringing telephone even if they are quite busy, simply because the telephone call may be even more important than the task in hand. Therefore good telephone callers will always ensure that the following are included in their calls:

A statement of who you are and where you are calling from.
A check on to whom you are speaking.
Finding out if it is a convenient time to call.
Stating clearly and precisely the nature of the call: social, business, etc.
Making sure phone numbers are exchanged if necessary.
Being clear who must do what as a result of the call.
Summarising at the end of the call.
Signing off courteously.

Finally, when transacting any business via the telephone it is good practice to keep a record of the call.

9.6 Managing meetings

Meetings are notorious for the extreme levels of communication effectiveness that can be experienced. Some meetings achieve vast amounts in short periods whilst others ramble on, decide no actions and might as well not have happened. In this section we will look at the meeting and at some simple steps which can be taken to avoid poor performance in meetings. Meetings have a wide variety of functions. Some are progress meetings where engineers report on progress to their colleagues and superiors, and receive information about the progress of others, or news that affects them. Others are briefing meetings where many people attend to benefit from the questions asked by colleagues and by the interactive nature of the meeting. There are also meetings with collaborators or customers where difficult issues may need to be discussed and agreed, such group conversations cannot easily be conducted by telephone or letter and the only way is to bring all concerned together.

The following paragraphs indicate some of the most important things to remember when setting up effective meetings.

Is a meeting appropriate?: Meetings are called to bring several people together and make the best use of their collective talents. The more a proposed meeting makes use of the assembled talents the more successful it will be. For instance, a project planning meeting conducted by a group of engineers who have a project to complete and the skills to do so between them, is likely to be productive. However, a design meeting in which one or two delegates have their own opinion about methods to be employed whilst the rest are excluded or superfluous is likely to waste time.

Meetings allow sharing of ideas, rapid communication, interactive reasoning and the ability to solve problems that span far more than the knowledge-base of one person. These are impressive attributes and they should be reserved for worthy problems.

The purpose of the meeting: Every meeting should have a purpose, there must be some desired result or decision that needs to be reached and the meeting must keep it in sight. It is the responsibility of all those present to maintain a focused view and not side-track the meeting; however, the chairperson has particular responsibility for this task.

Circulate agenda first: One cannot expect people to contribute their best if they are not forewarned of what is required. Apart from anything else delegates may be busy and the meeting will then have to be rearranged. Finding a time when all the required parties can attend can be a problem in some organisations. The agenda should contain the start time, location, details of subject matter, the chairperson should be named and the duration of the meeting. Having the agenda will also allow people who are unable to attend an opportunity to make a written contribution.

Have a chairperson: During meetings people often put forward creative ideas and intense debate can develop when important issues are at stake. Quiet personalities with much to offer can be lost beneath loud ones. If the meeting is to provide the benefits of many views and produce constructive ideas these pitfalls must be avoided. This is the purpose of the chairperson.

The chairperson has a special role in the meeting and is responsible for the process of the meeting rather than its content. It is the duty of the chairperson to ensure the meeting is kept on the subject and to alert the group when the meeting wanders. The chairperson must ensure all voices are heard and that the loud do not dominate the quiet. A chairperson must steer the discussion towards making conclusions at the appropriate time and ensure that the record of the meeting contains the important points made. The chairperson must not volunteer opinions on the matters at stake or enter into the technicalities of the discussion. Such a combination of skills does not come easily and people who are good at this task are not common.

Take minutes: Minutes record what was decided at the meeting. Some minutes are very brief such as those from a regular meeting and record no more than the actions that were agreed and who is responsible for them. Others, such as the minutes from court cases, are more comprehensive and record the arguments as they were put forward and how the discussion went. The style of minutes chosen must be appropriate for the meeting in question. A meeting to decide whether or not to take a particular action will not normally have the arguments recorded, only the result, whereas an appraisal meeting will normally record all the views and opinions of both the parties present.

It is always advisable to have someone taking minutes who is not involved in the discussion and decision-making, as this ensures that the minutes are objective and that someone who should be contributing is not distracted from doing so. After the meeting minutes should always be circulated promptly whilst the meeting is still fresh in the memory. It is usual to have three standing items on the agenda of regular meetings:

1. Apologies for absence.
2. Minutes of the last meeting.
3. Matters arising.

The second item is the ratification of the minutes where the group have the opportunity to point out any inaccuracies before the minutes are agreed as a true and accurate record of the previous meeting. The third item is where the actions that were agreed previously are followed up and reports are made to the meeting.

Ground rules: In meetings where tensions are likely to be high, time is of the essence or where interpersonal conflict is likely to occur ground rules are an excellent way of keeping things positive and constructive. Examples of ground rules that are normally taken for granted are: (i) only one person can speak at once; (ii) the chairperson's decision if final; (iii) all members are free to participate. In particular types of meeting other ground rules may be appropriate: for instance, in a brainstorming meeting it may be sensible to make the ground rule that seniority is not important. The ground rules should be established and agreed before the meeting starts.

Good meeting behaviour: Everyone at the meeting is responsible for their own contribution. It is a common mistake to think that the chairperson is responsible. People who come away from meetings criticising the decisions taken but have not contributed significantly themselves are just as much at fault as those who force their own points of view through at the expense of others. Incumbent upon every member is the responsibility for giving their own point of view, succinctly and unambiguously, and not wasting the time of the other delegates with ill-considered trivia. It is also no more than common courtesy to remember that everyone has their own point of view and is entitled to express it without censorship. There will always be people at different levels of understanding in a meeting and the people who ask questions that appear basic and obvious to some team members are to be respected — not only are they learning during the meeting but they often bring the more erudite members back to earth with basic questions. After all, if an idea is sound, it should be trivial for the idea's author to answer the question; if it is not, something is clearly wrong.

9.7 Summary of Chapter 9

In this chapter we have introduced the subject of communications and explained its importance from a management point of view. We have examined the need for communications and seen how communication occurs in the work-place. The majority of the chapter was consumed with an examination of written and oral communications. In these sections we described some of the most important methods used by engineers in their work. Finally, we looked at a special example of communication, the meeting.

It has been said that there are no problems in management, only problems in communication and it is certainly hard to think of a problem that has occurred that could not have been solved by a timely communication. The importance of being a good communicator is repeatedly emphasised by employers, by customers, by suppliers, and by our colleagues and friends. The individual who ignores this need does so at their own peril. It is possible to make great improvements in our communication abilities and there is no one who does not stand to gain by doing so.

Communication is self-evidently the most important skill we possess, not only from a management point of view, but from our need to communicate with other members of our own species. Communication underpins our abilities to perform any other management function. Those who disagree with this view would be well advised to steer clear of management for they will never be able to perform satisfactorily as managers of people.

9.8 Revision questions

1. Make a list of the communication techniques you have been exposed to over the last two weeks. For each describe the advantages and disadvantages as experienced by you.

2. What communication issues are important in the process of appraisals? Start by considering what an organisation needs from its appraisal process and then describe the communication issues associated with each area.

3. How do computers affect communications in your life? List all the communications that you have experienced that were computer assisted. What impact have computers had on communication techniques and communications within organisations? Use a library to do your research.

4. Why might memos be 'copied' to people other than the person to whom the communication is directed?

5. Obtain a copy of the 'mission statement' of your organisation, or an organisation you know well. (If no statement exists compose one that might be appropriate.) How do communications in your organisation exist to serve the statement? Are there communication needs that are not addressed? Is there more communication than is necessary? Can you explain the success of your organisation in terms of its communications?

6. Choose a set of instructions or directions you have found poorly written in the past. Try to use examples of domestic products, like a power drill or coffee percolator, with which you are familiar. Criticise the text and rewrite all or a portion of it so that it is perfect. Discuss your text with a friend and get them to criticise your text. Then answer the following questions. Why is writing such text so difficult? What issues must be considered in its preparation?

7. Explain the following communication terms:

   Feedback
   Redundancy
   Jargon
   Abstract
   Reference

8. Describe the communication issues that might be important within an engineering organisation between the manufacturing department and (i) the sales office, (ii) the personnel department.

9. Discuss the communication issues raised by the implementation of a quality system within an engineering company.

10. Discuss the communication issues raised by the implementation of a training programme within an engineering programme.

11. Discuss the communication issues that influence the effective operation of a team of engineers.

12. Prepare a letter of application to an engineering organisation for a post taken from a real advertisement.

13. Describe the interview process from both points of view. What communication techniques can be used to ensure that the interview goes well?

14. What is active listening? How does it help a conversation?

15. You are to chair a meeting in which the engineering department of a company will meet an important customer for a discussion about progress on a contract. Progress has not been good, the customer blames the company while the company feels that the selling price of the contract was competitive enough to begin with, but since the customer has started to insist on changes and additions to the original contract it has become impossible to meet the contract. How would you prepare for such a meeting? What communication issues would be addressed? How would you go about achieving a resolution between these parties?

16. Explain how information, which is intangible, can be of value to an engineering organisation. What things give a piece of information its value?

17. Why do customer complaints make a poor method of gaining market information? What else can an organisation do to gain market information from its customers?

18. What other organisations may be of use in providing useful information to a company? Describe them and the sort of information they can provide.

19. Examine a colleague's reading skills in the following way. Ask the colleague to read a page you have chosen at random from this book, in their own time. Ask them about the contents of the page afterwards and record your comments about the level of detail they remembered afterwards together with the length of time they took to read it. Ask them to read another page in only three-quarters of the time and record the same data. Finally, do the same but allow the colleague only a quarter of the original time. Repeat the test for several other colleagues. What conclusions or observations can you make about reading speed and its effect on recall?

9.9 Further reading

Fowler, H.W. (1965), *A Dictionary of Modern English Usage*, 2nd ed., Oxford: Oxford University Press.

Waldhorn, A., Zeiger, A. (advisory editors: South, R. MA, PhD and South, J. (1986), *English*, London: William Heinemann Ltd — Made Simple Books (ISSN 0265-0541).

Part 3

Engineering Management in Practice

Chapter 10
Case study

In the introduction to this book we explained in general terms the way in which management topics impact on the jobs of engineers. Throughout the book we have described the different areas of management and in fairly limited examples have shown their application in engineering situations. However, these examples cannot give an overall picture and they were specifically chosen to illustrate discrete areas. As engineers writing for engineers we will now redress the balance of the book.

In this chapter we have a case study. It is a complete engineering project, a job on which James Balkwill was employed for nearly two years. The type of situation described, that of a small project team, is quite typical. Many engineering graduates will wor in such teams. The technology may be unfamiliar but engineering is so varied that it would be impossible to choose a technology which would be widely known. We have, however, included the technical detail so that you have some understanding of the technical issues that were faced by James and his team.

The project is presented chronologically with the emphasis on the task. However, we can see the management issues and skills that are required to tackle an engineering job like this. After the case study it is debriefed. We have prepared a review of the case and the management issues raised. This is structured to follow the format of the book so that you can look back to the relevant section and reconsider the theory in practice. You will see that engineering management is not only important, it is an integral part of the job, and you will notice that none of the areas previously discussed works without affecting other areas.

10.1 The Cu100 project at Oxford Lasers Ltd

This case study is reproduced by kind permission of Oxford Lasers Ltd, 60/62 Magdalen Road, Oxford OX4 1RD, and of the Cu100 team members themselves: John Boaler, Richard Benfield, Steff Inns, Rhys Lewis and James Balkwill.

When I arrived at Oxford Lasers in my interview suit I realised that the interview was to be conducted by a panel. This was a first for me and it did little to calm my nerves. Two of the panel were directors and so at least I knew something of their role within the company, but the third person was an unknown quantity. We walked round from the reception area to the side entrance of some smart new industrial units from which the company was operating. At the top of a staircase was an empty room which seemed to serve as a library, perhaps a meeting room and, at this time, an interview room. I knew there was a post available on a large project to build a new and powerful laser, something that would break new ground but I was short of any real facts about it.

The interview began with a description of Oxford Lasers. The company had been trading for nearly ten years, they manufactured copper vapour lasers at this one site in Oxford, and sold them world-wide. There were about fifty employees, the turnover was in the millions and Oxford Lasers was the market leader by a large margin. The lasers were sold for various applications: high-speed photography, medical practice including cancer therapy, research, television projection systems and isotope separation. The post for which I had applied was indeed on a large project, the aim of which was to produce a new and more powerful copper laser than any previously built by the company. There was a customer with a special need and a contract was already in place. The project was to last ninety-six weeks, and it was already Week 4. Some specifications had already been completed; particularly, the diameter and power of the laser beam itself had now been agreed. The third person on the panel explained that he would have overall responsibility for the project. He then spoke of the importance placed on the project by the company commenting that he even had his diary marked out in project week numbers. We joked, I was used to doing the same.

I explained my reasons for wanting the advertised project engineer's appointment: I was already employed in a similar role elsewhere but wanted to tackle one very large project rather than many smaller ones. I wanted my own team and I wanted to increase my role with customers and be asked to produce equipment whose specification I had assisted in negotiating. I felt confident on the technical side and explained the sort of projects I had undertaken before which included many specialist one-off scientific instruments. I thought I'd lost it then, everything went quiet and they started asking lots of other questions — even technical things, which is unusual.

Much later I found out that the fit between my wants and their needs was so close that they were taken by surprise and just didn't want to seem too enthusiastic. About two weeks after the interview the papers came through and after a minor altercation over the salary I was hired. I started a month after the interview.

Project Week 8

I didn't know what the traffic would be like on the first day so I left a lot of time and arrived really early, but the place was locked and I spent the first hour of my new employment in the car park! The third person from the interview arrived and introduced himself again: he was Rhys Lewis, my boss. He looked at his watch and said, 'You're early.' Normally I'd been at work for over an hour by now. We walked upstairs into an office and I was shown my desk. There were five of us in the one office, the size of a large living room. They were clearly the desks of engineers. Papers were piled high, in-trays groaned under stacks of drawings and files, dotted about the room were several machined flanges, interestingly shaped components and some other proprietary parts I did not recognise. There were books on laser physics and electronics catalogues everywhere together with two computers. Anything that provided a flat surface was piled high with papers, journals, and handwritten design ideas; there was a general atmosphere of people doing things.

Over the first few days I started to get some idea of what was going on. There was a company induction programme that I went on, I was even introduced to every other employee. Rhys started giving me lessons in laser physics. We started with the energy levels in the copper atom and with a degree in engineering and a keen interest in physics I was able to follow the theory. The heart of a copper laser is a gas of copper through which a discharge current is passed, the electrons in the discharge then excite the copper atoms, by collisions, from their low-energy state, called the *ground state*, to a much higher energy level. The region in which all this occurs is called the *gain chamber* and it is contained within a tube of exotic material capable of withstanding the arduous conditions caused by the high temperature and corrosive action of the discharge. This tube is called the *plasma tube.*

Once excited, the atoms then start to shed energy in accordance with the probability distributions dictated by quantum theory. Before long many of them have fallen down to a lower energy level and emitted a photon in the process. The energy level to which the atoms fall back is not the ground state, they fall to another energy level just above it. The central principle of the laser is that if a photon emitted from one such decaying atom happens to collide with another atom still at the high energy level, it causes it to fall immediately to the next energy level down and therefore emit a photon as it does so. This is called stimulated emission, a photon 'stimulates' the excited atom to emit a photon. The initial photon is unchanged in this process and there are therefore now two photons travelling in the same direction, and, as it happens, the two photons are always in phase. This action continues until nearly all the atoms have fallen down to the lower energy level. During this phase there is a massive conversion of energy into light, and mirrors at each end of the lasing chamber channel the light created back into the gas of copper atoms and so increase the chances of a photon colliding with an excited atom and stimulating even more light.

It is an acronym for this process that gives the laser its name, this is '*L*ight *A*mplification by the *S*timulated *E*mission of *R*adiation'. After all the atoms have emitted a photon and are in the lower energy level one has to wait for a comparatively long time for all the atoms to relax to the ground state, and only then can the process be repeated. Because of this action, the copper laser is a pulsed laser and although the output beam appears to be a constant and intense beam of green light it is in fact made up of a series of javelin-shaped arrows of light each being only about one hundred feet long and each being separated from the next by about thirty miles (a distance that corresponds to tenths of a millisecond at light speed).

Week 9

The intensive induction programme and introduction to laser phsics continued, punctuated only by other introductory talks and visits. All the physics described by Rhys clearly posed great engineering problems such as: 'How does one achieve a contained gas of copper?', 'How can one achieve the massive rate of current rise required to make the process work?', 'What sort of materials and components can be used for the arduous conditions inside the laser?' Rhys explained how these issues were solved in the present designs of copper lasers produced by the company. This knowledge formed the basis of Oxford Lasers' success and some of it is a jealously guarded secret. Once I had developed a working knowledge of copper lasers we moved on to the project itself.

A large industrial concern in the UK had approached Oxford Lasers with a need for very large copper lasers. The customer wanted in excess of a hundred watts of light from a single machine yet the most powerful existing product gave only sixty watts after a lengthy and difficult production process. The product name for this machine was 'Cu60', 'Cu' for copper and '60' for the power in watts; it had therefore seemed logical to name the new, more powerful laser the 'Cu100' on the same basis.

A hundred watts of light does not sound much when compared to an ordinary light bulb of a hundred watts which may easily be bought at any hardware store. The reason a 100 watt copper laser is so different and so much more powerful is that the hundred watts used to describe a light bulb refers only to the power consumed by the light bulb, and in fact only a small fraction of this power is turned into light within the bulb, the vast majority being turned into heat. With the copper laser, however, the hundred watts refers to the power of the light and is vastly greater than the light output of a bulb. The second factor at play is that a light bulb emits its light radially outwards in all directions whereas the copper laser emits its light as a narrow beam, in our case 60mm in diameter. This factor also greatly increases the light density in the laser beam compared to that from a conventional bulb.

The customer for the Cu100 also wanted a very different style of machine: the need was for industrial use characterised by simple control, robust operation, easy servicing, long periods between routine services and large mean times between failures. The existing products by contrast were complicated to control since most customers were well acquainted with the operation of a copper laser and wanted to perform and control all the different sequences and modes of operation themselves. The copper laser is an intrinsically temperamental device with a number of unknown issues involved. A good design of copper laser comes at least in part from much empirical optimisation and Rhys explained there was a great danger that radical new geometries and ideas would bring a whole new range of problems. We would have to introduce a number of novel features in order to produce such a laser — 'We will have to be radical, but not too radical', he said.

In addition to these requirements the customer wished to place the lasers in chains, the light from one laser entering the next and being amplified in the process. In essence they wanted a chain of four or five lasers, and the first one in each chain was to be a special version of the laser, called an *oscillator*. The purpose of the oscillator was to produce the initial pulse of light. The pulse would be about a hundred feet long and be composed of very parallel rays so that they would not be too spread out by the time they reached the end of their flight path.

The other lasers in each chain existed to amplify this initial pulse. After a few nanoseconds, the pulse would enter the first amplifying laser in the chain. This laser would be discharged and start to amplify just as the front of the javelin-shaped pulse of light arrived. The pulse would grow on its passage through this laser and would then enter the next laser in the chain. This, in turn, would be discharged at the right moment to reach peak amplification as the light pulse entered its gain chamber. In this way the initial pulse would be boosted to several hundred watts of light.

The customer also intended to have many such chains operating simultaneously in a large rack containing many lasers. The rack would be built from four-inch square box girders of steel and sounded more like a shipyard project than an optical bench. It was imagined that each laser would weigh about a tonne and be placed in or out of the car-sized slots in the rack by fork-lift truck. The lasers were to be optically aligned with a dimensional error of less than a quarter of a millimetre over the 60 metre light path and each laser in each chain was to be synchronised to light speed and only start to amplify the pulse of light when it arrived. The amplification period of a copper laser is over so quickly that this additional complexity was necessary if the whole system was to stand any chance of working.

Week 11

The aim of the project was to produce the first two amplifying lasers, each capable of producing a hundred watts of light when run alone. At present we had no order for an oscillator because everybody had been too busy getting the amplifier contract going but we did expect that a similar contract for an oscillator would be negotiated.

The first amplifying laser was due for delivery in Week 68 and the total project duration would be ninety-six weeks. Eleven weeks had already passed in the preparation of plans and contracts and hiring project engineers. We also had a fixed budget. We were to use a team of five people or so and a new industrial unit was to be given over to the project. Rhys introduced me to some of the plans for the project that had already been draughted and it became clear that we already had things to do that were urgent even with more than a year and a half of the project to go. A deadline had been set for the mutual agreement of a design report in Week 32. As soon as this was formally signed, ordering and manufacture would commence and so the completion of the report became our first milestone. Although the project was for two lasers only, it was clearly our intention for our customer to be completely satisfied with the product and so to order the large number of lasers that would be necessary to fill the rack they were constructing. We felt that a considerable commitment to the remaining lasers was being expressed by our customer simply by building the rack, but on the other hand we were often reminded that an order for more lasers would not be made until the first two had proved our abilities.

The milestone report covered not just the design of the laser and its precise technical specification but included testing schedules, a quality assurance plan, and what inspection schedules would apply. Even measurement techniques, colours and test equipment were specified, while in one section the effect and operation of every single control and indicator were specified. The report was a complete specification for the project and it ran to 120 pages. The design report milestone had an associated stage payment and the money received at that time would cover the cost of much of the expensive materials used in the laser construction.

Week 12

The only piece of equipment that would have to be sorted out before the design report was finalised was the enclosure for the laser. This need arose from simple time pressure. The enclosure would take some months to design and build and this was too long to fit in the time allowed for laser manufacture. In any case the enclosure would be needed at the start of the manufacture period since all the laser components would have to be fitted into it. It was clear that some sort of box with considerable rigidity was required and no one yet knew how to achieve the design requirements. Many design constraints existed for the enclosure and it had about seven roles to play in the final laser performance.

My first real task was therefore to despatch the design of the enclosure and the next few weeks saw me write a specification for the enclosure and agree it with the customer. I then sent out a tender to about ten companies whose names I had taken from trade journals. I also had to re-learn my stress analysis and design a beam plinth to meet the specification and place an order. In the first days of my investigation exotic options involving such things as honeycombs of special light alloys with active temperature control looked like the only solution, and even the best advice I'd obtained made the design prohibitively expensive. In some of the options I considered it would be necessary for me to orchestrate the supply of components from many different suppliers and fit the components together for the first time at Oxford. Needless to say I was keen to avoid such undesirable commercial and technical arrangements. After a few weeks of hard work, an order was placed for an aluminium enclosure at about a fifth of the initial cost estimates with a supplier who could meet the stringent quality requirements of the customer. After collaborating with this supplier, and using some of their expertise, the design was now much simpler and more elegant. There had been a good deal of problem-solving to be done but between us we had the necessary knowledge. At one pont I also had to bring in a consultant from higher education for advice on the metallurgical aspects of the design.

The only technical problem we could not accurately predict came from welding so much aluminium and the possible distortion that might result. We identified a subcontractor specialising in aluminium welding who felt confident and arranged for a shipyard press to be used to bend the aluminium beam straight if it distorted too much. This was no mean feat since the aluminium beam was over four inches deep. We found a press in southern England of the correct specification and we booked time on it so that if needed, it would be available.

The commercial arrangements regarding the enclosure were now good. One supplier was responsible for ensuring that all the components fitted and met an acceptance specification rather than me trying to synchronise the production, delivery and assembly of the whole affair. In the end the aluminium did bend a little too much when welded but it was easily bent to a flatness well within the specification and on the subsequent enclosures we adopted the procedure as standard.

Week 13

Not every minute of every day over that first three-month period had been taken up with enclosure design and Rhys and I did a lot of work on the design report. I also spent some time with Richard Benfield, the technician who had been working with Rhys on research aimed at producing a 100W laser before I joined the team. Richard had a test bench in the same building which was furnished with the latest laser design on the goal of achieving reliable operation at 100W of light. Richard gave me many lessons on lasers and fast-pulse circuit design and the opportunity to spend some time working on a real laser gave me a much deeper understanding of the problems involved.

We had agreed the outside dimensions of the enclosure with the customer already and these were now fixed since they defined the dimensions of the optical rack that would house the lasers. The construction of this rack and its associated building would be a long job and could not wait until the lasers were built. We had been asked how large the lasers would be and had given our answer — this could not now be changed within the contract. The laser tube is a cylindrical object and is the longest item within the laser enclosure; consequently the length of the tube dictated the length of the box that contained all the laser components. Our reason for believing that the project was technically viable came from earlier work on a developmental 100W laser performed by Richard and Rhys. They had demonstrated a hundred watts for a short time from a tube of the proportions we were using prior to Oxford Lasers ever accepting the contract. This experiment, though difficult to do, had proved that the physical principles upon which the laser action depended, would not break down in some new and unforeseen way in a machine of this size and power. Of course there is a great distance between momentary demonstration and reliable operation of an industrial machine but at least we knew it was possible.

Week 14

Early in the new year, Rhys and I held interviews for the remaining two people we felt were required for the team. After the design report and with my increased knowledge of the project we felt able to be more accurate about personnel requirements and the skills we required. It was clear that another technician was required to look after the mechanical side of the laser and do much of the laser assembly. It could be that the same person would do some paperwork and ordering too — there would be a lot of this later on and a technician who could grow into the job would be preferable; certainly someone who could handle change was needed.

The second person was a draughtsperson: we had already had to use subcontract draughting services and the need for draughting was going to increase rapidly, especially to produce drawings for all the components that would finally be designed and detailed once the design report was out of the way. The new unit was available very soon and so with all these changes it seemed sensible to hire the two staff, set up in the new unit and officially change the reporting structure so that they reported to me, since I was to be their manager, while I still reported to Rhys.

Week 16

John Boaler joined almost immediately as junior technician. He was an internal applicant and certainly fitted the needs of the group very well. His old boss, the production manager, was somewhat vexed to lose a very good technician who had been trained by his department using his own budget. The view was that the production department spent all the time and money training technicians only to have the best ones appointed to other internal jobs. I could see his point of view, but then John wanted the job and Rhys and I wanted to give it to him.

Week 18

Hiring a draughtsperson took longer and in the first round of recruitment no one who fitted the needs of the team and who would fit in could be found. We therefore continued using subcontractor services and carried out another round of recruitment. To meet our customer's request that certain drawings be prepared only by Oxford Lasers personnel I had to prepare some of them myself. Eventually a draughtswoman from a local electronics company responded to our advertisement and interviewed well. She was older than the rest of us, had all the skills we needed and struck both Rhys and I as someone who would help to make the team more balanced in outlook. We offered her the job and three weeks later Steff Inns joined Oxford Lasers.

Week 19

Rhys and I often took an hour or so out to chat over progress so far and current events. This week we felt we should look at the totality of the project. On the one hand we were now nearly a third of the way through the project and the team of five would only come together for the first time in a fortnight. We hadn't yet even cleared out all the junk from the new industrial unit and we didn't have a final design for the laser tube itself. In addition there was a big technical question over how one could possibly design an alignment system that could hold the laser tube axis in the required position and yet remotely adjust this one-ton monster to a fraction of a millimetre. None of this sounded good.

On the other hand we were very close to final agreement with our customer over the design report. Once this was agreed everything would be defined and the rest of the project was, in some senses, simply a question of doing what we had written down and agreed. In addition, we also had the enclosure design solved and were shortly due to take delivery of the first one. We also had a good team. The manufacturing phase of the project had seemed impossibly far away at the start but now it was looking imminent. We concluded that we were doing well but had no reason for complacency.

Week 20

Steff visited Oxford Lasers during the week to attend the monthly meeting of the whole company. Often this informal meeting was held in the pub just fifty yards from the company and it always consisted of news from each of the departments followed by drinks. Rhys had made a short speech about our progress and other people gave other speeches about other departments. I introduced Steff and explained that she had come along to meet everyone and would be starting work the following week, then we all had a drink. A team meal was felt appropriate and curry was hailed as the only food with which to welcome a new member of the team. Consequently Steff's introduction to Oxford Lasers was beer and a curry. My only comment was that it was better than an hour in the car park!

Week 26

We visited our customer again for the last meeting before official signing of the design report. We still had a number of issues outstanding, some of them tricky, and had prepared extensively for the debate by considering our reaction to the alternatives that might come up. We also had a 'shopping list' of issues to be resolved and the way we wanted them to go. The customer must have had a similar shopping list and disagreement certainly occurred. The points were negotiated and agreement was reached with compromise on both sides. It was certainly true that even this far into the project, both parties failed to appreciate exactly why some issues were important to the other side. The delegation from the customer's side was quite large and people kept coming in and out of the meeting which was very disconcerting. Only Rhys and I represented Oxford Lasers. We stayed there all day. Fortunately nearly all the people involved were engineers or scientists and so a rational approach was taken with a cool examination of the points. With mutual trust established it was ultimately possible to resolve our differences but it was very hard work; the main lesson is to keep alternatives open and avoid either being cornered yourself or forcing your customer into a corner.

Week 31

We now had very little time until manufacture would start in earnest and the industrial unit was still filled with junk. The customer would soon visit to formally sign the design report. At that time we wanted to be able to show our readiness for the construction phase which would follow. John and I walked into the unit on Monday morning to confront the mess.

Richard had already been using some of the floor-space with a very experimental version of the power control system he wanted to use. A laser as large as the Cu100 uses a lot of power and the simple but bulky power control systems used on the other smaller lasers would not be practical for our purposes. We were concerned that there might be stability problems with the proposed solution and in any case we had no experimental evidence for believing that it would work. We, or rather Richard, therefore built a three-phase version of a new, and as yet unproven, power control system intended for use on another new product being developed elsewhere in Oxford Lasers. Apart from one flash-over when a loose wire arced to the concrete floor inside the safety enclosure and took out three fuses and blew a hole in the concrete floor that could accept a pencil, the system worked fine and Richard was confident of proposed design. I got him to write up the experiment — he hating writing things up but it had to be done. We now had many technical issues under way and the simplest way of losing the information you need is not to keep accurate documentation.

During the week the unit was cleaned out. John and I had hired a skip for the rubbish which included a timber mock-up of the enclosure which had been built to help us visualise what the situation would be like, with several large, one-ton lasers dotted about in what was rapidly looking like a very small industrial unit. We imagined that if the project was successful the remaining lasers that might be ordered would be built in the same unit.

Shelves were put up for stores, equipment brought in, and a small office area for Steff and me was put up. Richard's test rig was transformed into a very professional-looking 'power supply evaluation unit', which did just the same thing but looked good too. We even put up some pictures, and obtained a copy of the company safety book, a first aid kit and a company logo. John and I reflected on our jobs: 'After all' he said, 'you don't often start your job by having to spring clean your factory.' We agreed that whilst it is good to do such jobs from time to time there is only so much enthusiasm one can muster for cleaning on this scale. There were some real benefits from the clear-out, and in the process several useful things were sorted out. The layout for the shelves, the stores arrangement, how the lasers would actually be moved around — all things that John really needed to decide on before the lasers arrived — and there we were, sorting them out whilst giving a very good impression of being cleaners! By Friday the place was spotless and ready to be shown off to the customer whose delegation was arriving after the weekend for the official signing of the design report.

Week 32

Monday saw the jeans and trainers swapped for a suit and tie. The day went well and the signing meant that a stage payment was released and we had earned our first substantial payment for words alone. The report earned far more money per word than any other form of writing I have ever contributed to, either before or since.

Week 33

After the acceptance of the design report, our second milestone was to demonstrate the laser discharging in the presence of our customer. The discharge would not be at full power and the laser would not have copper in it, but it would be a continuous run of several hours. We were now working hard to achieve this milestone which was due in Week 52, only another twenty short weeks away. We were also giving thought to the milestone after that, which was due only four weeks later in Week 56. This was a full-power continuous demonstration of the laser at over a hundred watts.

A new copper laser requires much work to be done between the first stable discharge and finally running at full power with copper. Many of the circuits need optimisation, but we felt a month was ample for this — our real worries at that time were much more immediate. The first was that we still did not have a design for the final geometry of the water-cooling jacket and the flanges that mounted the laser tube itself. Rhys and Richard had this as a priority and with only twenty weeks to go before it needed to be running there was clearly a lot of work to do. The second worry was the alignment system for which there was no design at all. John and I had this as our priority, although we had other jobs too.

Week 34

Production was now in progress. Orders were going out to all the suppliers and a new filing system was set up to cope. We started to use some of the money we had generated from the design report. The whole project had budgets for each category of expenditure and a desired spend profile. As expenditure started in earnest financial control became an issue. There was a set of files, one for every major subassembly of the laser. Each file contained several sections: some covered the design of the subassembly, others covered the administration. One file was reserved for the bill of materials and another for the purchase orders. Because of the nature of the project we had to keep a separate store for the customer to inspect; however, we still wished to take advantage of the company's existing ordering and receipt of goods procedure. For this reason orders were processed by the company stores and finance department but as soon as they had been received they were sent to our unit for inspection. The store shelves filled up and we started manufacturing various subassemblies, those that could be built in advance of the complete laser construction. The first enclosure was looking likely to be be late and so a progress trip was made to try and prevent any further delay; even so the delivery date was delayed by ten days. We altered the manufacturing schedule to bring other jobs forward but ten days was about the limit. There was flexibility in the plans and although we had used up some float, progress was still on track. The enclosure finally came eight days late.

Week 35

At about this time we realised we had a new problem with the laser system design. One of the conditions agreed in the design report was that the laser would be safe under the simultaneous disconnection of all services. This in itself was true but it all depended on what your definition of 'safe' was.

As the design stood at the moment, personnel would be safe because the enclosure formed a safety shroud around the whole laser, and even an internal fire would be contained. However, we realised that some of the laser components might not be safe under disconnection. This was because, unlike any of the other lasers built by the company, the water lines had to be in-line connectors that sealed themselves off when disconnected. These were included so that when the customer took a laser in or out of the rack, the connectors (from which the water-cooling lines would first have to be disconnected) would not lead any of the thirty or forty litres of water held in the cooling circuit. There was even a specification in the design report on the acceptable 'weepage', which was to be less than six drops. We started to consider what would happen under the worst conditions.

If the cooling water was stopped at the same time as the power and the laser had been at full power for a long time there would be a lot of heat stored in the plasma tube. Upon disconnection, the heat from the plasma tube, at about 1400 degrees centigrade, would start to flow into the surrounding insulation and cooling vessels. After about half an hour (we estimated) the O-rings would probably melt as would any plastic near the metal cooling jacket. We started to think carefully about what else might happen. The cooling circuit might also be cut if the laser was being removed for some reason. The self-sealing water connections meant if the inlet and outlet connections were broken the water in the cooling system would be contained within a trapped volume. This water would certainly boil and something would have to give.

I called the customer there and then. I asked about the imagined operation of the lasers in the future when many lasers would be in service — the team hung on my every word. It took moments to establish that exactly the conditions we had discussed might apply. If a laser went down in service the plan was to remove the faulty unit and then fork-lift in a replacement. The operating crew would have their hands full for at least an hour before they could get round to servicing the faulty laser. Our previous conversation had shown that when they lifted the service panels they could expect to find melted or burnt plastic, a laser assembly that would need a complete strip-down to replace all the O-rings, which is a long job. As well as this there would be a ruptured cooling circuit as the trapped cooling water boiled and eventually burst its pipes, spraying scalding water over everything, and causing doubt that the control electronics in the laser would still be working. Steff spoke for all of us when she said: 'All that doesn't really have the ring of "safe under disconnection" about it, does it?'

We talked, all of us. None of us was prepared to tell the customer what would happen with the present design and that meant something had to be done. We all started on ideas for solution. If the heat was a problem, could we get rid of enough of it to remove the problem? Could we slow down the heat transfer so that peak temperatures were low? Could we somehow direct all the heat into the massive aluminium beam that formed the base of the laser? Could we cover the laser tube in cooling fins that would reduce the peak temperature? There were many good ideas and some non-starters; in fact we had a lot of fun thinking up ideas. After about forty minutes the team was flagging and it was clear that answers to some thermodynamic calculations were needed before any rational selection of options could proceed. Rhys was the best at that so he went off to do the sums. We all carried on with the building programme.

Week 36

One of my responsibilities as a manager was to conduct appraisals for my staff. The company held annual appraisals and the time for them had come round. The appraisal scheme was two-tier, each person being appraised firstly by their immediate boss. This appraisal considered the facts of performance over the last year and existed to give objective feedback to the individual and offer ways in which the individual might develop. The second appraisal was with the boss's boss and covered longer-term issues such as career development and company developments. I appraised the team individually. Each had their own particular points to make and some problems, although no one had anything particularly difficult to deal with. The appraisals acted as a chance to re-examine my feelings about how the team was working and whether everybody within it was being fulfilled by their work. The second appraisals were with Rhys and he and I met to go through the appraisal forms for each team member before he held the second appraisals. Appraisals are time-consuming and are pointless if the appraiser does not know the appraisee's work very well. At the same time they can be extremely effective for the individual if done well and from my point of view that particular round worked very well in providing me with an objective look at each team member through a long conversation with just the individual and no interruptions. This was something I never normally got the chance to do.

The general pattern of the next few weeks was hasty construction according to the project plans defined in the design report. We would all gather round the laser to sort out particular issues; otherwise we just carried on with the tasks in hand.

Since we first got together as a group I had insisted on a weekly meeting of the whole group. The format was always the same. I went through general progress, developments with the customer or any other issues that affected the group. Each individual in turn then compared last week's achievements against the minuted aims from that week's meeting. With only five this is not too time-consuming if the meeting is well run and it does have good communication advantages. Even with five people, however, such a meeting can become too long and the meeting needed to be run carefully in order to work. However, the great motivation and team spirit that resulted from having the meeting was very important. Each Monday we set new targets for the following week that fitted with the overall goals of the project. Some weeks I would take extra time to talk through the major plans for the whole project and explain the work involved over the next two or three months.

Week 39

We met to finalise our reaction to the simultaneous disconnection problem. After some time away from the problem the options seemed simpler. In fact the only solution that really had any chance of success on technical grounds was surrounding the laser tube with enough water in the cooling circuit to absorb the heat without boiling. The diameter of the water-cooling jacket could be increased sufficiently with a little reorganisation of the internal components.

Rhys had calculated the required amount of water for safety, accounting for the duration of the process and the heat lost by convection. There were two consequences of the proposal: firstly, the laser tube would look rather fatter than we had imagined and we felt that we would lose the slim 'hi-tech' appearance. In fact it wasn't long before we nicknamed it 'the boiler'. The second consequence was an increase in the weight of the whole laser. The only real difficulty with the first consequence was that Rhys, a man skilled in the art of laser design, was anxious for his latest creation to look absolutely right and didn't much care for the aesthetics of this 'boiler'. The second consequence, however, was rather more serious.

The customer was very concerned about weight since the rack of lasers was a considerable structure in its own right and the precise alignment required for the operation's success would be upset by even a little too much extra weight. We already knew that various other parts of the laser had come out rather overweight and although we could not yet be accurate about the final weight of the laser it was clear that it would greatly exceed the 750kg specified in the contract. We did the sensible thing and had a technical meeting with the customer almost immediately.

There was, in fact, some spare capacity in the design of the rack and the new weight was acceptable. The rack would still work up to a weight of 1050kg. Even with the new 'boiler' we didn't think we would get anywhere near that much. Ultimately, the final consequence of the problem was that the customer would have to buy a larger fork-lift truck than had been anticipated. We felt that the weight should not really have been specified in the contract: after all, we had not increased the weight intentionally. The weight of the laser would simply be a consequence of the other constraints imposed by the design. The customer had been insistent on including a weight in the specification and in the end Oxford had been forced to agree. Paradoxically, it appeared that whilst Oxford Lasers was able to supply the highest-specification copper lasers commercially available anywhere on the planet, estimating their weight accurately was proving a good deal more difficult!

Week 42

We were visited by the customer for one of the many inspection meetings that occurred during the construction phase. A few days before the visit we again had the test rig running at over 100 watts for some experiments on component life and proudly showed our visitors the awesome sight of such a beam. The technical excellence and importance of this achievement was lost on the inspectors and they were only interested in the paperwork. They tested our ability to trace various components they identified in the laser to their original suppliers. The paperwork was in order and the components were traced. Various other aspects of the project's administration and some quality issues were also scrutinised. The inspectors went away happy. We were happy that they were happy but incredulous that the test rig didn't impress them.

Week 50

In this week the whole laser was together and would be operational. It was time to test our designs. We had tested most of the subsystems separately and had even been required to do so by the inspection and test procedures described in the design report. It was a nervous time for us and we tested several of the systems again for safety. About mid-morning on Wednesday there was nothing else left to delay us and we really did have to try the whole laser together.

Richard turned the whole laser on for the first time and left it going for about five seconds and then turned it off again very quickly. The laser had instrumentation on all the vital measuring points. John was on duty looking down the bore through a special protective attenuating plastic and reported on the behaviour of the discharge. The lights were dim so that we could quickly spot any unwanted discharges underneath the perspex covers used for testing. Richard and Rhys paced up and down alongside the laser, talking technical things. Steff came to watch as well, the lights were too dim for her to work and anyway this was an important event. Rhys and Richard finished discussing their technical things and announced that we were satisfied and that we were going to switch it on again and, as Richard said, 'go for it'.

There is a safety protocol used for switching on a copper laser and the team automatically went into it again. 'Thyratron warm,' announced Richard. 'Ready?' he then asked. Two 'readies' came from myself and Rhys, followed by one from John when he was in position. Everyone looked at Steff who wasn't used to all this laser talk. 'Yeah, fine — "ready", I mean,' she said to all the waiting faces. 'Switching on' called Richard. The loud thump of a chain of contactors switching in confirmed we had started. 'Volts on' came next as the power controller started to deliver power into the laser tube. By now the whole machine was buzzing. 'Breakdown,' called John, as the discharge started to flow properly: '14kV.' Rhys looked at one of the oscilloscopes and paced round the laser. He and Richard compared thoughts then he looked up and smiled.

'Piece of cake,' he said.

'So what was all the fuss about?' I asked. We walked off to carry on working.

Steff looked at the laser, which to her seemed the same as normal apart from the messy test equipment and terrible noise, and asked, 'Is that good then?' Rhys explained that it *was* good. 'Oh', she commented. 'Good' and went off to switch the lights on again, we didn't need the darkness any more.

The world's first commercially available 100W copper laser was running smoothly in our industrial unit before our very eyes, but one cannot be in awe too long and after about ten minutes we had all settled down to other jobs. About half an hour later John went to the toilet at the back of the unit and something strange happened. The moment he flushed it, the laser tripped out. We all flew from our other jobs to see what had happened. We soon found that the laser had shut itself down because it detected a drop in water pressure. The laser was interlocked to the water pressure so that if the supply slowed down and there was a risk of overheating it switched itself off. The water supply to the unit was not enormous and it turned out that there was not enough for the laser and the loo. The laser obviously took priority and so the loo was closed and John is the last person ever to have used that particular convenience. Over the next fortnight we became more and more confident with the laser and planned to demonstrate it at the next milestone meeting with copper in the tube, ahead of the original plan.

Week 52

The second milestone of a laser continuously running went ahead on time, a year to the day after the project started. We ran the laser at about three-quarters of full power and the beam of green light given at this power looked rigid enough to sit on. Our customer was delighted with the sight. A scientific delegation had been sent this time including their most senior scientist, a man who appreciated what he was looking at on occasions like this. The delegation accepted without reservation that if we were able to achieve this demonstration ahead of schedule we wouldn't have to much difficulty with the few outstanding circuit boards and control systems that needed completion. In the team afterwards we joked that there was a month in the plan between this milestone and the next of demonstrating full power yet we were confident now that all we had to do was turn up the power control knob to the full power and we would be fine. 'Not a bad month's work,' we said.

In reality, of course, we used the month of time we had clawed back to help in the areas where we were behind, particularly the alignment system which John and I were able to sort out completely in just two weeks now that we didn't have interruptions taking us away from it every hour or so.

Week 56

In Week 56 the demonstration of the laser running at full power went ahead without incident although behind the scenes there had been more problems than expected. Laser output power had not been as steady as we would have liked and in the week just before the test Richard began to suspect that the plasma tube had cracked. We discussed it and, on his recommendation, decided to fit a replacement. This was not a popular decision since there is a great deal of work involved and Richard and John particularly would have a lot of overtime to do in order to meet the deadline. In the event the tube was cracked, copper had leaked out and a new tube was the right decision. On the day of the test the tube was hardly cool from all the installation processes and running that accompanies the fitting of a new tube, before we were running it again for the milestone.

Week 57

After the full-power milestone meeting we regrouped and listed all the 'little jobs' that needed to be completed before the pre-shipment inspection. The laser was due for delivery in Week 68 and this was our next and fourth milestone. The tasks required a large amount of time finalising the first laser. It's not just the tasks themselves — in a sense that's the easy bit — it's the documentation that goes with them so that you can make many more the same. Steff had a long list of drawings to prepare, many of which were associated with quite small changes made recently to components that had been modified. Many such changes involved modification to the general assembly drawings higher up the bill along with the relevant component drawings. The drawing set was now up to about 250 and still had some way to go before it would form the minimum statement of the designs with which we would be happy. There were also other drawings required for the operation and service manual that had to be prepared.

I wrote the manual, a task which took many weeks and which involved further consultation with and approval from the customer. The major block of work for the team, however, was to complete the characterisation tests of the new product. A laser needs an extensive set of tests to determine the operating envelope and these would take many days of running time to complete; in addition there was a need for a continuous test of a hundred hours of unbroken running — one stoppage for only a few moments would negate the whole test and it would have to be started again. The other daunting task was the continuous construction of the second laser. In Week 70 that was due to be discharging stably. Work continued with the regular weekly meetings forming the basis for planning. John constructed the second laser and Richard and Rhys finished the testing and construction of the first one. Steff continued drawing whilst I wrote the manual and other support documentation. I also arranged shipping, including insurance and all the other details that came with freighting out a large laser by surface to the customer's site.

Week 67

The customer arrived in a delegation. As usual our jeans and trainers were swapped for suits and ties and the acceptance papers for the first laser were signed. We were getting very practised at hosting visitations from our customer and the initial detailed planning and effort that had gone into preparation for the early meetings had given way to a more relaxed and informal feel. However, each meeting still contained a formal recognition of the state of the project and minutes were always taken recording actions and decisions. We all knew each other by now, though, and somehow it seemed easier to resolve differences of opinion.

Week 68

On the day of the delivery we visited the customer's site, the laser arrived by lorry and was lifted by crane into a second-floor laboratory. It is a heart-stopping spectacle to see the culmination of such a large proportion of one's time and effort covered with polythene and dangling from a crane jib in the breeze and rain. We breathed again when it was safely inside the laboratory. Rhys and Richard stayed to commission the laser whist John and I returned to Oxford to continue with the next laser.

When we got back Steff was really sad: she had wanted to go too and somehow, unbelievably, no one had said anything about it, not me, not Rhys, not Richard or John, not even Steff herself. I wished she had said, of course she could have gone — I felt awful. I promised her the installation trip next time, but as it happened she wouldn't be going on that trip either.

Week 69

Rhys and I had one of our reflective sessions on project progress. We had spent about a third of the project time thinking about what we were going to do, about a third of the time actually doing it, and about a third measuring what we did and confirming it worked. This was a ratio that served us well, and we planned to repeat it when we were next hired to build a 200 watt laser! Our minds were turning more and more to the future. It was clear that things would change: there wasn't enough work for the whole group between now and the arrival of an order for lasers to fill the rest of the customer's rack. Even if we were successful, it was some way off. In addition, much as we would have loved it, there was no 200 watt copper laser programme anywhere in the world and we had no prospect whatsoever of getting work that way.

Week 70

The discharge test on the second laser went ahead on time and was accepted.

Week 74

The group finalised a new contract for the production of an oscillator, a special variant on the laser design to be used by the customer to inject light at the start of each chain of lasers in the rack. We had anticipated this order but it was only for one machine, and it involved much less work than one of the big amplifying lasers. It would, however, provide work for a few more months. It was becoming more certain that there would be a considerable time-lag between the customer receiving the first two lasers and the first oscillator and then testing the lasers in the rack to prove in principle that their proposed system would work. Only then would they be in a position to order the future amplifiers and oscillators. Such an order would clearly be a very attractive proposition to Oxford Lasers and stood to be their biggest order to date, probably for quite some time to come.

Week 80

Sadly Steff resigned from Oxford Lasers and returned to her original employers. I was concerned by this development — the team had worked well and I had thought everyone was happy. We had a chat and I asked why she wanted to leave. She explained that there was little of the electronics draughting that she had liked so much in her previous job. There has also been a number of changes at her previous employer and the reason for her leaving in the first place had been removed. The company had contacted Steff and made her an offer that was difficult for her to refuse. We chatted some more, partly about Oxford Lasers, partly about our own lives and partly about the rest of the group. She would be sad to leave the team and would miss us all but she felt the move was best and we left it at that.

'James,' she said, 'you do realise that this means I never will get to make an installation trip to the customer after all.'

'Well, that's your fault,' I joked, 'you don't *have* to go.'

She smiled. 'I do really,' she said. 'It's time to move on.'

She was right. Things were moving on at Oxford Lasers too.

Week 83

Oxford Lasers was undertaking a second major product development programme to completely redesign the existing range of lasers and bring them into new markets. I had been asked to assist in this venture and after discussion with my superiors an appointment was made. The following Monday I left the Cu100 group to run a new project as Head of Engineering.

Week 86

The group delivered the second laser as scheduled.

Week 96

I met with Rhys for a drink to mark the passage of the project. Both the lasers were now delivered and the customer had been trained in their operation. Commissioning was over and the project was completed. The order for the oscillator would keep the three of us busy for a while but after that the future was rather open — 'Until the order for loads more big lasers comes in,' I said.

Richard's particular skills were needed on another product development programme to develop a new design of copper laser altogether using compounds of copper, instead of elemental, which would operate at much lower temperatures but had the disadvantage of only producing very low powers. However, there was the possibility of commercial exploitation and Oxford Lasers decided to pursue it. Richard started this work straight away and spent some time working in both groups before leaving the Cu100 group altogether.

A year later, when Oxford Lasers did receive an order for more amplifiers and oscillators, only John and Rhys remained from the original team of five. The receipt of the order was a massive commercial achievement and vindication of all our original time and effort.

Finally, it only remains for me to say how saddened I was to leave the group. In Oxford Lasers there is a rare combination of excellence, technical humour and *esprit de corps* but within that particular project team these properties were especially strong. We did cover new and very difficult technical ground but we thoroughly enjoyed doing it. It was my privilege to experience the special brand of engineering that was the Cu100 group. I miss them.

10.2 The Cu100 project debrief

In the case study many management issues were raised. We will now consider these following the structure of the book.

Nature of organisations

In Chapter 2 we considered the structure of organisations and the role of corporate objectives. Reading through this case study it is clear that one of the objectives of Oxford Lasers was to secure that very large order, for Cu100 lasers. This project was the first phase in their strategy to meet that objective. What is clear is that the objective caused the company to consider the type of organisation required to achieve their goal. The project team was established although it functioned within a company that was organised by function (references to production department, stores, finance). You will notice, however, that at the end of the Cu100 project both James and Richard moved on to other projects. It may be that owing to the success of operating with the project team the company decided to move to that method for other work.

Functions of organisations

In Chapter 3 we looked at the different functions that were required in an engineering organisation. Although the Cu100 team was multi-functional we still see many references in the case to the discrete functions. Parts were ordered by the team, to specifications set by them. We even saw an example of James visiting suppliers, working with them and using their specialist knowledge (the distortion due to welding problem). The company had established ordering and goods receipt procedures and these were used by the team, whilst retaining responsibility for inspection and storage, as required by the customer. This is a prime example of how it is important to draw on resources that are available when they meet the criteria. It would have been expensive and time-consuming to operate a separate procedure.

In this project we see little interaction of the team with the manufacturing function, apart from recruiting one of their technicians. However, throughout the project the plan was always that the Cu100 laser would go into production by receiving the order for more lasers. Because of this the team were developing the product with manufacture in mind, defining bills of material and providing a complete drawing set.

James and Rhys were involved in the 'sales' function, in negotiating the contract with the customer. Throughout the project they also retained close contact with the customer and were finally responsible for installation and commissioning on the customer's premises.

The company's finance department was used for the processing of orders, although it is clear from the case study that responsibility for managing the project budget rested with the team.

The whole case study describes the product development process for the Cu100. The project team were wholly responsible for all aspects of the design, from producing the design report to building the laser. Drawings were originally subcontracted but part-way through the project the team had its own draughtswoman.

Finally, we consider the quality function. It was clear in the case study that there was a stringent quality system requirement from the customer. References implying this relate to the checking of inspection schedules, the requirement to trace parts to source, and the specific reference to having to use an enclosure supplier who could meet the quality requirements of the customer.

Personnel management

In Chapter 4 we considered some of the issues related to employing people. These included legislation governing employment and recruitment, motivation, appraisal, training and development, and job design. There are many examples of the first four of these topics in the case study, and we shall consider these in turn.

In the case study we see an example of the recruitment and selection process from the candidate's point of view, when James applied for the job. We also see the other side of this when James and Rhys were recruiting for other members of the team. You will notice from the case study that they had a clear idea of not only the skills they were looking for but also the type of person they needed for the team. There was also an example of the problems that can arise if you are not able to make a satisfactory appointment (James having to prepare some drawings, subcontractors doing others).

Unfortunately we also saw another aspect of employing people: that was 'poaching' of staff (when Steff's old company went directly to her with a new job offer). Following this Steff gave notice to terminate her employment with Oxford Lasers.

In the case study the high level of motivation of the team is obvious. Not so obvious are the many techniques employed to keep that motivation high. Working in a small group usually works well, but there were other things like the weekly meeting. The team spirit was engendered by being a bit different from the rest of the company, in their own industrial unit, and socialising together (beer and curry). In particular there was the total involvement of everyone with the whole project and with the customer, which only broke down when Steff wasn't asked to go when the laser was delivered.

We saw that Oxford Lasers operate a two-tier appraisal system for everyone in the company. We also saw that it took a considerable amount of James' and Rhys' time. But as James said, he needed to establish how things were really going and he didn't normally get time for that. The appraisal system provided that opportunity.

Throughout the text we see many references to training and development. These start with James learning about laser physics and fast electronics, and him having to relearn his stress analysis. This is not unusual — again, as engineering is so vast and is constantly changing, we can't be experts in everything. What we have as engineers is a general knowledge, such as management, thermodynamic principles etc. that is generally applicable. However, we also have the ability to acquire more knowledge; in particular we have the ability to define what knowledge we need. We also have a reference to John who was trained by the production department. It is likely that as a technician John acquired much of his formal training whilst working in the company, although this may have been coupled with some form of tertiary education.

Teamworking and creativity

In Chapter 5 we looked at the selection and use of teams and the process of planning motivation, as exemplified by the Cu100 product development. This whole case study is about teamworking and no further discussion of that is necessary. However, we saw an example of problem-solving by teamwork and the way in which the whole group contributed ideas to the brainstorming session to solve the disconnection problem.

Personnel management

In Chapter 6 we considered some techniques that can be used to achieve good personnel management. In the case study we see few references that we could point to as examples of personnel management. However, it is clear that in the early weeks of the project James was very much on his own, designing the enslosure and learning about laser physics. He would not have been able to do these tasks and meet the deadline if he had not had an effective time-management system, well-organised filing system and generally good personal organisation.

Finance

In Chapter 7 we saw a general need for monetary control and then looked at some specific accounting techniques including the preparation of balance sheets, product costing, and the use of budgets.

We see in the case study that the project has its own budget and also that the customer makes stage payments when milestones are achieved. The first of these means that there is a difficult financial task for the team, the second means that the team has responsibility for producing significant cash in-flows to the company.

Managing a budget means that you have to plan recruitment, purchases, equipment hire, and allow for the costs of operating an industrial unit. James wouldn't have needed to know about balance sheets and profit and loss accounts for this job but he must have had to cost the lasers and ensure that there was sufficient cash flow to cover necessary expenditure.

Project planning and management

In Chapter 8 we looked at the requirements for project management and at some specific planning techniques. This was one large project and it was obviously well managed — all the deadlines were met. The customer was so pleased that more orders were placed.

The project constraints, the milestones and deadlines were all defined by the customer. However, the way in which these requirements were met was determined by James and Rhys. Plans were developed during the first eleven weeks of the project and must have been costed as they led to a fixed project budget. Further detailed plans, in the form of quality plans, were agreed with the customer in the design report.

Monitoring of the project by Rhys and James seems to have been done on a fairly informal basis, although there was a much more formal review with the customer. The day-to-day management of the project and scheduling of tasks was done by James during the weekly meeting. There is an indication that there may also have been a very informal review of progress at the monthly company meeting.

Communication skills

In Chapter 9 we saw that the job of an engineer is very largely concerned with gathering and using information, and we looked at some techniques for doing these things. Throughout the case study it is obvious that information control was of the utmost importance. This was a big project and they could not afford to repeat mistakes or lose information. Experiments were recorded, design files were kept. All the drawings were cross-referenced via the general assembly drawings. There must have been a document control procedure because the customer was insisting on quality plans and on traceability. In addition there were inspection and test schedules that had to be met.

Information-gathering was done in a variety of ways. James talks of receiving instruction from his boss, Rhys, and from his subordinate, Richard, who had specific technical expertise. The customer provided information, particularly about the system into which the lasers would fit, and then later at technical meetings. The suppliers of the enclosure also provided technical expertise, and further specialist help was provided by an institution of higher education. Without all of these inputs to the product development process it is unlikely that the target would have been met. We now live in a world of such complexity that no one person can have an adequate grasp of all the knowledge required to do a job like this. The effective engineer uses all the resources available to provide the extra knowledge required.

In Chapter 9 we also looked at the requirement for engineers to be able to communicate well both in written work and orally. We looked at some specific techniques that would be of use here and at some methods of communicating in the workplace. Communication is such a big issue in engineering that the case study is full of examples. Throughout there are constant references to communication issues; in fact every single paragraph contains a communication issue. The major written communications were James' job application, the design report, and the operating manuals for the lasers. However, all the design was documented, with experiments being recorded. There was obviously a significant requirement for all the team members to communicate in writing.

Examples of oral communications range from the very formal communciations with the customer to the weekly meetings with the team. These are equally important. You have to be able to identify the customer need but you can only meet that need if you can then translate it into tasks and activities by communicating with the people who work with you, and for you. Another aspect of communciation was that of wider communication with the other people in the company and in fact we saw that Oxford Lasers put so much emphasis on this that a monthly company meeting was held.

Case study

In Chapter 10 we have given you a case study, a real engineer working on a real engineering project. The project would not have been achievable without that engineer's management skill. Every single management issue we have described as separate issues within the body of the book can be seen to be in use in this case study. The work described in the case study is clearly an engineering undertaking and yet as we have seen the work is filled with management issues.

In the introduction to this book we asked whether it was worthwhile for engineers to study the contents of the book and so develop their management skills. In this last case study we have conclusively proved that engineers use management and technical skills. We have shown that engineers are capable of being excellent at both of these. We have shown that management can indeed be done well, when it is done by engineers.

Index